Advances in Pattern Recognition

Springer

London
Berlin
Heidelberg
New York
Barcelona
Hong Kong
Milan
Paris
Singapore
Tokyo

Advances in Pattern Recognition is a series of books which brings together current developments in all areas of this multi-disciplinary topic. It covers both theoretical and applied aspects of pattern recognition, and provides texts for students and senior researchers.

Springer also publishes a related journal, **Pattern Analysis and Applications**. For more details see: http://link.springer.de

The book series and journal are both edited by Professor Sameer Singh of Exeter University, UK.

Also in this series:

Michael S. Lew (Ed.)

Principles of Visual Information Retrieval

With 93 Figures

 Springer

Michael S. Lew, PhD
Leiden Institute of Advanced Computer Science, Leiden University,
Niels Bohrweg 1, 2333 CA Leiden, The Netherlands

Series editor
Professor Sameer Singh, PhD
Department of Computer Science, University of Exeter, Exeter, EX4 4PT, UK

British Library Cataloguing in Publication Data
Principles of visual information retrieval. – (Advances in
    pattern recognition)
    1. Image processing – Digital techniques 2. Information
    retrieval
    I. Lew, Michael S.
    621.3'67

ISBN 978-1-84996-868-3
Library of Congress Cataloging-in-Publication Data
Principles of visual information retrieval / Michael S. Lew (ed.).
    p. cm. – (Advances in pattern recognition)
    Includes bibliographical references and index.

    1. Image processing – Digital techniques. 2. Optical scanners. I. Lew, Michael S.,
    1965- II. Series.
    TA1637.P77   2001
    621.36'7—dc21   00-063529

© Springer-Verlag London Limited 2010
Printed in Great Britain

34/3830-543210 Printed on acid-free paper

To my parents

Preface

Digital visual information is here to stay. The major world nations are converting their cultural heritage into digital libraries. The era of digital video has also arrived for television, movies, and even home video. Tying it together, the World Wide Web gives us the ability to communicate and share the information. Visual information permeates every facet of our life in the 21st century. Finding it is the next step.

The goal of this book is to describe and illuminate the fundamental principles of visual information retrieval. Consequently, the intention is to introduce basic concepts and techniques, and develop a foundation, which can be used as a cornerstone for later study and research. Chapter 1 introduces the reader to the paradigms, issues, and important applications. In multimedia search, the three prevalent features are color, texture, and shape. Therefore, Chapters 2, 3, and 4 cover each of these, respectively, and discuss both models and techniques for matching in the scope of content based retrieval. Methods for measuring similarity between features are given in Chapter 5, and the state of the art in feature selection methods in the context of visual learning is given in Chapter 6.

Beyond the features of color, texture, and shape, the next logical step is to include time-based image sequences or video. In Chapter 7, the intent is to cover the important issues in video indexing which include shot break detection, automatic storyboarding, finding smooth transitions such as fades and dissolves, and generating visual summaries of the video. The coverage does not only compare different video indexing methods, but also introduces the basic models and mathematical formulations.

Regardless of the features which are used in the multimedia search system, queries need to be specified. Chapter 8 discusses the state of the art in multimedia query languages, and Chapter 9 explains how to improve the search process using multiple queries based on continual user feedback (relevance feedback). Having introduced the basic features, the next logical questions would be how to combine the basic features and integrate context into the search process. These questions are directly addressed in Chapters 10 and 11, respectively.

Perhaps the ultimate goal of a search system is for it to understand concepts beyond combinations of basic features. Instead of only knowing primitive features, it would be preferable if the system understood higher level semantic visual concepts such as car, child, violin, etc. Therefore, Chapter 12 describes how to integrate semantics towards improving multimedia search. A representative example of applying content-based retrieval to solve a real world problem is the area of trademark retrieval, which is covered in Chapter 13.

In summary, this book begins by describing the fundamental principles underlying multimedia search, which are color, texture, and shape-based search. The discussion shifts to indexing sequences of images, and then to natural extensions of the fundamentals. The extensions include methods for integrating and selecting features, specifying multimedia queries, and semantic-based search.

Acknowledgments

I would like to thank Nicu Sebe for his tireless efforts and dedication to the project. Erwin Bakker was very helpful in checking and reviewing each chapter. Daniel Lewart solved numerous formatting problems. Moreover, I would like to thank the authors for their outstanding contributions. I also want to thank Hyowon Kim for contributions which are too deep to be expressed.

M.L.

December, 2000
Leiden, The Netherlands

Contents

xvi   Contents

List of Contributors

Vassilis Athitsos
Computer Science Department, Boston University
111 Cummington Street, Boston, MA 02215, USA

Shi-Kuo Chang
Computer Science Department, University of Pittsburgh
Pittsburgh, PA 15260, USA

John P. Eakins
Institute for Image Data Research, University of Northumbria at Newcastle
Newcastle, NE1 8ST, UK

Charles Frankel
Electric Knowledge LLC
Cambridge, MA 02139, USA

Paul C. Gardner
Excalibur Technologies Corporation
1959 Palomar Oaks Way
Carlsbad, California, 92009, USA

Theo Gevers
ISIS, Faculty of WINS, University of Amsterdam
Kruislaan 403
1098 SJ, Amsterdam, The Netherlands

Michiel Hagedoorn
Department of Computer Science, Utrecht University
Centrumgebouw Noord, office A210, Padualaan 14, De Uithof
3584 CH Utrecht, The Netherlands

Thomas S. Huang
Beckman Institute, University of Illinois at Urbana-Champaign
Urbana, IL 61801, USA

Jean-Michel Jolion
Pattern Recognition and Vision Lab
INSA Lyon, Bat. 403
F-69621 Villeurbanne Cedex, France

Erland Jungert
Swedish Defence Research Institute
172 90 Stockholm, Sweden

Marco La Cascia
OffNet S.p.A.
Via Aurelia km 8.7
00165 Roma, Italy

Clement Leung
Visual Information Systems Research Group
Communications & Informatics, Victoria University
P.O. Box 14428, Melbourne CMC
Victoria 8001, Australia

Michael S. Lew
Leiden Institute of Advanced Computer Science, Leiden University
Niels Bohrweg 1
2333 CA, Leiden, The Netherlands

Yong Rui
Microsoft Research
Redmond, WA 98052, USA

Stan Sclaroff
Computer Science Department, Boston University
111 Cummington Street
Boston, MA 02215, USA

Nicu Sebe
Leiden Institute of Advanced Computer Science, Leiden University
Niels Bohrweg 1
2333 CA, Leiden, The Netherlands

Saratendu Sethi
Computer Science Department, Boston University
111 Cummington Street
Boston, MA 02215, USA

Simon So
Visual Information Systems Research Group
Communications & Informatics, Victoria University
P.O. Box 14428, Melbourne CMC
Victoria 8001, Australia

Dwi Sutanto
Visual Information Systems Research Group
Communications & Informatics, Victoria University
P.O. Box 14428, Melbourne CMC
Victoria 8001, Australia

Michael J. Swain
Cambridge Research Laboratory
Compaq Computer Corp.
Cambridge, MA 02139, USA

Audrey Tam
Visual Information Systems Research Group
Communications & Informatics, Victoria University
P.O. Box 14428, Melbourne CMC
Victoria 8001, Australia

Leonid Taycher
Artificial Intelligence Laboratory, Massachusetts Institute of Technology
545 Technology Square (MIT NE43)
Cambridge, MA 02139, USA

Philip Tse
Department of Computing, School of MPCE, Macquarie University
North Ryde NSW 2109, Australia

Remco C. Veltkamp
Department of Computer Science, Utrecht University
Centrumgebouw Noord, office A210, Padualaan 14, De Uithof
3584 CH, Utrecht, The Netherlands

PART I
Fundamental Principles

1. Visual Information Retrieval: Paradigms, Applications, and Research Issues

Michael S. Lew and Thomas S. Huang

1.1 Introduction

Imagine you are a designer working on the next Star Wars movie. You have seen thousands of images, graphics, and photos pass by your monitor. However, you can only recall a few characteristics of the images – perhaps it had a gorgeous night sky scene, or lonely sand dunes; or maybe it had a Gothic feeling. How do you find the visual imagery? Instead of being a designer, perhaps you are a news journalist who needs to quickly make a compilation of the millennium celebrations from around the world. How do you find the right video shots? Visual information retrieval (VIR) is focussed on paradigms for finding visual imagery: i.e., photos, graphics, and video from large collections which are spread over a wide variety of media such as DVDs, the WWW, or wordprocessor documents.

Visual information retrieval lies at the crossroads of multiple disciplines such as databases, artificial intelligence, image processing, statistics, computer vision, high performance computing, and human-computer intelligent interaction. Moreover, it is precisely at these crossroads and curious intersections where VIR makes contributions which are not covered by any of the other disciplines. The goal of this chapter is to introduce the reader to the major paradigms in VIR, applications, and research issues.

1.2 Retrieval Paradigms

There are several retrieval paradigms used in visual information retrieval. When text annotation is available, it can be directly used for keyword-based searching [3, 19]. However, in many situations, text annotation does not exist or it is incomplete. In fact, it is quite rare for complete text annotation to be available because it would entail describing every color, texture, shape, and object within the visual media.

When text annotation is unavailable or incomplete, we must turn to content-based retrieval methods (see Gudivada and Raghavan [7] for a system based overview and a more detailed research summarization can be found in Gupta and Jain [8], Chang et al. [2], and Rui et al. [17]). In content based retrieval methods, the search is performed on features derived from the raw visual media such as the color or texture in an image. The dominant VIR paradigms include querying for similar images, sketch queries, and iconic queries as shown in Fig. 1.1. In the similar images query paradigm, the user selects an image, and then the system responds with a list of images which are similar to the user-selected image. In the sketch search paradigm, the user manually draws a sketch of the desired image and then the system finds images which similar features (i.e., shape) to the user sketch. In the iconic query methods, the user places symbolic icons where the visual features should be.

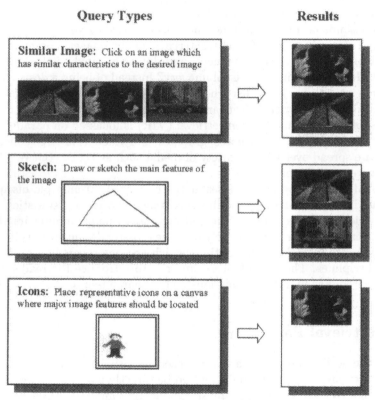

Fig. 1.1 Retrieval paradigms.

In all of the methods, the results of the search are usually shown as a list which is sorted by the VIR system's estimate of the relevance of the image to the user query. None of the VIR paradigms is appropriate for every application. For example, the similar images query paradigm has the disadvantage that a list of relevant images must first be found through another mechanism. The sketch search paradigm allows the user to compose a query for a wide set of images, but it also requires that the user has skill in sketching. The iconic query paradigm is probably the most intuitive query method, but it is restricted to the vocabulary of visual concepts which can be reliably understood by the VIR system. Furthermore, the different retrieval paradigms can be mixed. For example, one could first retrieve by keywords, and then use low level similarity-based search. Thus, one could ask for sunset images and receive 20 results, among which a few are appealing in color and layout. Then, one could ask for more images which are similar to these with respect to low level features.

In designing a VIR system, it is important to select appropriate features. In current systems, these features are typically related to color, texture, and/or shape. The process of automatically or interactively choosing the best features for a particular application is denoted as *feature selection*. When the features have been selected, a similarity (or distance) measure is chosen to rank the candidates. An ideal measure of feature similarity should be correlated to the user's intuitive sense of similarity. Roughly speaking, the steps are as follows:

Feature selection. From the set of low level features (i.e., color, texture, and shape) and high level features (i.e., probability of being a face, car, building, or street), select a subset (either automatically or interactively) which has good or optimal discriminatory power in the particular application.

Similarity measure selection. Choose a measure of similarity or feature distance. This similarity measure is used to give a distance between the user query and a candidate match from the media collection.

Ranking. Rank the candidates by the similarity measure.

Note that the above steps are meant as a rough guideline, not a definition of VIR. Methods such as relevance feedback [14–17, 20] improve the similarity measure and the usage of the features by learning from the previous user selections. It is also possible to dynamically change the feature sets [1]. Other interesting advances include reducing the images to perceptually significant coefficients [12], or directly learning the similarity measure [4, 9], from a training set. Some methods [10] do not perform ranking at all. High performance system aspects such as efficient high dimensional indexing have also been studied [5, 11, 21].

1.3 Applications

There are many applications where visual information retrieval is important. Some of these are:

- Architecture, real estate, and interior design – finding the right appearance
- Biochemical applications – indexing molecules
- Cultural services - exploring museums and art galleries
- Digital catalog shopping – browsing for clothes
- Education – preparing lectures and exploring the past

- Entertainment – browsing the WWW
- Film and video archives – video editing
- Identification – who is he/she?
- Geographical information systems – where are the local attractions?
- Journalism – background research and preparing a story
- Medicine – is this a tumor?
- Remote sensing – which satellite images contain tanks?
- Surveillance – which cars were speeding?
- Tourist information – browse for a fun place to go

In the following sections, a sampling of the applications are described in more detail.

1.3.1 Architecture, Real Estate, and Interior Design

Describing the shape of buildings is difficult because there is no common terminology which perfectly describes the subtle aspects of architecture. Methods for finding similar buildings allows users to find blueprints which correspond to more appealing structures. In interior design, there is an abstract intuition of what colors, textures, and shapes make good combinations. However, this information is rarely found in text annotation. For example, suppose the user is interested in finding a painting to match a room with wood floors, yellow carpet, and black chairs. By finding paintings which have similar colors to the room, it is possible to quickly find matching candidates.

1.3.2 Biochemical

Each day more molecules are found and cataloged by biochemical researchers. Methods for indexing these molecules would be helpful in the design of drugs. For example, the user could search for molecules with shapes similar to a candidate drug to get an idea of possible side effects. Biochemical indexing would naturally be conducive to combined text- and shape-based search because the quantity of each constituent atomic element is known in the molecule.

1.3.3 Digital Catalog Shopping

Many shoppers only have a vague idea of what they are interested in, which prompted the phrase, "I'll know it when I see it." A shopper might see something that is almost right, but not perfect. Consequently, he or she will want to fine tune the search. There are a wide variety of catalogs where the search process is visual. These would include topics such as clothes, cars, vacation spots, posters, etc. Furthermore, in the case of furniture such as sofas, there can also be the option of mixing different covers with a particular frame.

1.3.4 Education

In education, the role of visual information retrieval varies depending on the area. For example, in history, it is useful to have immediate access to images and short video sequences of relevant events and people. In giving a lecture on modern production methods, it is useful to be able to include images of the early mass-production factories. In the performing arts, images and video shots can be used as learning tools or as backgrounds for productions. In foreign languages, visual media can be used to place the new words and concepts into the right context.

1.3.5 Film and Video Archives

When television stations such as Discovery shoot a story, they typically record over 50 times as much material as they use. The rest of the video footage is stored for later use. Until recently, the only option available for finding a particular video shot was to manually fast-forward through the video tapes until the video segment was found. Using VIR methods, the video movie can be reduced to the constituent shots which are then represented using a single frame from each shot, which are known as keyframes. The keyframes can be searched automatically for particular characteristics such as color, texture, shape or even high level concepts such as particular people, places, or objects. The ability to find video shots quickly is particularly important to news stations because they often only have minutes to put together the late-breaking news story.

1.3.6 Medicine

Diagnosis in medicine is often performed using visual recognition of abnormal conditions which are either directly viewed by the physician or scanned into an image from X-rays, magnetic resonance imaging, or computed tomography. The medical literature contains volumes of photographs of normal vs. pathological conditions in every part of the body. Diagnosis may require recalling that the current condition resembles a condition from the literature. Since medical treatment is often more effective when given early, it is important to search the literature quickly and accurately.

1.4 Research Issues

Currently, there are several important issues in the visual information retrieval area: (1) low level features vs. semantics; (2) fusion of different modalities; (3) one-shot versus navigational queries; (4) evolution or revolution for multimedia databases, and (5) performance evaluation.

In practice it is easier to compute low level features than high level features which may be linked with semantics. Consequently, many systems rely entirely on low level features. The problem comes from trying to transform the user's language to the specification of low level features. For example, how does one express the concept of a chair in terms of color histograms? Should we expect the user to be able to perform this transformation? Petkovic [13] suggests trying to automatically extract simple semantics such as whether the image is B&W or color; detecting faces; and whether it is an indoor or outdoor scene.

In many visual retrieval and multimedia search applications, diverse modalities exist which can potentially improve the accuracy of the search. For example, in video we have audio and sometimes closed caption text and even transcripts. In video retrieval, these other modalities could contribute significantly. Furthermore, the automatic inference of high level (semantic) concepts can be made easier by the availability of multiple modalities. Fusing the different modalities is an important research area.

A one-shot query is thought to be the classical equivalent of being able to write a single query, for example an SQL SELECT, and finding the desired results immediately. Navigational queries are different from one-shot queries in the sense that the expectation changes from finding the correct result immediately to taking part in an interactive journey to find the correct result. An example of a navigational query is relevance feedback. In relevance feedback, the user is presented with a set of images. For each image, the user decides whether or not the image is relevant. After the user has sent the relevance information to the system, the system uses the relevant images toward improving the next set of results. This cycle continues until the user has found the desired image. At each cycle, the system gains more information as to the relevance of different images and is usually able to make better suggestions in future cycles.

As with scientific paradigms, there is the question of evolution or revolution. This refers to making incremental enhancements to the current paradigm versus throwing out the current paradigm for a completely new paradigm. In visual databases, this refers to the issue of whether it is better to enhance classical text databases with visual queries or abandon the text methods. Martinez and Guillaume [10], tried to extend an object-oriented database with image-oriented queries. However, Santini and Jain [18] assert that image databases are not databases with images. Their primary argument is that image database queries must be patterned after similarity search, which are then ranked using a human-oriented similarity metric.

Currently there are no established standards, nor widely recognized test sets for benchmarking visual retrieval systems. For example, the most frequently published and referenced query by visual content system is arguably the QBIC [6] system, and yet there are no widely accepted findings on its effectiveness in finding relevant images. We think that this has to do with the inherent ambiguity in defining a relevant image. Each user may have a different perception of whether an image is relevant.

1.5 Summary

We are at the beginning of the digital age of information, a digital Renaissance. Worldwide networking allows us to communicate, share, and learn information in a global manner. However, having access to all of the information in the world is pointless without a means to search for it. Visual information retrieval is poised to give access to the myriad forms of images and video which have a diverse set of applications in education, science, and business.

References

1. Buijs, J and Lew, M, "Learning Visual Concepts," ACM Multimedia'99, 2, pp. 5–8, 1999.
2. Chang, SF, Smith, JR, Beigi, M, and Benitez, A, "Visual Information Retrieval from Large Distributed Online Repositories," Commun ACM, Special Issue on Visual Information Retrieval, 40(12), pp. 12–20, December, 1997.
3. Chang, SK and Hsu, A, "Image Information Systems: Where Do We Go From Here?" IEEE Trans Knowl Data Eng, 4(5), pp. 431–442, October, 1992.
4. Del Bimbo, A and Pala, P, "Visual Image Retrieval by Elastic Matching of User Sketches," IEEE Trans Patt Anal Mach Intell, 19(2), pp. 121–132, February, 1997.
5. Egas, R, Huijsmans, DP, Lew, M, and Sebe, N, "Adapting K-D Trees for Image Retrieval," VISUAL'99, pp. 131–138, 1999.
6. Flickner, M, Sawhney, M, Niblack, W, Ashley, J, Huang, Q, Dom, B, Gorkani, M, Hafner, J, Lee, D, Petkovic, D, Steele, D, and Yanker, P, "Query by Image and Video Content: The QBIC System," Computer, IEEE Computer Society, pp. 23–32, September, 1995.
7. Gudivada, VN and Raghavan, VV, "Finding the Right Image, Content-Based Image Retrieval Systems," Computer, IEEE Computer Society, pp. 18–62, September, 1995.
8. Gupta, A and Jain, R, "Visual Information Retrieval," Commun ACM, 40(5), pp. 71–79, May, 1997.
9. Lew, M, Sebe, N, and Huang, TS, "Improving Visual Matching," Proc. IEEE Conf on Computer Vision and Pattern Recognition, Hilton Head Island, 2, pp. 58–65, June, 2000.
10. Martinez, J and Guillaume, S, "Color Image Retrieval Fitted to 'Classical' Querying," Proc. of the Int. Conf. on Image Analysis and Processing, Florence, 2, pp. 14–21, September, 1997.
11. Ng, R and Sedighian, A, "Evaluating Multi-Dimensional Indexing Structures for Images Transformed by Principal Component Analysis," Proc. SPIE Storage and Retrieval for Image and Video Databases, 1996.
12. Pentland, A, Picard, R, and Sclaroff, S, "Photobook: Content-Based Manipulation of Image Databases," Int J Computer Vision, 18, pp. 233–254, 1996.

13. Petkovic, D, "Challenges and Opportunities for Pattern Recognition and Computer Vision Research in Year 2000 and Beyond," Proc. of the Int. Conf. on Image Analysis and Processing, Florence, 2, pp. 1–5, September, 1997.
14. Picard, R, "A Society of Models for Video and Image Libraries," IBM Syst, pp. 292–312, 1996.
15. Picard, R, Minka, T, and Szummer, M, "Modeling User Subjectivity in Image Libraries," Proc. IEEE Int. Conf. on Image Processing, Lausanne, September, 1996.
16. Rui, Y, Huang, TS, Mehrotra, S, and Ortega, M, "Automatic Matching Tool Selection via Relevance Feedback in MARS," Proc. of the 2nd Int. Conf. on Visual Information Systems, San Diego, California, pp. 109–116, December 15-17, 1997.
17. Rui, Y, Huang, TS and Chang, SF, "Image Retrieval: Current Techniques, Promising Directions and Open Issues," Visual Commun and Image Represent, 10, pp. 39–62, March, 1999.
18. Santini, S and Jain, R, "Image Databases Are Not Databases With Images," Proc. of the Int. Conf. on Image Analysis and Processing, Florence, 2, pp. 38–45, September, 1997.
19. Tamura, H and Yokoya, N, "Image Database Systems: A Survey," Patt Recogn, 17(1), pp. 29–43, 1984.
20. Taycher, L, Cascia, M, and Sclaroff, S, "Image Digestion and Relevance Feedback in the ImageRover WWW Search Engine," VISUAL'97, San Diego, pp. 85–91, December, 1997.
21. White, D and Jain, R, "Similarity Indexing: Algorithms and Performance," Proc. SPIE Storage and Retrieval for Image and Video Databases, 1996.

2. Color-Based Retrieval

Theo Gevers

2.1 Introduction

Can you imagine an existence without color? Indeed, color does not only add beauty to objects but does give more information about objects as well. Further, color information often facilitates your life, like in traffic, or in sport to identify your favorite team when both teams wear dark shirts.

After more than three hundred years since Newton established the fundamentals of color in his "Opticks" (1704) [17], color has been involved in many fields ranging from purely scientific, to abstract art, and applied areas. For example, the scientific work on light and color resulted in the quantum mechanics started and elaborated by Max Planck, Albert Einstein, and Niels Bohr. In painting, Albert Munsell provided the theoretical basis in his "A Color Notation" (1905) [15] on which most painters derived their notions about color ordering. The emotional and psychological influence of color on humans has been studied by Goethe in his famous book "Farbenlehre" (1840) [10]. Further, the value of the biological and therapeutic effect of light and color has been analyzed, and views on color from folklore, philosophy and language have been articulated by Descartes, Schopenhauer, Hegel, and Wittgenstein.

Today, with the growth and popularity of the World Wide Web, a new application field is born through the tremendous amount of visual information, such as images and videos, which has been made publicly accessible. With this new application area, color has returned to the center of interest of a growing number of scientists, artists, and companies. Aside from decorating and advertising potentials for Web design, color information has already been used as a powerful tool in content-based image and video retrieval. Various color-based image search schemes have been proposed, based on various representation schemes such as color histograms, color moments, color edge orientation, color texture, and color correlograms, [3, 19, 24–28]. These image representation schemes have been created on the basis of RGB, and other color systems such as HSI and $L^*a^*b^*$ [3, 19, 25–27]. In particular, the Picasso [1] and ImageRover [20] system use the $L^*u^*v^*$ color space for image indexing and retrieval. The QBIC system [3] evaluates similarity of global color properties using histograms based on a linear combination of the RGB color space. MARS [21] is based on the $L^*a^*b^*$ color space which is (like $L^*u^*v^*$) a perceptual uniform color space. The PicToSeek system [9] is based on color models robust to a change in viewing direction, object geometry and illumination. Hence, the choice of color systems is of great importance for the purpose of proper image retrieval. It induces the equivalence classes to the actual retrieval algorithm. However, no color system can be considered as universal, because color can be interpreted and modeled in different ways. Each color system has its own set of color models, which are the parameters of the color system. Color systems have been developed for different purposes: (1) display and printing processes: RGB, CMY; (2) television and video transmission efficiency: YIQ, YUV; (3) color standardization: XYZ; (4) color uncorrelation: $I_1I_2I_3$; (5) color normalization and representation: rgb, xyz; (6) perceptual uniformity: $U^*V^*W^*$, $L^*a^*b^*$, $L^*u^*v^*$, and (7) intuitive description: HSI, HSV. With this large variety of color systems, the inevitable question arises as to which color system to use for which kind of image retrieval application. To this end, criteria are required to classify the various color systems for the purpose of content-based image retrieval. Firstly, an important criterion is that the color system is independent of the underlying imaging device. This is required when images in the image database are recorded by different imaging devices such as scanners, cameras, and camcorders (e.g., images on the Internet). Another prerequisite is that the color system should exhibit perceptual uniformity, meaning that numerical distances within the color space can be related to human perceptual differences. This is important when images are to be retrieved which should be visually similar (e.g., stamp, trademark and painting databases). Also, the transformation needed to compute the color system should be linear. A non-linear transformation may introduce instabilities with respect to noise, causing poor retrieval accuracy. Further, the color system should be composed of color models which are understandable and intuitive to the user. Moreover, to achieve robust and discriminative image retrieval, color invariance is an important criterion. In general, images and videos are taken from objects from different viewpoints. Two recordings made of the same object from different viewpoints will yield different shadowing, shading, and highlighting cues changing the intensity data fields considerably. Moreover, large differences in the illumination color will drastically change the photometric content of images even when they are taken from the same object. Hence, a proper retrieval scheme should be robust to imaging conditions discounting the disturbing influences of a change in viewpoint, object pose, and/or illumination.

In this chapter, the aim is to provide a survey on the basics of color, color models and ordering systems, and the state of the art on color invariance. For the purpose of color-based image retrieval, our aim is to provide a taxonomy on color systems composed according to the following criteria:

– Device independence
– Perceptual uniformity
– Linearity
– Intuitivity
– Robustness against varying imaging conditions

- Invariance to a change in viewing direction
- Invariance to a change in object geometry
- Invariance to a change in the direction of the illumination
- Invariance to a change in the intensity of the illumination
- Invariance to a change in the spectral power distribution (SPD) of the illumination

The color system taxonomy can be used to select the proper color system for a specific application. For example, consider an image database of textile printing samples (e.g., curtains). The application is to search for samples with similar color appearances. When the samples have been recorded under the same imaging conditions (i.e., camera, illumination and sample pose), a perceptual uniform color system (e.g., $L^*a^*b^*$) is most suitable. When the lighting conditions are different between the recordings, a color invariant system is most appropriate, eliminating the disturbing influences such as shading, shadows, and highlights.

This chapter is organized as follows. Firstly, the fundamentals of color will be given in Section 2.2. As color appearance depends on the light, object and observer, we will study and analyze this triplet in detail in Section 2.3. Further, in Section 2.4, the standardization of the light–object–observer triplet will be presented. A survey on color invariance is presented in Section 2.5. The taxonomy of color systems is given in Section 2.6. Color and image search engines are discussed in Section 2.7. Conclusions will be drawn in Section 2.8. Further, the various color systems and their performance can be experienced within the PicToSeek and Pic2Seek systems on-line at: http://www.wins.uva.nl/research/isis/zomax/.

2.2 Color Fundamentals

Fundamentally, color is part of the electromagnetic spectrum with energy in the wavelength range from 380 to 780 nm. This is the part of the spectrum to which the human eyes are sensitive. For color measurements, it is often restricted to 400–700 nm. This visible part of the spectrum is perceived as the colors from violet through indigo, blue, green, yellow, and orange to red. This continuous spectrum of colors is obtained when a beam of sunlight is split through a glass prism. Wavelength is the physical difference between the various regions of the spectrum. The unit length of the wavelength is the nanometer (nm). Each wavelength value within the visible band corresponds to a distinct color. Most of the colors that we see do not correspond to one single wavelength but are a mixture of wavelengths from the electromagnetic spectrum, where the amount of energy at each wavelength is represented by a spectral energy distribution. For example, the energy emitted by a white-light source contains a nearly equal amount of all wavelengths, see Fig. 2.1(a). When white light shines upon an object, some wavelengths are reflected and some are absorbed. For example, a green object reflects light with wavelengths primarily around the 500 nm range. The other wavelengths are absorbed, see Fig. 2.1(b).

Color can be described in different ways than only physically by its wavelength characteristics. A system that describes color is called a color system. Color can be defined and modeled in different ways and each color system has its own set of color models (usually three). The three color features which are usually taken to describe the visual (intuitive) sensation of color are hue, saturation, and lightness. The *hue* corresponds with the dominant wavelength of the spectral energy distribution and is the color we see when viewing the light. For instance, if one sees a color with predominant high wavelengths then the color is interpreted as red. Thus, the hue is the kind of color, like red, green, blue, yellow, cyan, etc. *Saturation* corresponds to the excitation purity of the color and is defined as the proportion of pure light with respect to white light needed to produce the color. High saturation denotes little dilution. In other words, saturation is the richness of a hue and therefore denotes how pure a hue is. *Lightness* is related to the intensity (or energy) of the light reflected from objects. Further, *brightness* corresponds to the intensity of the light coming from light sources (i.e., self- luminous objects). The higher the emitted intensity, the brighter the color appears. The description of spectral information in terms of these three color models is derived as follows. Consider the energy emitted by a white light source as shown in Fig. 2.1(a). All wavelengths contribute more or less equally to the total energy. When the color of the light source has a dominant wavelength around the 506 nm part of the spectrum, as shown in Fig. 2.1(b), then the perceived color will be perceived as greenish. Let the power energy of the dominant wavelength be denoted by E_H and the wavelengths contributing to produce white light of intensity by E_W, see Fig. 2.1(b). Hence, the hue is green corresponding to the dominant wavelength E_H of 506 nm where the energy power is given by E_H. Further, saturation depends on the difference between E_H and E_W: the larger the difference the more pure the color is. The lightness is equal to the area under the curve of the spectral power distribution (i.e., total energy).

In conclusion, the relative spectral power (light) gives complete information about the color (i.e., color fingerprints). However, the human eye is incapable of analyzing color into its spectral components. An intuitive approximation of the spectral information of color by humans is in terms of hue, saturation, and lightness. Although, the approximation of the spectral curves by these three visual attributes suffice for a large number of problems, there are still a number of problems for which spectral information is essential and where human interpretation fails such as standards for color measurement, color matches under different illumination, and color mixture analysis.

The spectral power distribution of the light from objects (Fig. 2.1(b)) is the result of different complex factors such as the light source and material characteristics. In the next section, these complex factors will be discussed.

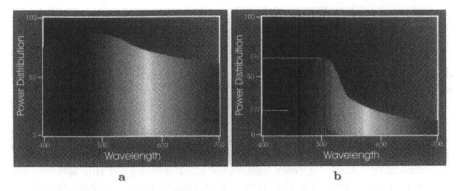

Fig. 2.1 a Relative spectral power distribution of a white light source. **b** Spectral power distribution of the light reflected from a green sample.

2.3 Color Appearance

Let's follow the light emitted by a light source. Through gas discharge (e.g., fluorescent lamps) or by heating up a material (the filament of the lamp), the light source generates light illuminating an object. Some part of the spectrum of the light is reflected from the object and is subsequently measured by an observer such as our light-sensitive eyes or by a color camera. The measured light is then sent to our brain (or computer) where the color of the light is observed (interpreted). The light–object–observer triplet depends on very complex factors: the observed color may vary with a change in the intensity and energy distribution of the light source, material characteristics, mode of viewing, the observer and so on.

Let's take a closer look at the triplet light–object–observer. As stated above, the traveling of light starts with the *light* source. It is known that the color of the light source has a large influence on the observed color. Examples are the spotlights installed in theaters or disco's. A substantial change in spotlight color will subsequently change the color of the dresses of the dancers. In contrast, people have a large degree of color constancy: the ability to observe the same color under different lightning conditions. For example, an object illuminated by daylight is perceived the same as when the object is illuminated by candlelight.

The second major component is the *object* or *sample* itself. The object color determines the observed color by reflecting (opaque or reflecting materials), absorbing (transparent materials), or absorbing and transmitting (translucent materials) parts of the incident light. Also the viewing mode influences the observed color. A change in viewing position of the camera will change the color and intensity field of the object color with respect to shading and highlighting cues. Further, material characteristics of the object, in terms of structure, gloss, and metallic effects, influence the observed color.

Thirdly, the light-sensitive detectors of the *observer*, such as the human eyes or color (CCD) cameras, determine which color will be observed. It often occurs that different human observers will judge the color of an object differently even when the object is seen under the same viewing conditions. Similarly, two recordings made from the same object with two different cameras (i.e., with different color filters) but under the same imaging conditions, may yield different pictures.

Hence, the observed color of an object depends on a complex set of imaging conditions. To get more insight in the imaging process, it is essential to understand the basics of color measurement. Therefore, in this section, we will subsequently focus on the three major components: the light source in Section 2.3.1, the object characteristics and viewing mode in Section 2.3.2, and the observer (eye and camera) in Section 2.3.3.

2.3.1 The Light Source

The main light source is the sun. Further, artificial light sources exist generating light by gas discharge (neon, argon or xenon) such as fluorescent lamps, or by heating up material (e.g., the filament of a lamp). Light produced by different light sources may vary with respect to their spectral power distribution (SPD), which is the amount of radiant power at each wavelength of the visible spectrum. Instruments exist for measuring the radiation power at each wavelength. To make the SPD independent of the amount of light, the spectral power distribution is computed with respect to the radiation of light at a single wavelength (usually 560 nm which is given the value 100). The relative spectral power distribution of a light source is denoted by $E(\lambda)$.

Color temperature is commonly used to express the relative SPD of light sources. Color temperatures correspond to the temperature of a heated black body radiator. The color of the black body radiator changes with temperature. For example, the radiator changes from black at 0 K (Kelvin), to red at about 1,000 K, white at 4,500 K to bluish white at about 6,500 K. An incandescent lamp would give a color temperature of 2,900 K and a fluorescent lamp about 4,100 K. Historically, color temperature has been introduced to match and describe (old fashioned) light sources such as candles and incandescent lamps, generating light in a similar way to the black body due to their thermal radiation. However, modern light sources such as fluorescent lamps often do not properly match a color temperature. These light sources have so-called correlated color temperatures instead of exact color temperatures.

The most important light source is the *sun*. The color temperature of the sun may vary during the time of the day (e.g., reddish at sunrise and bluish at noon). Further, the weather plays also an important role in the determination of the color of the sun. For example, on a clear day the color temperature is about 5,500 K. The color of the sun changes with the day and month of the year due to latitude and atmospheric conditions. Hence, the color of the sun may vary between color temperatures of 2,000 K to 10,000

K (indoor north sky daylight). For standardization of colorimetric measurements, in 1931, the International Lighting Commission (CIE), recommended that the average daylight has the color temperature of 6,500 K and this has been specified as the standard illuminant D65. Also other standard illuminants have been provided such as D50, D55, and D75. Note the difference between light sources and illuminants. In contrast to real light sources, illuminants may not be physically realizable but are given by numerical values determining the SPD. Before illuminant D65, standard illuminant C was used as the average daylight. In contrast to the D series, illuminant C can be reproduced by filtering a tungsten filament lamp. However, real daylight has more ultraviolet then illuminant C. This ultraviolet band (380–400 nm) has been specified in D65. In Fig. 2.2, the relative spectral power distribution are given for illuminant A, C, and D65. The spectral power distribution of illuminant A (black body with a color temperature of 2,856 K) is similar to that of a incandescent lamp.

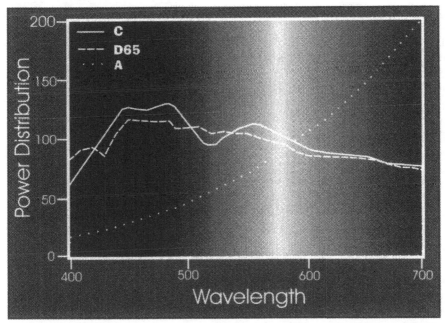

Fig. 2.2 Relative spectral power distribution of illuminant A, C, and D65.

Artificial light sources have been used in a wide variety of places for illuminating, for example, building interiors, and roads. The most important light sources are the fluorescent lamps based on gas discharge of rare gases such as neon and xenon, or metallic vapors such as mercury and sodium vapor. The filtered xenon arc lamp approximates the D65 illuminant (i.e., average daylight) the closest from all existing light sources. The xenon arc lamp has a high amount of continuum in contrast to other discharge lamps which have short peaks in their SPD corresponding to the characteristics of the excited molecules (gas dependent). Finally, standard illuminants have been recommended by the CIE denoted by F1–F12 corresponding to SPDs of different types of fluorescent lamps.

In conclusion, color temperatures are used to describe the relative spectral power distribution of light sources. The CIE recommended different SPDs as standards such as illuminant A (incandescent light), D65 and C (average daylight), and F1–F12 (fluorescent lamps). Manufacturers are in search of lamps having an SPD similar to that of average daylight D65. Filtered xenon light is the closest approximation but is very expensive. Fluorescent lamps are cheap and mostly used for the lighting in stores. However, their SPDs differ significantly from daylight. This causes the problem that the color of clothes may differ significantly if you wear them indoors or outdoors (under daylight).

2.3.2 The Object

In this chapter, colored materials are called objects or samples. Objects may consist of different materials such as plastics, glass, wood, textile, paper, coating, metals, brick, etc. Objects can be divided into three major groups: (1) transparent objects, where some part of the light is absorbed and some part goes unscattered through the sample (e.g., sunglasses, windows, bottles etc.), (2) translucent objects, which absorb, transmit, and scatter parts of the light (e.g., lamp panels and shades), and (3) opaque objects, where objects absorb and reflect a portion of the light and no light is transmitted (e.g., paper, textiles, and walls). Because most of the materials that surround us are opaque, we will focus on opaque materials.

Objects are characterized by the percentage of light that they reflect (or transmit) at each wavelength of the visible spectrum. The amount of reflected light from an object depends on the color of the material. The amount of light reflected from an object is computed with respect to that of a white standard. The ratio of the two reflected values is the reflectance factor determining the so-called relative spectral reflectance curve, a number from 0 to 1 (or 0 to 100 in percentage). Thus the relative spectral reflectance curve is not absolute but relative to a white reference and is denoted by $S(\lambda)$. Spectral reflectance curves can be obtained by measurement instruments such as spectrophotometers. The reflectance curve can be seen as the fingerprint of the object's color. In Fig. 2.3(a) the spectral curves are shown for green and orange paint samples. The green paint sample absorbs the blue and red parts of the visible spectrum. Further, the orange paint sample absorbs the green and blue part of the visible spectrum. The SPD of a reflecting object is calculated by adding the product of the SPD of the illuminant and the reflectance (transmittance) of the object at each wavelength of the visible spectrum:

$$P(\lambda) = E(\lambda)S(\lambda) \tag{2.1}$$

where $P(\lambda)$ is the spectral power in the light reflected by the object with reflectance $S(\lambda)$ under illuminant $E(\lambda)$.

Figure 2.3 illustrates this concept. The SPD reflected from a green sample with reflectance given in Fig. 2.3(a) when the light illuminant has the SPD of Fig. 2.3(b) is shown in Fig. 2.3(c).

In conclusion, the color of an object can be measured objectively (i.e., independent of the illuminant) and is given by the relative spectral reflectance curve. The color reflected from an object is the product of the SPD of the illuminant and the spectral reflectance of the object and is computed by $P(\lambda) = E(\lambda)S(\lambda)$. In the next section, the measured SPD of $P(\lambda)$ will be transformed into three numbers corresponding to the numbers we use to describe the color of the object.

2.3.3 The Observer

The third component of the triplet is the observer. The observer measures light coming directly from a light source $E(\lambda)$ or light which has been reflected (or transmitted) from objects in the scene $P(\lambda)$. The observer can be a color camera or the human eyes. For the human eye, the retina contains two different types of light-sensitive receptors, called *rods* and *cones*. Rods are more sensitive to light and are responsible for vision in twilight. There are theories stating that rods also play a small role in color vision. The cones are responsible for color vision and consist of three types of receptors sensitive to long (red), middle (green), and short (blue) wavelengths. Human eyes have been studied by many scientists. There are two theories which have been widely accepted and used in practice: the trichromacy and opponent theories. In fact, the two theories are completive. It has been recognized that both views have their value.

In this section, the trichromacy and opponent theories are outlined. Further, the spectral sensitivities of the different receptors (cones) of the eye are given.

Trichromacy Theory. The comprehension of color perception started with Newton in 1666, the year in which he dispersed (white) light with a glass prism. The result was the amazing discovery that white light is composed of a full range of colors. In 1801, the English scientist Thomas Young suggested the theory that just three primary additive colors were sufficient to produce all colors. The trichromacy theory has been expanded by the German scientist Hermann von Helmholtz and is usually called the Young–Helmholtz approach. They suggested that the human eye perceives color by the stimulation of three pigments in the cones of the retina. So the color perceived by the human eye has three degrees of freedom. By taking three arbitrary colors from the electromagnetic spectrum, the range of perceived colors which is

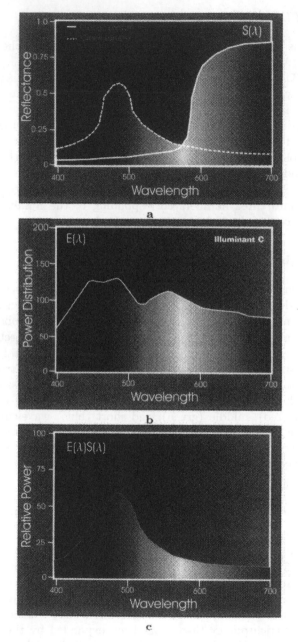

Fig. 2.3 a Spectral reflectance curves of a green and orange sample. **b** Relative spectral distribution of the power from illuminant C. **c** Spectral power distribution of the light reflected from a green sample under illuminant C.

obtained by the mixture of these three colors with corresponding intensities is called the *color gamut* and the colors themselves are called *primary colors*. The trichromacy theory was confirmed only in 1960, when three types of receptors were identified in the retina. The maximum responses of these receptors correspond to blue (near 440 nm), green (near 540 nm), and red (near 590 nm).

The trichromacy theory has found its way into different applications such as color cameras, camcorders, and video monitors. For example, a TV tube is composed of a large array of triangular dot patterns of electron-sensitive phosphor. Each dot in a triad is capable of producing light in one of the primary colors. When each dot in a triad is activated by electrons it emits light, and the amount of light depends upon the strength of the electron stream. The three primary colors from each phosphor triad are added together and the result will be a color because the phosphor dots are very small. The color of a triangular dot pattern depends upon the ratio of the primary color intensities.

Opponent Theory. The opponent color theory started in about 1500 when Leonardo da Vinci came to the conclusion that colors are produced by the mixture of yellow and green, blue and red, white and black. Arthur Shopenhauer noted the same opposition of red–green, yellow–blue, and white–black. This opponent color theory has been completed by Edwald Hering who concluded that the working of the eye is based on the three kinds of opposite colors. An example of opponent color theory is the so-called after-image. Looking for a while at a green sample will cause a red after-image (excluding yellow and blue). Focussing on the chromatic channels (i.e., red–green and blue–yellow), they are opponent in two different ways. Firstly, as mentioned above, no color seems to be a mixture of both members of any opponent pair. Secondly, each member of an opponent pair exhibits the other. In other words, by adding a balanced proportion of green and red, a hue will be produced which is neither greenish nor reddish. The opponent color theory was confirmed in 1950 when opponent color signals were detected in the optical connection between eye and brain.

Modern theories combine the trichromacy and opponent color theory; the process starts by light entering the eye, which is detected by trichromacy cones in the retina, and is further processed into three opponent signals on their way to the brain. More recent developments are in the Retinex theory proposed by Edwin Land [13]. Experiments show that people have a considerable amount of color constancy (i.e., colors are perceived the same even under different illumination). Unfortunately, all theories discussed so far are constrained and are incapable of explaining the effects of, for example, influences of intensities/colors adjacent to the sample, intensity and color distribution of the light source, mode of viewing and so on.

Color Response of the Eye. We have seen that the light from a light source $E(\lambda)$ as well as light reflected by an object $S(\lambda)$ can be determined with the aid of physical measurement instruments. However, the sensation of a human observer cannot be measured by an objective instrument. Therefore, the spectral sensitivities of the human eyes are measured only indirectly as follows.

Experiments have been conducted on human observers without any vision anomalies. The observers were asked to match a test light, consisting of only one wavelength, by adjusting the energy level of three separate primary lights. The three primary lights were additively mixed to match the test light of one wavelength. At each wavelength the amount of energy was recorded for the three primary colors yielding the so-called color matching functions. Different triplets of primaries can be used to get different color matching functions matching the visible spectrum. Color matching functions can be transformed to those which are obtained by other primaries. Historically, the above experiments were performed by W.D. Wright (7 observers) and J. Guild (10 observers) using different primary colors: 460, 530, and 650 nm, and 460, 543, and 630 nm, respectively. The viewing angle was 2°, corresponding to the viewing of a sample of about 0.4 inches at reading distance of 10 inches, in which the light illuminates only the fovea. Later experiments have been conducted with a viewing angle of 10°, corresponding to the viewing of a sample of about 1.9 inches at reading distance of 10 inches, by Stiles and Burch (about 50 observers). The tricolor functions of the human eye are shown in Fig. 2.4(a).

The CIE 2° Standard Observer. The complication of the color matching functions so far is that a negative amount of at least one of the primaries was necessary to produce the full set of spectral colors, see Fig. 2.4(a). A negative amount was accomplished by adding one of the primaries to the test spectral light. However, a negative amount is inconvenient for computational reasons. To this end, the CIE recommended mathematical transformations based on three primary standards X, Y, and Z. These primary standards are not real but imaginary. The reason to choose these imaginary primary standards are: (1) no negative values in the color matching functions, (2) the primaries X, Y, and Z enclose all possible colors, and (3) the Y primary color corresponds to intensity.

This resulted in the CIE color matching functions denoted by $\bar{x}$, $\bar{y}$, and $\bar{z}$ giving the amount of each of the primary colors required to match a color of one watt of radiant power for each wavelength, see Fig. 2.4(b). The curves correspond to the spectral sensitivity curves of the three receptors in the human eye. Note that $\bar{x}$ corresponds to the spectral luminance factor of the eye. Depending on the viewing angle, again two versions exist: the standard observer – CIE 1931, 2° (shortened to the 2° standard observer), and the standard observer – CIE 1931 10° (cf. 10° standard observer). The standard observer – CIE 1931 10° has been introduced to correspond better to the visual matching of larger samples.

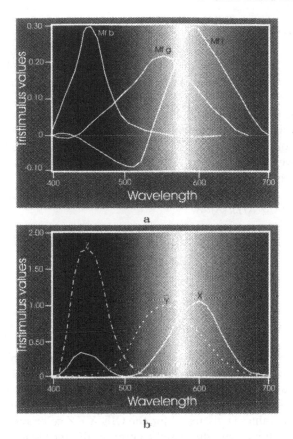

Fig. 2.4 a Color matching functions $\bar{r}$, $\bar{g}$ and $\bar{b}$ of the human observer. **b** Color matching functions $\bar{x}$, $\bar{y}$ and $\bar{z}$ from the standard observer 2° (CIE 1931).

Summary. In conclusion, the spectral sensitivities of the different cone types of the human retina have been determined indirectly by testing on the additive mixture of three primary colors. The CIE recommends two sets of standard spectral curves (also called color matching functions) depending on the angle an observer views a sample: 2° standard observer (corresponding to the viewing of a sample of about 0.4 inches at reading distance of 10 inches) and 10° standard observer (corresponding to the viewing of a sample of about 1.9 inches at reading distance of 10 inches). In the next section, the matching functions are used to express a color spectrum as three numbers.

2.4 Colorimetry

Standardization of the light–object–observer triplet is required to measure objectively the imaging process. As discussed in Section 2.3.1, different light sources exist with different spectral power distributions $E(\lambda)$. To get comparable results, the CIE recommended various illuminants. Further, in Section 2.3.2, the light reflected by objects $S(\lambda)$ can be measured. Then, the reflected color of the object can be determined by $P(\lambda) = E(\lambda)S(\lambda)$. Further, the observer perceives color in terms of three color signals based on the trichromacy theory, which can be modeled by:

$$R = \int_\lambda E(\lambda)S(\lambda)f_R(\lambda)d\lambda, \tag{2.2}$$

$$G = \int_\lambda E(\lambda)S(\lambda)f_G(\lambda)d\lambda, \tag{2.3}$$

$$B = \int_\lambda E(\lambda)S(\lambda)f_B(\lambda)d\lambda \tag{2.4}$$

where the tristimulus values are obtained by adding the product of the SPD of the light source $E(\lambda)$, the reflectance (or transmittance) factor of the object $S(\lambda)$ and the color matching functions $f_C(\lambda)$ for $C \in \{R, G, B\}$ of the observer (eye or camera) at each wavelength of the visible spectrum. Depending on the color matching functions ($\bar{r}$, $\bar{g}$, and $\bar{b}$ or $\bar{x}$, $\bar{y}$, and $\bar{z}$), the sum of each equation is equal to the amount of primaries to match the sample.

This section is outlined as follows. Firstly, in Section 2.4.1 the calculation of the tristimulus values is discussed for the CIE XYZ color system. Further, chromaticity coordinates are discussed to graphically represent color. In Section 2.4.2, calculation of the tristimulus values is derived from color matching functions from a RGB system (e.g., color camera). Different color systems are then discussed and defined in terms of R, G, and B coordinates. In Section 2.4.5, color order systems are presented, which compute color differences approximating human perception.

2.4.1 XYZ System

Having three color matching functions of the standard observer, we are now able to compute three numbers (called the tristimulus values) equivalent to what a standard observer perceives:

$$X = \int_\lambda E(\lambda)S(\lambda)\bar{x}(\lambda)d\lambda, \tag{2.5}$$

$$Y = \int_\lambda E(\lambda)S(\lambda)\bar{y}(\lambda)d\lambda, \tag{2.6}$$

$$Z = \int_\lambda E(\lambda)S(\lambda)\bar{z}(\lambda)d\lambda \tag{2.7}$$

where $\overline{x}(\lambda)$, $\overline{y}(\lambda)$, and $\overline{z}(\lambda)$ are the CIE color matching functions of the 2° CIE standard observer, see Fig. 2.4(b).

In this way, we are able to compute the color tristimuli values equivalent to a CIE 1931 2° standard observer viewing an object composed of green paint illuminated by light source D65. Consider the reflectance graph of a green sample. Further, assume that the light source is illuminant D65, then the X, Y, and Z tristimulus values is computed by adding up the product of light, object, and matching functions at each wavelength.

The tristimulus values do not have a rather comprehensive meaning for human beings. For example, if we have two colors $X_1 = 34$, $Y_1 = 45$, and $Z_1 = 102$, and $X_2 = 64$, $Y_2 = 90$, and $Z_2 = 202$ respectively, then it can be deduced that the first color is twice as intense as the second color. However, it is difficult to describe the chromaticity of the color. Therefore, the notion of the chromaticity of the color is accomplished by defining the so-called chromaticity coordinates:

$$x = \frac{X}{X + Y + Z},\tag{2.8}$$

$$y = \frac{Y}{X + Y + Z},\tag{2.9}$$

$$z = \frac{Z}{X + Y + Z}.\tag{2.10}$$

It is obvious that the intensity information is factored out of the system, because the chromaticity coordinates reflect only the ratio of the three standard primary colors.

Since the sum of the chromaticity coordinates equals unity, two of these three quantities are sufficient to describe a color. When the x and y values are represented in a plane, the CIE 1931 *chromaticity diagram* is obtained, see Fig. 2.5. Spectral colors lie on the tongue-shaped curve according to their wavelengths. In the middle of the diagram, we find neutral white at $x = 0.3333$ and $y = 0.3333$ with corresponding color temperature of $6,000$ K. For illuminant D65, we obtain $x = 0.3127$ and $y = 0.3290$. Illuminant C has chromaticity coordinates $x = 0.3101$ and $y = 0.3163$, see Fig. 2.5(a). A third dimension can be imagined related to the brightness which is perpendicular to the chromaticity diagram. Instead of using X, Y, and Z, a color is more easily understood when it is specified in terms of x, y, and Y (i.e., intensity).

Hue and saturation are defined in the chromaticity diagram as follows, see Fig. 2.5(b). First a reference white-light point should be defined. This point is defined as a point in the chromaticity diagram that approximately represents the average daylight, for example illuminant C or D65. For a color $G1$, the hue is defined as the wavelength on the spectral curve that intersects the line from reference white through $G1$ to the spectral curve, which is $G2$ at 523 nm. When $\|G1\|$ is the distance from $G1$ to the white point, then the saturation is the relative distance to the white point, given by $\frac{\|G1\|}{\|G2\|}$, where

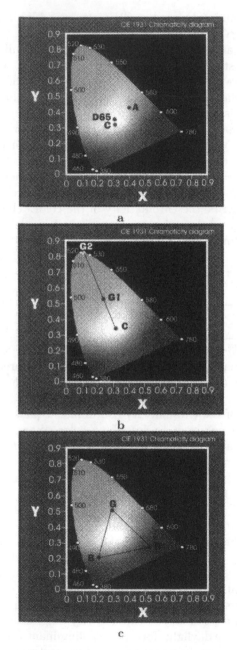

Fig. 2.5 a Illuminant A, C, and D65 plotted in the xy-plane. **b** Definition of hue and saturation under illuminants A and D65. **c** Color gamuts.

$||G2||$ is the distance from $G2$ to the white point. Hence the most saturated colors are the spectral colors.

Color gamuts are represented in the chromaticity diagram by lines joining the points defining the color gamut, see Fig. 2.5(c). All colors along the line joining R and G can be obtained by mixing the amounts, corresponding to the distances from R and G, of the colors represented by the endpoints. The color gamut for three points R, G, and B is the triangle formed by the three vertices of the three colors. Any color within the triangle can be produced by a weighted sum of the three colors of the triangle, but no colors represented by points outside the triangle could be produced by these colors. It is now obvious why no set of three primary colors can produce all colors, since no triangle within the diagram can encompass all colors. It also explains why there are many different standards possible for the primary colors, because any triangle in the diagram, defined by its vertices, could be taken as a standard.

In conclusion, we are now able to assign precise numerical values to the color sensation of a standard observer in terms of intuitive attributes hue, saturation, and brightness in a totally objective manner. In fact, the XYZ system introduced by CIE is the scientific basis of objective color measurement. The XYZ system allows us to compute tristimulus values describing the sensation of a human being, given by matching functions $\overline{x}(\lambda)$, $\overline{y}(\lambda)$ and $\overline{z}(\lambda)$, viewing an object $S(\lambda)$, illuminated by a light source $E(\lambda)$. Although any set of primary colors can be taken, in the next section, the RGB primaries are given for a color camera. The purpose of the RGB system is to derive mathematical formulae to define color systems in terms of RGB coordinates coming directly from a color camera.

2.4.2 RGB System

As discussed above, a linear function of the tristimulus values converts a set of color primaries into another set. The standard RGB established by the CIE in 1931 with three monochromatic primaries at wavelengths 700 nm (R)ed, 546.1 nm (G)reen, and 435.8 nm (B)lue is the RGB Spectral Primary Color Coordinate System corresponding to the 2° standard observer. The color matching functions have already been shown in Fig. 2.4(b). Another set of primaries have been recommended by the National Television Systems Committee (NTSC) for phosphor standardization. As the NTSC primaries are not pure monochromatic sources of radiation, the color gamut produced by the NTSC primaries is smaller than available from the spectral primaries. Digital cameras and videos mostly use RGB NTSC.

To represent the RGB color space, a cube can be defined on the R, G, and B axes, see Fig. 2.6(a). White is produced when all three primary colors are at M, where M is the maximum light intensity, say $M = 255$. The axis connecting the black and white corners defines the intensity:

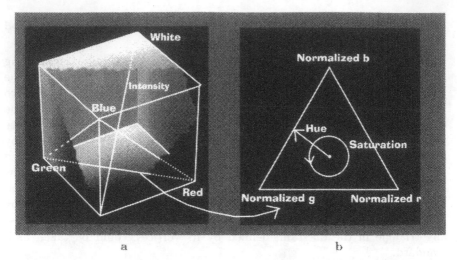

a   b

Fig. 2.6 a RGB color space **b** Definition of hue and saturation in the chromaticity plane. The circle has equal saturation and the line has equal hue.

$$I(R, G, B) = R + G + B. \tag{2.11}$$

All points in a plane perpendicular to the gray axis of the color cube have the same intensity. The plane through the color cube at points $R = G = B = M$ is one such plane. The projection of RGB points on the rgb chromaticity triangle is defined by:

$$r(R, G, B) = \frac{R}{R + G + B}, \tag{2.12}$$

$$g(R, G, B) = \frac{G}{R + G + B}, \tag{2.13}$$

$$b(R, G, B) = \frac{B}{R + G + B}, \tag{2.14}$$

yielding the rgb color space which is normalized with respect to intensity and graphically represented by Fig. 2.6(a) and 2.6(b) where the intensity axis I in RGB space is projected onto $r = g = b = 1/3$ in the chromaticity plane.

2.4.3 HSI System

Hue and saturation are defined in the chromaticity triangle in the standard way as follows. Similar to Section 2.4.1 for the xy diagram, the reference white point is first defined in the rg chromaticity diagram. Let's assume a white light source and hence the reference point is represented by $r = g = b = 1/3$. Then saturation S_{rgb} is defined as the radial distance of a point from the reference white-light point mathematically specified as:

$$S_{rgb}(r,g,b) = \sqrt{(r-1/3)^2 + (g-1/3)^2 + (b-1/3)^2} \qquad (2.15)$$

and graphically represented by Fig. 2.6(b).

H_{rgb} is a function of the angle between a reference line (e.g., horizontal axis) and the color point:

$$H_{rgb}(r,g,b) = \arctan\left(\frac{r-1/3}{g-1/3}\right). \qquad (2.16)$$

See Fig. 2.6(b) as well.

The transformation from RGB used here to compute H in terms of R, G, and B is given by [14]:

$$H(R,G,B) = \arctan\left(\frac{\sqrt{3}(G-B)}{(R-G)+(R-B)}\right), \qquad (2.17)$$

and S measuring the relative white content of a color as having a particular hue by:

$$S(R,G,B) = 1 - \frac{\min(R,G,B)}{R+G+B}. \qquad (2.18)$$

Note that rgb, H, and S are undefined for achromatic colors (i.e., $R = G = B = 0$).

In conclusion, hue, saturation, and intensity are calculated from the original R, G, B values from the corresponding red, green, and blue images provided by the color camera.

2.4.4 *YIQ* and *YUV* System

Various cameras provide images in terms of the YIQ NTSC transmission color coordinate system. The NTSC developed the three color attributes Y, I, and Q for transmission efficiency. The tristimulus value Y corresponds to the luminance of a color. I and Q correspond closely to the hue and saturation of a color. By reducing the spatial bandwidth of I and Q without noticeable image degradation, efficient color transmission is obtained. For the PAL and SECAM standards used in Europe, the Y, U, and V tristimulus values are used. The I and Q color attributes are related to U and V by a simple rotation of the color coordinates in the color space. The conversion matrix to compute the YIQ values from the original RGB NTSC tristimulus values is given by:
$Y = 0.299 * R + 0.587 * G + 0.114 * B$, $I = 0.596 * R - 0.274 * G - 0.312 * B$, and $Q = 0.211 * R - 0.523 * G + 0.312 * B$.

2.4.5 Color Order Systems

A color system is visually uniform when numerical distances can be related to human perceptual differences: the closer a color is to another color in the color space, the more similar they are. It is known that XYZ and RGB are not visually uniform. To achieve visual uniformity, color order systems have been proposed. A color order system is a multi-dimensional (usually three) space arranging the gamut of colors of visual sensation. This geometric ordering relates a notation to each color corresponding to its position. Each color is given a number yielding an objective classification criterion. Hence, the goal of a color order system is to represent and classify colors objectively, and to provide color differences. Color order systems are often accompanied by atlases and catalogs containing real samples from specific position within the space. Color systems are usually based on perceptual uniformity where the Euclidean distance between two position in the space closely corresponds to the difference between the two colors as perceived by a human being. In this section, Munsell's color system, the CIELAB colorimetric space, and the color difference will be discussed.

Munsell's Color Order System. Albert H. Munsell proposed in 1905 [15], in his "A Color Notation," a geometric ordering space to classify colors. The publication was accompanied with an atlas containing charts, derived from psychological and experimental measurements, designed to produce a uniform color space. The Munsell color space is based on three perceptual attributes. First, he developed the notation of *value* ranging from white to black with perceptual equal steps. Further, a *hue* circle has been designed corresponding to a certain *chroma* value. The hue circle contains the property that steps between two different hues closely correspond to their perceptual differences. To graphically illustrate the Munsell color space, the color order system can be represented by a cylinder from which the center axis relates to value. Further, chroma corresponds to the distance from this central axis of constant hue forming circles around the value axis. In this way, Munsell provided a visually balanced ordering of colors to yield an objective color identification scheme. The Munsell system has been updated by modern instruments. Today, it is possible to compute the Munsell notations from the CIE XYZ tristimulus values.

CIELAB Color Order Systems. As stated above, a color system is visual uniform when numerical distances can be related to human perceptual differences. Hence, the closer a color is to another color in the color space, the more similar they are. MacAdam proved the visual non-uniformity of the XYZ color system by doing experiments. These experiments are based on the *just noticeable difference* (JND) concept. The JND is defined as the minimal visual difference of two colors: let Q_0 be a color, then color Q_1 is just noticeably different from Q_0, if there is no color Q_2, lying on the line from Q_0 through Q_1, which is closer to and noticeably different from Q_0. The results of the experiments made by MacAdam is that colors with coordinates (x_i, y_i) in the xy chromaticity diagram, which are just noticeably different from color center (x_0, y_0), are lying on an ellipsis with center (x_0, y_0) and all colors which are not (just) noticeably different from (x_0, y_0) are lying outside the ellipsis. The ellipses of an arbitrary number of colors were calculated and analyzed. MacAdam proved that the differences between the axis diameters of the ellipses were rather (too) large and therefore non-uniformity of the XYZ color system can easily be seen.

Over the last decades various attempts have been made to develop perceptual uniform spaces and color difference equations. To this end, in 1976, the CIE recommended a new system $L^*a^*b^*$, computed from the XYZ color system, having the property that the closer a point (representing a color) is to another point (also representing a color –possibly the same), the more visually similar the colors (represented by the points) are. In other words, the magnitude of the perceived color difference of two colors corresponds to the Euclidean distance between the two colors in the color system.

The $L^*a^*b^*$ system is based on the three-dimensional coordinate system based on the opponent theory using black–white L^*, red–green a^*, and yellow–blue b^* components. The L^* axis corresponds to the lightness where $L^* = 100$ is white and $L^* = 0$ is black. Further, a^* ranges from red $+a^*$ to green $-a^*$ while b^* ranges from yellow $+b^*$ to blue $-b^*$.

The L^*, a^*, and b^* coordinates are computed from the X, Y, and Z tristimulus values as follows:

$$L^* = 116f\left(\frac{Y}{Y_n}\right), \tag{2.19}$$

$$a^* = 500f\left(\frac{X}{X_n} - \frac{Y}{Y_n}\right), \tag{2.20}$$

$$b^* = 500f\left(\frac{Y}{Y_n} - \frac{Z}{Z_n}\right), \tag{2.21}$$

where X_n, Y_n, and Z_n are the tristimulus values of the nominal white stimulus. For example, for D65/10°: $X_n = 94.81$, $Y_n = 100.00$, and $Z_n = 107.304$. Further, $f(\frac{X}{X_n}) = (\frac{X}{X_n})^{1/3}$ for values $\frac{X}{X_n}$ greater than 0.008856, and $f(\frac{X}{X_n}) = 7.787(\frac{X}{X_n}) + 16/116$ for values $\frac{X}{X_n}$ less or equal to 0.008856. Equal arguments hold for Y and Z.

The intuitive visual attributes hue (h) and chroma (saturation) (C^*) can be expressed by:

$$C^* = (a^{*2} + b^{*2})^{1/2}, \tag{2.22}$$

$$h = \arctan\left(\frac{b^*}{a^*}\right). \tag{2.23}$$

Note that the CIE $L^*a^*b^*$ chroma is not the same as the Munsell chroma. Further, the CIE $L^*a^*b^*$ is still not perfectly visually uniform. However, the system allows us to define a color difference by measuring the geometric distance between two colors in the color space. The color difference ruler is discussed in the following section.

Color Difference. Many image retrieval applications must provide the ability to retrieve visually similar images satisfying the expectation of the operator. Therefore, images should be retrieved within the limits of acceptable variations of color. For example, consider a database of images taken from paper samples. The operator searches for samples with a specific color. Then, color differences are essential for the retrieval task at hand. Note the difference between perceptibility and acceptability. The color difference is perceptible if the difference can be seen by a human operator. Even when the difference is perceptible for different tasks, the color difference can still be acceptable. As discussed above, the $L^*a^*b^*$ is approximately visually uniform and provides us the ability to define a perceptual color difference in the Euclidean way.

First, the differences are subtracted from the L^*, a^*, and b^* components of the trial from the standard:

$$\Delta L^* = L^*_{trial} - L^*_{standard}, \tag{2.24}$$

$$\Delta a^* = a^*_{trial} - a^*_{standard}, \tag{2.25}$$

$$\Delta b^* = b^*_{trial} - b^*_{standard}, \tag{2.26}$$

where a positive value of ΔL^* implies that the sample is lighter than the standard one. Further, a positive value of Δa^* and Δb^* indicates that the sample is more red and more yellow than the standard, respectively.

The total CIE $L^*a^*b^*$ is given by:

$$\Delta E^*_{ab} = \sqrt{(\Delta L^*)^2 + (\Delta a^*)^2 + (\Delta b^*)^2}. \tag{2.27}$$

The difference in saturation is given by:

$$\Delta C^* = C^*_{trial} - C^*_{standard}. \tag{2.28}$$

Due to the circular nature of hue, the hue difference is given by:

$$\Delta H^*_{ab} = \sqrt{(\Delta E^*_{ab})^2 + (\Delta C^*)^2 + (\Delta L^*)^2}. \tag{2.29}$$

Summary. Color order systems are required to identify colors into a geometric structure. The Munsell space was one of the first color order systems with the aim of representing color objectively. Due to the perceptual nonuniformity of the CIE XYZ system, the CIE recommended the $L^*a^*b^*$ color order system where the Euclidean distance between two colors closely corresponds to the visual color difference.

2.5 Color Invariance

In the previous section, the scientific measurement of color has been outlined to measure objectively the light–object–observer process yielding basic notions on color, color models, and ordering systems. The advantage of using, for example, the $L^*a^*b^*$ system, is that it corresponds to human perception, which is useful when retrieving images which are visually similar. However, it is known that the $L^*a^*b^*$ and RGB color system are dependent on the imaging conditions. Therefore, for general image retrieval, an important color property is color invariance. A color invariant system contains color invariant models which are more or less insensitive to the varying imaging conditions such as variations in illumination (e.g., lamps having different spectral power distribution) and object pose (changing shadows, shading, and highlighting cues). Color invariants for the purpose of viewpoint invariant image retrieval are given by Gevers and Smeulders [9]. In this section, an overview is given of color invariants.

This section is outlined as follows. First, in Section 2.5.1, a general reflection model is discussed to model the interaction between light and material. Then, in Section 2.5.2, assuming white illumination, basic color systems are analyzed with respect to their invariance. In Section 2.5.3, color constant color systems are given. Color constant color systems are independent of the relative spectral power distribution of the light source.

2.5.1 Reflection from Inhomogeneous Dielectric Materials

To arrive at a uniform description it is common to divide opaque materials into two classes on the basis of their optical properties: optically homogeneous and inhomogeneous materials.

Optically homogeneous materials have a constant index of refraction throughout the material. This means that when light is incident upon the surface of an homogeneous materials, some fraction of it is reflected. A smooth surface reflects light only in the direction such that the angle of incidence equals the angle of reflection. The properties of this reflected light are determined by the optical and geometric properties of the surface. Metals are homogeneous materials having a larger specular component than other materials.

Optically inhomogeneous materials are composed of a vehicle with many embedded colorant particles that differ optically from the vehicle. The fraction of the incident light that is not reflected from the surface enters the body of the material. The body is composed of a vehicle and many colorant particles. When light encounters a colorant particle, some portion of it is reflected. After many reflections, the light is diffused, and a significant fraction can exit back through the surface in a wide range of directions. Some examples of inhomogeneous materials are plastics, paper, textiles, and paints.

The dichromatic reflection model describes the light which is reflected from a point on a dielectric, inhomogeneous material as a mixture of the light reflected at the material surface and the light reflected from the material body.

Let $E(x, \lambda)$ be the spectral power distribution of the incident (ambient) light at the object surface at x, and let $L(x, \lambda)$ be the spectral reflectance function of the object at x. The spectral sensitivity of the kth sensor is given by $F_k(\lambda)$. Then ρ_k, the sensor response of the kth channel, is given by:

$$\rho_k(x) = \int_\lambda E(x, \lambda) L(x, \lambda) F_k(\lambda) d\lambda \tag{2.30}$$

where λ denotes the wavelength, and $L(x, \lambda)$ is a complex function based on the geometric and spectral properties of the object surface. The integral is taken from the visible spectrum (e.g., 380–700 nm).

Further, consider an opaque inhomogeneous dielectric object, then the geometric and surface reflection component of function $L(x, \lambda)$ can be decomposed in a body and surface reflection component as described by Shafer [22]:

$$\phi_k(x) = G_B(x, n, s) \int_\lambda E(x, \lambda) B(x, \lambda) F_k(\lambda) d\lambda$$
$$+ G_S(x, n, s, v) \int_\lambda E(x, \lambda) S(x, \lambda) F_k(\lambda) d\lambda \tag{2.31}$$

giving the kth sensor response. Further, $B(x, \lambda)$ and $S(x, \lambda)$ are the surface albedo and Fresnel reflectance at x, respectively. n is the surface patch normal, s is the direction of the illumination source, and v is the direction of the viewer. Geometric terms G_B and G_S denote the geometric dependencies on the body and surface reflection component, respectively.

2.5.2 Reflectance with White Illumination

Considering the neutral interface reflection (NIR) model (assuming that $S(x, \lambda)$ has a nearly constant value independent of the wavelength) and approximately white illumination, then $S(x, \lambda) = S(x)$, and $E(x, \lambda) = E(x)$. Then, the measured sensor values are given by [8]:

$$\omega_k(x) = G_B(x, n, s) E(x) \int_\lambda B(x, \lambda) F_k(\lambda) d\lambda$$
$$+ G_S(x, n, s, v) E(x) S(x) \int_\lambda F_k(\lambda) d\lambda \tag{2.32}$$

giving the kth sensor response of an infinitesimal surface patch under the assumption of a white light source.

If the integrated white condition holds (i.e., the area under the sensor spectral functions is approximately the same):

$$\int_\lambda F_i(\lambda) d\lambda = \int_\lambda F_j(\lambda) d\lambda, \tag{2.33}$$

the reflection from inhomogeneous dielectric materials under white illumination is given by [8]:

$$\omega_k(x) = G_B(x, n, s)E(x) \int_\lambda B(x, \lambda)F_k(\lambda)d\lambda$$
$$+ G_S(x, n, s, v)E(x)S(x)F. \tag{2.34}$$

If $\omega_k(x)$ is not dependent on x, we obtain:

$$\omega_k = G_B(n, s)E \int_\lambda B(\lambda)F_k(\lambda)d\lambda + G_S(n, s, v)ESF. \tag{2.35}$$

Invariance for Matte Surfaces. Consider the body reflection term of Eq. 2.34:

$$\beta_k(x) = G_B(x, n, s)E(x) \int_\lambda B(x, \lambda)F_k(\lambda)d\lambda \tag{2.36}$$

giving the kth sensor response of an infinitesimal *matte* surface patch under the assumption of a white light source.

The body reflection component describes the way light interacts with a dull surface. The light spectrum E falls on a surface B. The geometric and photometric properties of the body reflection depends on many factors. If we assume a random distribution of the pigments, the light exits in random directions from the body. In this case, the distribution of exiting light can be described by Lambert's law. Lambertian reflection models dull, matte surfaces which appear equally bright regardless from the angle they are viewed. They reflect light with equal intensity in all directions.

As a consequence, a uniformly colored surface which is curved (i.e., varying surface orientation) gives rise to a broad variance of RGB values. The same argument holds for intensity I.

In contrast, rgb is insensitive to surface orientation, illumination direction, and illumination intensity. This is mathematically specified by substituting Eq. 2.36 in Eqs 2.12 - 2.14:

$$r(R_b, G_b, B_b) = \frac{G_B(x, n, s)E(x)k_R}{G_B(x, n, s)E(x)(k_R + k_G + k_B)} = \frac{k_R}{k_R + k_G + k_B}, \tag{2.37}$$

$$g(R_b, G_b, B_b) = \frac{G_B(x, n, s)E(x)k_G}{G_B(x, n, s)E(x)(k_R + k_G + k_B)} = \frac{k_G}{k_R + k_G + k_B}, \tag{2.38}$$

$$r(R_b, G_b, B_b) = \frac{G_B(x, n, s)E(x)k_B}{G_B(x, n, s)E(x)(k_R + k_G + k_B)} = \frac{k_B}{k_R + k_G + k_B}, \tag{2.39}$$

thereby, factoring out dependencies on illumination and object geometry. Hence rgb is only dependent on the sensors and the surface albedo.

Because S corresponds to the radial distance from the color to the main diagonal in the RGB-color space, S is an invariant for matte, dull surfaces illuminated by white light (cf. Eq. 2.36 and Eq. 2.18):

$$S(R_b, G_b, B_b) =$$

$$1 - \frac{\min(G_B(x, n, s)E(x)k_R, G_B(x, n, s)E(x)k_G, G_B(x, n, s)E(x)k_B)}{G_B(x, n, s)E(x)(k_R + k_G + k_B)}$$

$$= 1 - \frac{\min(k_R, k_G, k_B)}{k_R + k_G + k_B} \tag{2.40}$$

and hence is only dependent on the sensors and the surface albedo.

Similarly, H is an invariant for matte, dull surfaces illuminated by white light (cf. Eq. 2.36 and Eq. 2.17):

$$H(R_b, G_b, B_b) = \arctan\left(\frac{\sqrt{3}G_B(x, n, s)E(x)(k_G - k_B)}{G_B(x, n, s)E(x)((k_R - k_G) + (k_R - k_B))}\right)$$

$$= \arctan\left(\frac{\sqrt{3}(k_G - k_B)}{(k_R - k_G) + (k_R - k_B)}\right). \tag{2.41}$$

Obviously, in practice, the assumption of objects composed of matte, dull surfaces is not always realistic. To that end, the effect of surface reflection (highlights) is discussed in the following section.

Invariance for Shiny Surfaces. Consider the surface reflection term of Eq. 2.34:

$$C_s = G_S(x, n, s, v)E(x)S(x)F \tag{2.42}$$

where $C_s \in \{R_s, G_s, B_s\}$ and R_s, G_s, and B_s are the red, green, and blue sensor response for a highlighted infinitesimal surface patch with white illumination, respectively.

When light hits the surface of a dielectric inhomogeneous material, it must first pass through the interface between the surrounding medium (e.g., air) and the material. Since the refraction index of the material is generally different from that of the surrounding medium, some percentage of the incident light is reflected at the surface of the material.

Several models have been developed in the physics and computer graphics communities to describe the geometric properties of the light reflected from rough surfaces. Issues in modeling this process involve the roughness scale of the surface, compared to the wavelengths of the incident light, and self-shadowing effects on the surface which depend on the viewing direction and the direction of the incident light. The commonly used surface reflection model is described by a function with a sharp peak around the angle of perfect mirror reflection. For smooth surfaces, surface (specular) reflections are in the direction of r, which is s mirrored about n. For imperfect rough reflectors, the maximum surface reflection occurs when α, the angle between r and e is zero; and falls off rapidly as α increases. The falloff is approximated by $\cos^n \alpha$, where n is the surface-reflection exponent or degree of shininess.

Because H is a function of the angle between the main diagonal and the color point in RGB-sensor space, all possible colors of the same (shiny) surface region (i.e., with fixed albedo) have to be of the same hue as follows from substituting Eq. 2.42 in Eq. 2.17:

$$H(R_w, G_w, G_w) = \arctan\left(\frac{\sqrt{3}(G_w - B_w)}{(R_w - G_w) + (R_w - B_w)}\right)$$

$$= \arctan\left(\frac{\sqrt{3}G_B(x, n, s)E(x)(k_G - k_B)}{G_B(x, n, s)E(x)((k_R - k_G) + (k_R - k_B))}\right)$$

$$= \arctan\left(\frac{\sqrt{3}(k_G - k_B)}{(k_R - k_G) + (k_R - k_B)}\right) \tag{2.43}$$

factoring out dependencies on illumination, object geometry, viewpoint, and specular reflection coefficient and hence are only dependent on the sensors and the surface albedo. Note that $C_w = G_B(x, n, s)E(x)\int_\lambda B(x, \lambda)F_C(\lambda)d\lambda + G_S(x, n, s, v)E(x)S(x)F$ for $C \in \{R, G, B\}$.

Obviously other color features depend on the contribution of the surface reflection component and hence are sensitive to highlights.

2.5.3 Color Constancy

Existing color constancy methods require specific a priori information about the observed scene (e.g., the placement of calibration patches of known spectral reflectance in the scene) which will not be feasible in practical situations (e.g. [4, 6, 13]). To circumvent these problems, simple and effective illumination-independent color ratios have been proposed by Funt and Finlayson [7] and Nayar and Bolle [16]. In fact, these color models are based on the ratio of surface albedos rather then the recovering of the actual surface albedo itself. However, these color models assume that the variation in spectral power distribution of the illumination can be modeled by the coefficient rule or von Kries model, where the change in the illumination color is approximated by a 3×3 diagonal matrix among the sensor bands and is equal to the multiplication of each RGB-color band by an independent scalar factor. The diagonal model of illumination change holds exactly in the case of narrow-band sensors. Consider the the body reflection term of the dichromatic reflection model:

$$C_c = G_B(x, n, s)E(x)\int_\lambda B(x, \lambda)F_k(\lambda)d\lambda \tag{2.44}$$

where $C_c = \{R_c, G_c, B_c\}$ and R_c, G_c, and B_c are the red, green, and blue sensor response of a matte infinitesimal surface patch of an inhomogeneous dielectric object under unknown spectral power distribution of the illumination, respectively.

Suppose that the sensor sensitivities of the color camera are narrow-band with spectral response be approximated by delta functions $f_K(\lambda) = \delta(\lambda - \lambda_K)$, then the measured sensor values are:

$$C_K = G_B(\boldsymbol{x}, \boldsymbol{n}, \boldsymbol{s})E(\boldsymbol{x}, \lambda_K)B(\boldsymbol{x}, \lambda_K). \tag{2.45}$$

The color ratios are proposed by Nayar and Bolle [16]:

$$N(C^{\boldsymbol{x}_1}, C^{\boldsymbol{x}_2}) = \frac{C^{\boldsymbol{x}_1} - C^{\boldsymbol{x}_2}}{C^{\boldsymbol{x}_2} + C^{\boldsymbol{x}_1}} \tag{2.46}$$

and by Funt and Finlayson [7]:

$$F(C^{\boldsymbol{x}_1}, C^{\boldsymbol{x}_2}) = \frac{C^{\boldsymbol{x}_1}}{C^{\boldsymbol{x}_2}} \tag{2.47}$$

where $C \in \{R, G, B\}$ and $\boldsymbol{x}_1$ and $\boldsymbol{x}_2$ denote the image locations of the two neighboring pixels. Note that the set $\{R, G, B\}$ must be colors from narrow-band sensor filters and that they are used in defining the color ratio because they are immediately available from a color camera, but any other set of narrow-band colors derived from the visible spectrum will do as well. Although standard video cameras are not equipped with narrow-band filters, spectral sharpening could be applied [2] to achieve this to a large extent.

Assuming that the color of the illumination is locally constant (i.e., $E(\boldsymbol{x}_1, \lambda_K) = E(\boldsymbol{x}_2, \lambda_K)$) and that neighboring points have the same surface orientation (i.e., $G_B(\boldsymbol{x}_1, \boldsymbol{n}, \boldsymbol{s}) = G_B(\boldsymbol{x}_2, \boldsymbol{n}, \boldsymbol{s})$), then the color ratio N is independent of the illumination intensity and color as shown by substituting Eq. 2.46 in Eq. 2.44:

$$\frac{G_B(\boldsymbol{x}_1, \boldsymbol{n}, \boldsymbol{s})E(\boldsymbol{x}_1, \lambda_K)B(\boldsymbol{x}_1, \lambda_K) - G_B(\boldsymbol{x}_2, \boldsymbol{n}, \boldsymbol{s})E(\boldsymbol{x}_2, \lambda_K)B(\boldsymbol{x}_2, \lambda_K)}{G_B(\boldsymbol{x}_1, \boldsymbol{n}, \boldsymbol{s})E(\boldsymbol{x}_1, \lambda_K)B(\boldsymbol{x}_1, \lambda_K) + G_B(\boldsymbol{x}_2, \boldsymbol{n}, \boldsymbol{s})E(\boldsymbol{x}_2, \lambda_K)B(\boldsymbol{x}_2, \lambda_K)}$$
$$= \frac{B(\boldsymbol{x}_1, \lambda_K) - B(\boldsymbol{x}_2, \lambda_K)}{B(\boldsymbol{x}_1, \lambda_K) + B(\boldsymbol{x}_2, \lambda_K)}. \tag{2.48}$$

Equal arguments hold for the color ratio F by substituting Eq. 2.47 in Eq. 2.44:

$$\frac{G_B(\boldsymbol{x}_1, \boldsymbol{n}, \boldsymbol{s})E(\boldsymbol{x}_1, \lambda_K)B(\boldsymbol{x}_1, \lambda_K)}{G_B(\boldsymbol{x}_2, \boldsymbol{n}, \boldsymbol{s})E(\boldsymbol{x}_2, \lambda_K)B(\boldsymbol{x}_2, \lambda_K)} = \frac{B(\boldsymbol{x}_1, \lambda_K)}{B(\boldsymbol{x}_2, \lambda_K)}. \tag{2.49}$$

However, it is assumed that the neighboring points, from which the color ratios are computed, have the same surface normal. Therefore, the method depends on varying surface orientation of the object (i.e., the geometry of the objects) affecting negatively the recognition performance. To this end, a color constant color ratio has been proposed not only independent of the illumination color but also discounting the object's geometry [8]:

$$m(C_1^{\boldsymbol{x}_1}, C_1^{\boldsymbol{x}_2}, C_2^{\boldsymbol{x}_1}, C_2^{\boldsymbol{x}_2}) = \frac{C_1^{\boldsymbol{x}_1} C_2^{\boldsymbol{x}_2}}{C_1^{\boldsymbol{x}_2} C_2^{\boldsymbol{x}_1}} \tag{2.50}$$

where $\boldsymbol{x}_1$ and $\boldsymbol{x}_2$ denote the image locations of the two neighboring pixels, $C_1, C_2 \in \{R, G, B\}$, and $C_1 \neq C_2$.

The color ratio is independent of the illumination intensity and color, and also to a change in viewpoint, object geometry, and illumination direction as shown by substituting Eq. 2.50 in Eq. 2.44:

$$\frac{(G_B(x_1,n,s)E(x_1,\lambda_K)B(x_1,\lambda_{C_1}))(G_B(x_2,n,s)E(x_2,\lambda_K)B(x_2,\lambda_{C_2}))}{(G_B(x_2,n,s)E(x_2,\lambda_K)B(x_2,\lambda_{C_1}))(G_B(x_1,n,s)E(x_1,\lambda_K)B(x_1,\lambda_{C_2}))}$$

$$= \frac{B(x_1,\lambda_{C_1})B(x_2,\lambda_{C_2})}{B(x_2,\lambda_{C_1})B(x_1,\lambda_{C_2})} \tag{2.51}$$

factoring out dependencies on object geometry and illumination direction $G_B(x_1,n,s)$ and $G_B(x_2,n,s)$, and illumination for $E(x_1,\lambda_{C_2}) = E(x_2,\lambda_{C_2})$, and hence is only dependent on the ratio of surface albedos, where x_1 and x_2 are two neighboring locations on the object's surface not necessarily of the same orientation.

Note that the color ratios do not require any specific a priori information about the observed scene, as the color model is an illumination-invariant surface descriptor based on the ratio of surface albedos rather then the recovering of the actual surface albedo itself. Also the intensity and spectral power distribution of the illumination is allowed to vary across the scene (e.g., multiple light sources with different SPDs), and a certain amount of object occlusion and cluttering is tolerated due to the local computation of the color ratio.

2.6 Color System Taxonomy

The purpose of this section is to give a compact formulation on the relevance of the color systems for the purpose of color-based image retrieval. Therefore, a survey is given on the color models. Each color system is briefly discussed and transformations are given in terms of RGB NTSC tristimulus color coordinate system. Further, the following criteria are used to classify the color systems: (1) Is the color system device independent? (2) Is the color system perceptually uniform? (3) Is the color system non-linear? (4) Is the color system intuitive? (5) Is the color system robust against varying imaging conditions? The section will end up with a table summarizing the main and important characteristics of the color systems.

2.6.1 Gray-Value System

The color model $GRAY$ or $INTENSITY$ is calculated from the original R, G, B NTSC tristimulus values from the corresponding red, green, and blue images provided by a CCD color camera and hence is dependent on the imaging device. Gray is not perceptually uniform as a just noticeably brighter gray-value does not necessarily correspond to a difference between two successive gray-values. Gray is heavily influenced by the imaging conditions.

- Color model: $GRAY$
- Transformation:

$$GRAY = 0.299R + 0.587G + 0.144B. \tag{2.52}$$

- Characteristics:
  - Device dependent
  - Not perceptually uniform
  - Linear
  - Intuitive
  - Dependent on viewing direction, object geometry, direction of the illumination, intensity and, color of the illumination
- Remarks: Gray-value information.

2.6.2 RGB Color System

The RGB color system represents the colors of (R)ed, (G)reen, and (B)lue. In this chapter, the R, G, and B color features correspond to the primary colors where red = 700 nm, green = 546.1 nm, and blue = 435.8 nm. Similar to gray-value, the RGB color system is not perceptually uniform and dependent on the imaging conditions. Using RGB values for image retrieval cause problems when the query and target image are recorded under different imaging conditions.

- Color models: R, G, and B
- Transformation: Identity
- Characteristics:
  - Device dependent
  - Not perceptually uniform
  - Linear
  - Not intuitive
  - Dependent on viewing direction, object geometry, direction of the illumination, intensity and, color of the illumination
- Remarks: no transformation is required.

2.6.3 rgb Color System

The rgb color system has three color features: r, g, and b. These color models are called normalized colors, because each of them is calculated by dividing the values of respectively R, G, and B by their total sum. Because the r, g, and b coordinates depend only on the ratio of the R, G, and B coordinates (i.e., factoring luminance out of the system), they have the important property that they are not sensitive to surface orientation, illumination direction and illumination intensity (cf. Eqs 2.37-2.39). Another important property is that it is convenient to represent these features in the chromaticity diagram. Normalized colors become unstable and meaningless when the intensity is small [12].

- Color models: r, g, and b
- Transformation:

$$r = \frac{R}{(R+G+B)}, \tag{2.53}$$

$$g = \frac{G}{(R+G+B)}, \tag{2.54}$$

$$b = \frac{B}{(R+G+B)}. \tag{2.55}$$

- Characteristics:
  - Device dependent
  - Not perceptually uniform
  - Non-linear: become unstable when intensity is small
  - Not intuitive
  - Dependent on highlights and a change in the color of the illumination
- Remarks: the model is conveniently represented in the chromaticity diagram.

2.6.4 XYZ Color System

This color system is based on the additive mixture of three imaginary primaries X, Y, and Z introduced by the CIE. These primaries cannot be seen by a human eye or produced, because they are too saturated. The fact that these primaries are imaginary colors is not important, since any perceived color can be described mathematically by the amounts of these primaries. Another important property is that the luminance is determined only by the Y-value. Because XYZ system is a linear combination of R, G, and B values, the XYZ color system inherits all the dependencies on the imaging conditions from the RGB color system. Note that the color system is device independent as the X, Y, and Z values are objective in their interpretation. Further, the conversion matrix given below is based on the RGB NTSC color coordinates system.

- Color models: X, Y, Z
- Transformation:

$$X = 0.607R + 0.174G + 0.200B, \tag{2.56}$$

$$Y = 0.299R + 0.587G + 0.114B, \tag{2.57}$$

$$Z = 0.000R + 0.066G + 1.116B. \tag{2.58}$$

- Characteristics:
  - Device independent
  - Not perceptually uniform
  - Linear transformation
  - Not intuitive

- Dependent on viewing direction, object geometry, highlights, direction of the illumination, intensity, and color of the illumination
- Remarks: all colors are described mathematically by three color features, luminance is based on the Y-value alone.

2.6.5 xyz Color System

The color features of this system are the chromaticity coordinates: x, y, and z. Two of those chromaticity coordinates are enough to provide a complete specification and can be represented in the chromaticity diagram. Similar to rgb, this system cancels intensity out, yielding independence of surface orientation, illumination direction, and illumination intensity. Again problems arise when intensity is low.

- Color models: x, y, z
- Transformation:

$$x = \frac{X}{X + Y + Z},$$ (2.59)

$$y = \frac{Y}{X + Y + Z},$$ (2.60)

$$z = \frac{Z}{X + Y + Z}.$$ (2.61)

- Characteristics:
  - Device independent
  - Not perceptually uniform
  - Non-linear transformation: become unstable when intensity is small
  - Not intuitive
  - Dependent highlights and a change in the color of the illumination
- Remarks: the model is conveniently represented in the chromaticity diagram.

2.6.6 $U^*V^*W^*$ Color System

The CIE introduced the $U^*V^*W^*$ color system which has three color features: U^*, V^*, and W^*. The color model W^* is based on the scaling of luminance. The luminance of a color is determined only by its Y value. Scaling luminance between 0 (black) and 100 (white), the scaling method starts with black and selects a *just noticeably brighter* gray-value. Taking this just noticeably brighter gray-value the next just noticeably brighter gray-value is selected. This process continues until white is reached.

The other two color features solve the problem of large difference of the axis diameters of the ellipses in the chromaticity diagram, where colors which are not noticeably different for a particular color are lying on the ellipses and all colors which are (just) noticeably different are lying outside the ellipses.

The system is visually uniform, because a luminance difference corresponds to the same noticed luminance difference and the ellipses in the adjusted chromaticity diagram have constant axis diameters. The U^* and V^* color models become unstable and meaningless when intensity is small.

– Color models: U^*, V^*, W^*
– Transformation:

$$U^* = 13W^*(u - u_0), \tag{2.62}$$

$$V^* = 13W^*(v - v_0), \tag{2.63}$$

$$W^* = \begin{cases} 116\left(\dfrac{Y}{Y_0}\right)^{\frac{1}{3}} - 16 & \text{if } \dfrac{Y}{Y_0} > 0.008856, \\[2ex] 903.3\left(\dfrac{Y}{Y_0}\right) & \text{if } \dfrac{Y}{Y_0} \leq 0.008856i, \end{cases} \tag{2.64}$$

$$u = \frac{4X}{X + 15Y + 3Z}, \tag{2.65}$$

$$v = \frac{6Y}{X + 15Y + 3Z}, \tag{2.66}$$

$$u_0 = \frac{4X_0}{X_0 + 15Y_0 + 3Z_0}, \tag{2.67}$$

$$v_0 = \frac{6Y_0}{X_0 + 15Y_0 + 3Z_0}. \tag{2.68}$$

X_0, Y_0, Z_0 are values of a nominally white object-color stimulus.
– Characteristics:
  – Device independent
  – Perceptually uniform
  – Nonlinear transformation: become unstable when intensity is small
  – Not intuitive
  – Dependent on viewing direction, object geometry, highlights, direction of the illumination, intensity, and color of the illumination
– Remarks: the model is visually uniform.

2.6.7 $L^*a^*b^*$ Color System

Another kind of visually uniform color system proposed by the CIE is the $L^*a^*b^*$ color system. The color feature L^* correlates with the perceived luminance and corresponds to W^* of the $U^*V^*W^*$ color system. Color feature a^* correlates with the red–green content of a color and b^* reflects the yellow–blue content.

- Color models: L^*, a^*, b^*
- Transformation:

$$L^* = \begin{cases} 116 \left(\dfrac{Y}{Y_0} \right)^{\frac{1}{3}} - 16 & \text{if } \dfrac{Y}{Y_0} > 0.008856, \\ 903.3 \left(\dfrac{Y}{Y_0} \right) & \text{if } \dfrac{Y}{Y_0} \leq 0.008856, \end{cases} \tag{2.69}$$

$$a^* = 500 \left[\left(\dfrac{X}{X_0} \right)^{\frac{1}{3}} - \left(\dfrac{Y}{Y_0} \right)^{\frac{1}{3}} \right], \tag{2.70}$$

$$b^* = 200 \left[\left(\dfrac{Y}{Y_0} \right)^{\frac{1}{3}} - \left(\dfrac{Z}{Z_0} \right)^{\frac{1}{3}} \right]. \tag{2.71}$$

- Characteristics:
  - Device independent
  - Perceptually uniform
  - Non-linear transformation: become unstable when intensity is small
  - Not intuitive
  - Dependent on viewing direction, object geometry, highlights, direction of the illumination, intensity, and color of the illumination

2.6.8 $I1I2I3$ Color System

When RGB images are highly correlated, it might be desirable to down-weight this correlation. This can be achieved by computing the Karhunen–Loève transformation. This transformation is calculated from the covariance matrix and calculates an uncorrelated basis.

Three color models $I1$, $I2$, and $I3$ have been presented by doing experiments and by analyzing the results of eight color scenes by Ohta [18]. The color scenes were digitalized with 256×256 spatial resolution and 6-bit intensity resolution for each R, G, and B. Ohta calculated the eigenvectors of each of the eight color scenes.

- Color models: $I1$, $I2$, $I3$
- Transformation:

$$I1 = \frac{R + G + B}{3}, \tag{2.72}$$

$$I2 = \frac{R - B}{2}, \tag{2.73}$$

$$I3 = \frac{2G - R - B}{4}. \tag{2.74}$$

- Characteristics:
  - Device dependent
  - Not perceptually uniform

- Linear
- Intuitive
- Dependent on viewing direction, object geometry, highlights, direction of the illumination, intensity and, color of the illumination
- Remarks: uncorrelation based on the Karhunen–Loève transformation of eight different color images.

2.6.9 YIQ and YUV Color Systems

The National Television Systems Committee (NTSC) developed the three color attributes Y, I, and Q for transmission efficiency. The tristimulus value Y corresponds to the luminance of a color. I and Q correspond closely to the hue and saturation of a color. By reducing the spatial bandwidth of I and Q without noticeable image degradation, efficient color transmission is obtained. The PAL and SECAM standards used in Europe, the Y, U, and V tristimulus values are used. The I and Q color attributes are related to U and V by a simple rotation of the color coordinates in color space.

- Color models: Y, I, Q
- Transformation:

$$Y = 0.299R + 0.587G + 0.114B, \tag{2.75}$$

$$I = 0.596R - 0.274G - 0.312B, \tag{2.76}$$

$$Q = 0.211R - 0.523G + 0.312B. \tag{2.77}$$

- Characteristics:
  - Device independent
  - Not perceptually uniform
  - Linear transformation
  - Not intuitive
  - Dependent on viewing direction, object geometry, highlights, direction of the illumination, intensity and, color of the illumination
- Remarks: Y is the luminance of a color.

2.6.10 HSI Color System

The human color perception is conveniently represented by the following set of color features: I(ntensity), S(aturation), and H(ue). I is an attribute in terms of which a light or surface color may be ordered on a scale from dim to bright. S denotes the relative white content of a color and H is the color aspect of a visual impression.

The problem of hue is that it becomes unstable when S is near zero due to the non-removable singularities in the non-linear transformation, which a small perturbation of the input can cause a large jump in the transformed values.

- Color models: H, S, I
- Transformation:

$$I = \frac{R+G+B}{3},$$ (2.78)

$$H = \arctan\left(\frac{\sqrt{3}(G-B)}{(R-G)+(R-B)}\right),$$ (2.79)

$$S = 1 - 3\min(r,g,b).$$ (2.80)

- Characteristics:
  - Device dependent
  - Not perceptually uniform
  - I is linear. Saturation is non-linear: becomes unstable when intensity is near zero. Hue is non-linear: becomes unstable when intensity and saturation are near zero
  - Intuitive
  - Intensity I: dependent on viewing direction, object geometry, direction of the illumination, intensity, and color of the illumination
  - Saturation S: dependent on highlights and a change in the color of the illumination
  - Hue H: dependent on the color of the illumination

2.6.11 Color Ratios

Color constant color ratios have been proposed by Funt and Finlayson [7], Nayar and Bolle [16], and Gevers and Smeulders [8]. These color constant models are based on the ratio of surface albedos rather then the recovering of the actual surface albedo itself. The constraint is that the illumination can be modeled by the coefficient rule. The coefficient model of illumination change holds exactly in the case of narrow-band sensors. Although standard video cameras are not equipped with narrow-band filters, spectral sharpening could be applied [2] to achieve this to a large extent.

- Color models: color ratios N, F, m
- Transformation: Nayar and Bolle [16]:

$$N(C^{x_1}, C^{x_2}) = \frac{C^{x_1} - C^{x_2}}{C^{x_2} + C^{x_1}}.$$ (2.81)

Funt and Finlayson [7]:

$$F(C^{x_1}, C^{x_2}) = \frac{C^{x_1}}{C^{x_2}}.$$ (2.82)

Gevers and Smeulders [8]:

$$m(C_1^{x_1}, C_1^{x_2}, C_2^{x_1}, C_2^{x_2}) = \frac{C_1^{x_1} C_2^{x_2}}{C_1^{x_2} C_2^{x_1}}.$$ (2.83)

where $C \in \{R, G, B\}$ and x_1 and x_2 denote the image locations of the two neighboring pixels.
- Characteristics:
  - Device dependent
  - Not perceptually uniform
  - Non-linear: becomes unstable when intensity is near zero
  - Not intuitive
  - N and F: dependent on the object geometry, m no dependencies

2.7 Color and Image Search Engines

Very large digital image archives have been created and used in a number of applications including archives of images of postal stamps, textile patterns, museum objects, trademarks and logos, and views from everyday life as it appears in home videos and consumer photography. Moreover, with the growth and popularity of the World Wide Web, a tremendous amount of visual information is made accessible publicly. As a consequence, there is a growing demand for search methods retrieving pictorial entities from large image archives. Attempts have been made to develop general purpose image retrieval systems based on multiple features (e.g., color, shape, and texture) describing the image content (e.g. [3, 19, 25–27]). Further, a number of systems are available for retrieving images from the World Wide Web (e.g. [5, 11, 20, 23]). Aside from different representation schemes, these systems retrieve images on the basis color. In the Picasso [1] and ImageRover [20] systems, the $L^*u^*v^*$ color space has been used for image retrieval. The QBIC system [3] evaluates similarity of global color properties using histograms based on a linear combination of the RGB. MARS [21] uses the the $L^*a^*b^*$ color space because the color space consists of perceptually uniform colors, which better matches the human perception of color. The PicToSeek system [9] is based on color models robust to a change in viewing direction, object geometry, and illumination.

2.8 Conclusion

We have seen that each color system has its own characteristics. A number of systems are linear combinations of the R, G, and B values, such as the XYZ and the $I1I2I3$ color system, or normalized with respect to intensity, such as the rgb and the xyz color system. The $U^*V^*W^*$ and the $L^*a^*b^*$ color systems have distances which reflect the perceived similarity. Each image retrieval application demands a specific color system. In Table 2.1, a taxonomy is given of the different color systems/models. Note that only Hue is invariant to highlights. Further, the various color systems and their performance can be experienced within the PicToSeek and Pic2Seek systems on-line at: http://www.wins.uva.nl/research/isis/zomax/.

Table 2.1 Overview of the dependencies differentiated for the various color systems. + denotes that the condition is satisfied; – denotes that the condition is not satisfied

Color system/model	Dev. indep.	Perc. unif.	Linear	Intuit.	View-point	Object shape	Illum. intens.	Illum. SPD
RGB	–	–	+	–	–	–	–	–
XYZ	+	–	+	–	–	–	–	–
Norm. rgb	–	–	–	–	+	+	+	–
Norm. xyz	+	–	–	–	+	+	+	–
$L^*a^*b^*$	+	+	–	–	–	–	–	–
$U^*V^*W^*$	+	+	–	–	–	–	–	–
$I1I2I3$	–	–	+	–	–	–	–	–
YIQ	–	–	+	–	–	–	–	–
YUV	–	–	+	–	–	–	–	–
Intens.	–	–	+	+	–	–	–	–
Hue	–	–	–	+	+	+	+	–
Sat.	–	–	–	+	+	+	+	–
N, F	–	–	–	–	+	–	+	+
m	–	–	–	–	+	+	+	+

In this chapter, a survey of the basics of color has been given. Further, color models and ordering systems were discussed, and the state of the art of color invariance has been presented. For the purpose of color-based image retrieval, a taxonomy on color systems has been provided. The color system taxonomy can be used to select the proper color system for a specific application.

References

1. Del Bimbo, A, Mugnaini, M, Pala, P, and Turco, F, "Visual Querying by Color Perceptive Regions," Patt Recogn, 31(9), pp. 1241–1253, 1998.
2. Finlayson, GD, Drew, MS, and Funt, BV, "Spectral Sharpening: Sensor Transformation for Improved Color Constancy," JOSA, 11, pp. 1553–1563, May, 1994.
3. Flicker, M, Sawhney, H, Niblack, W, Ashley, J, Huang, Q, Dom, B, Gorkani, M, Hafner, J, Lee, D, Petkovic, D, Steele, D, and Yanker, P, "Query by Image and Video Content: The QBIC System," IEEE Computer, 28(9), pp. 23–32, 1995.
4. Forsyth, D, "A Novel Algorithm for Color Constancy," Int. J Computer Vision, 5, pp. 5–36, 1990.
5. Frankel, C, Swain, M, and Athitsos, V, "Webseer: An Image Search Engine for the World Wide Web," TR-96-14, U Chicago, 1996.
6. Funt, BV and Drew, MS, "Color Constancy Computation in Near-Mondrian Scenes," In Proceedings of the CVPR, IEEE Computer Society Press, pp. 544–549, 1988.
7. Funt, BV and Finlayson, GD, "Color Constant Color Indexing," IEEE PAMI, 17(5), pp. 522–529, 1995.
8. Gevers, T and Smeulders, A, "Content-Based Image Retrieval by Viewpoint-invariant Image Indexing," Image Vision Comput, (17)7, 1999.
9. Gevers, T and Smeulders, A, "Color Based Object Recognition," Patt Recogn, 32, pp. 453–464, March, 1999.
10. Goethe, "Farbenlehre," 1840.

11. Gupta, A, "Visual Information Retrieval Technology: A Virage Perspective," TR 3A, Virage Inc., 1996.
12. Kender, JR, "Saturation, Hue, and Normalized Colors: Calculation, Digitization Effects, and Use," Technical Report, Department of Computer Science, Carnegie-Mellon University, 1976.
13. Land, EH, "The Retinex Theory of Color Vision," Scient Am, 218(6), pp. 108–128, 1977.
14. Levkowitz, H and Herman, GT, "GLHS: A Generalized Lightness, Hue, and Saturation Color Model," CVGIP: Graph Models Image Process, 55(4), pp. 271–285, 1993.
15. Munsell, AH, A Color Notation, 1st edn (Geo. H. Ellis Co., Boston), 11th edn. (edited and rearranged) (Munsell Color Company, Baltimore, 1946); Now available from the Munsell Color Group, New Windsor, New York, 1905.
16. Nayar, SK and Bolle, RM, "Reflectance Based Object Recognition," Int J Computer Vision, 17(3), pp. 219–240, 1996.
17. Newton, I, Opticks (W Innys, London; reprinted: Dover, New York, 1952), 1704.
18. Ohta, Y, "Knowledge-Based Interpretation of Outdoor Natural Scenes," Pitman Publishing, London, 1985.
19. Pentland, A, Picard, RW, and Sclaroff, S, "Photobook: Tools for Content-based Manipulation of Image Databases," Proceedings Storage and Retrieval for Image and Video Databases II, 2, 185, SPIE, Bellingham, Wash. pp. 34–47, 1994.
20. Sclaroff, S, Taycher, L, and La Cascia, M, "ImageRover: A Content-based Image Browser for the World Wide Web," Proceedings of IEEE Workshop on Content-based Access and Video Libraries, 1997.
21. Servetto, S, Rui, Y, Ramchandran, K, and Huang, TS, "A Region-Based Representation of Images in MARS," J on VLSI Sig Process Systs, 20(2), pp. 137–150, October, 1998.
22. Shafer, SA, "Using Color to Separate Reflection Components," Color Res Appl, 10(4), pp 210–218, 1985.
23. Smith, JR and Chang, SF, "Visualseek: a Fully Automated Content-based Image Query System," Proceedings of ACM Multimedia, 1996.
24. Swain, MJ and Ballard, DH, "Color Indexing," Int J Computer Vision, 7(1), pp. 11–32, 1991.
25. Proceedings of First International Workshop on Image Databases and Multi Media Search, IDB-MMS '96, Amsterdam, The Netherlands, 1996.
26. Proceedings of IEEE Workshop on Content-Based Access and Video Libraries, CVPR, 1997.
27. Proceedings of Visual97 Information Systems: The Second International Conference on Visual Information Systems, San Diego, USA, 1997.
28. Proceedings of Visual99 Information Systems: The Third International Conference on Visual Information Systems, Amsterdam, The Netherlands, 1999.

3. Texture Features for Content-Based Retrieval

Nicu Sebe and Michael S. Lew

3.1 Introduction

Texture is an intuitive concept. Every child knows that leopards have spots but tigers have stripes, that curly hair looks different from straight hair, etc. In all these examples there are variations of intensity and color which form certain repeated patterns called *visual texture*. The patterns can be the result of physical surface properties such as roughness or oriented strands which often have a tactile quality, or they could be the result of reflectance differences such as the color on a surface. Even though the concept of texture is intuitive (we recognize texture when we see it), a precise definition of texture has proven difficult to formulate. This difficulty is demonstrated by the number of different texture definitions attempted in the literature [7, 12, 38, 65, 70].

Despite the lack of a universally accepted definition of texture, all researchers agree on two points: (1) within a texture there is significant variation in intensity levels between nearby pixels; that is, at the limit of resolution, there is non-homogeneity and (2) texture is a homogeneous property at some spatial scale larger than the resolution of the image. It is implicit in these properties of texture that an image has a given resolution. A single physical scene may contain different textures at varying scales. For example, at a large scale the dominant pattern in a floral cloth may be a pattern of flowers against a white background, yet at a finer scale the dominant pattern may be the weave of the cloth. The process of photographing a scene, and digitally recording it, creates an image in which the pixel resolution implicitly defines a finest scale. It is conventional in the texture analysis literature to investigate texture at the pixel resolution scale; that is, the texture which has significant variation at the pixel level of resolution, but which is homogeneous at a level of resolution about an order of magnitude coarser.

Some researchers finesse the problem of formally defining texture by describing it in terms of the human visual system: textures do not have uniform intensity, but are nonetheless perceived as homogeneous regions by a human observer. Other researchers are completely driven in defining texture by the application in which the definition is used. Some examples are given here:

- "An image texture may be defined as a local arrangement of image irradiances projected from a surface patch of perceptually homogeneous irradiances." [7]

– "Texture is defined for our purposes as an attribute of a field having no components that appear enumerable. The phase relations between the components are thus not apparent enumerable. The phase relations between the components are thus not apparent. Nor should the field contain an obvious gradient. The intent of this definition is to direct attention of the observer to the global properties of the display, i.e., its overall 'coarseness,' 'bumpiness,' or 'fineness.' Physically, nonenumerable (aperiodic) patterns are generated by stochastic as opposed to deterministic processes. Perceptually, however, the set of all patterns without obvious enumerable components will include many deterministic (and even periodic) textures." [65]

– "An image texture is described by the number and types of its (tonal) primitives and the spatial organization or layout of its (tonal) primitives ... A fundamental characteristic of texture: it cannot be analyzed without a frame of reference of tonal primitive being stated or implied. For any smooth gray tone surface, there exists a scale such that when the surface is examined, it has no texture. Then as resolution increases, it takes on a fine texture and then a coarse texture." [38]

– "Texture regions give different interpretations at different distances and at different degrees of visual attention. At a standard distance with normal attention, it gives the notion of macroregularity that is characteristic of the particular texture." [12]

A definition of texture based on human perception is suitable for psychometric studies, and for discussion on the nature of texture. However, such a definition poses problems when used as the theoretical basis for a texture analysis algorithm. Consider the three images in Fig. 3.1. All three images are constructed by the same method, differing in only one parameter. Figs 3.1(a) and (b) contain perceptually different textures, whereas Figs 3.1(b) and (c) are perceptually similar. Any definition of texture, intended as the theoretical foundation for an algorithm and based on human perception, has to address the problem that a family of textures, as generated by a parameterized method, can vary smoothly between perceptually distinct and perceptually similar pairs of textures.

3.1.1 Human Perception of Texture

Julesz has studied texture perception extensively in the context of texture discrimination [42, 43]. The question he posed was "When is a texture pair discriminable, given that the textures have the same brightness, contrast and color?" His approach was to embed one texture in the other. If the embedded patch of texture visually stood out from the surrounding texture, then the two textures were considered to be dissimilar. In order to analyze if two textures are discriminable, he compared their first and second order statistics.

First order statistics measure the likelihood of observing a gray value at a randomly chosen location in the image. These statistics can be computed

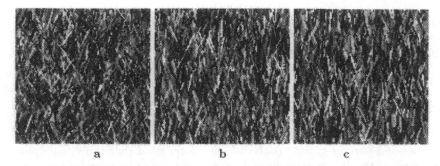

a   b   c

Fig. 3.1 Visibility of texture distinctions; Each of the images is composed of lines of the same length having their intensity drawn from the same distribution and their orientations drawn from different distributions. The lines in **a** are drawn from the uniform distribution, with a maximum deviation from the vertical of 45°. The orientation of lines in **b** is at most 30° from the vertical and in **c** at most 28° from the vertical.

from the histogram of pixel intensities in the image. These depend only on individual pixel values and not on the interaction or co-occurrence of neighboring pixel values. The average intensity in an image is an example of a first order statistic. Second order statistics are defined as the likelihood of observing a pair of gray values occurring at the endpoints of a dipole of random length placed in the image at a random location and orientation. These are properties of pairs of pixel values.

Julesz found that textures with similar first order statistics, but different second order statistics were easily discriminated. However, he could not find any textures with the same first and second order statistics that could be discriminated. This led him to the conjecture that "iso-second-order textures are indistinguishable." [42]

Later Caelli et al. [9] did produce iso-second-order textures that could be discriminated with pre-attentive human visual perception. Further work by Julesz [44, 45] revealed that his original conjecture was wrong. Instead, he found that the human visual perception mechanism did not necessarily use third order statistics for the discrimination of these iso-second-order textures, but rather use the second order statistics of features he called *textons*. These textons are described as being the fundamentals of texture. Three classes of textons were found: *color*, *elongated blobs*, and the *terminators* (endpoints) of these elongated blobs. The original conjecture was revised to state that "the pre-attentive human visual system cannot compute statistical parameters higher than second order." Furthermore Julesz stated that the pre-attentive human visual system actually uses only the first order statistics of these textons.

Since these pre-attentive studies into the human visual perception, psychophysical research has focussed on developing physiologically plausible models of texture discrimination. These models involved determining which measurements of textural variations humans are most sensitive to. Textons were not found to be the plausible textural discriminating measures as envisaged by Julesz [5, 76]. Beck et al. [4] argued that the perception of texture segmentation in certain types of patterns is primarily a function of spatial frequency analysis and not the result of a higher level symbolic grouping process. Psychophysical research suggested that the brain performs a multichannel, frequency, and orientation analysis of the visual image formed on the retina [10, 25]. Campbell and Robson [10] conducted psychophysical experiments using various grating patterns. They suggested that the visual system decomposes the image into filtered images of various frequencies and orientations. De Valois et al. [25] have studied the brain of the macaque monkey which is assumed to be close to the human brain in its visual processing. They recorded the response of the simple cells in the visual cortex of the monkey to sinusoidal gratings of various frequencies and orientations and concluded that these cells are tuned to narrow ranges of frequency and orientation. These studies have motivated vision researchers to apply multi-channel filtering approaches to texture analysis. Tamura et al. [70] and Laws [48] identified the following properties as playing an important role in describing texture: uniformity, density, coarseness, roughness, regularity, linearity, directionality, direction, frequency, and phase. Some of these perceived qualities are not independent. For example, frequency is not independent of density, and the property of direction only applies to directional textures. The fact that the perception of texture has so many different dimensions is an important reason why there is no single method of texture representation which is adequate for a variety of textures.

3.1.2 Approaches for Analyzing Textures

The vague definition of texture leads to a variety of different ways to analyze texture. The literature suggests three approaches for analyzing textures [1, 26, 38, 73, 79].

Statistical texture measures. A set of features is used to represent the characteristics of a textured image. These features measure properties such as contrast, correlation, and entropy. They are usually derived from the *gray value run length*, *gray value difference*, or Haralick's *co-occurrence matrix* [37]. Features are selected heuristically and the image cannot be re-created from the measured feature set. A survey of statistical approaches for texture is given by Haralick [38].

Stochastic texture modeling. A texture is assumed to be the realization of a stochastic process which is governed by some parameters. Analysis is performed by defining a model and estimating the parameters so that the stochastic process can be reproduced from the model and associated parameters. The estimated parameters can serve as features for texture classification and segmentation problems. A difficulty with this texture modeling is that many natural textures do not conform to the restrictions of a particular model. An overview of some of the models used in this type of texture analysis is given by Haindl [35].

Structural texture measures. Some textures can be viewed as two-dimensional patterns consisting of a set of primitives or subpatterns which are arranged according to certain placement rules. These primitives may be of varying or deterministic shape, such as circles, hexagons or even dot patterns. Macrotextures have large primitives, whereas microtextures are composed of small primitives. These terms are relative to the image resolution. The textured image is formed from the primitives by placement rules which specify how the primitives are oriented, both on the image field and with respect to each other. Examples of such textures include tilings of the plane, cellular structures such as tissue samples, and a picture of a brick wall. Identification of these primitives is a difficult problem. A survey of structural approaches for texture is given by Haralick [38]. Haindl [35] also covers some models used for structural texture analysis.

Francos et al. [29] describe a texture model which unifies the stochastic and structural approaches to defining texture. The authors assume that a texture is a realization of a 2D homogeneous random field, which may have a strong regular component. They show how such a field can be decomposed into orthogonal components (2D Wold-like decomposition). Research on this model was carried out by Liu and Picard [49, 50, 60].

3.2 Texture Models

The objective of modeling in image analysis is to capture the intrinsic character of images in a few parameters so as to understand the nature of the phenomenon generating the images. Image models are also useful in quantitatively specifying natural constraints and general assumptions about the physical world and the imaging process. Research into texture models seeks to find a compact, and if possible a complete, representation of the textures commonly seen in images. The objective is to use these models for such tasks as texture classification, segmenting the parts of an image with different textures, or detecting flaws or anomalies in textures.

There are surveys in the literature which describe several texture models [27, 38, 73, 75, 79, 80]. As described before, the literature distinguishes between stochastic/statistical and structural models of texture. An example of a taxonomy of image models is given by Dubes and Jain [26].

We divide stochastic texture models into three major groups: probability density function (PDF) models, gross shape models and partial models, as suggested by Smith [68] (see Fig. 3.2). The PDF methods model a texture as a random field and a statistical PDF model is fitted to the spatial distribution of intensities in the texture. Typically, these methods measure the interactions of small numbers of pixels. For example, the Gauss–Markov random field (GRMF) and gray-level co-occurrence (GLC) methods measure the interaction of pairs of pixels.

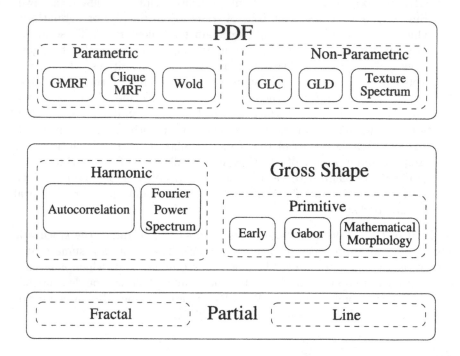

Fig. 3.2 Taxonomy of stochastic texture models.

Gross shape methods model a texture as a surface. They measure features which a human would consciously perceive, such as the presence of edges, lines, intensity extrema, waveforms, and orientation. These methods measure the interactions of larger numbers of pixels over a larger area than is typical in PDF methods. The subgroup of harmonic methods measures periodicity in the texture. These methods look for perceptual features which recur at regular intervals, such as waveforms. Primitive methods detect a set of spatially compact perceptual features, such as lines, edges, and intensity extrema and output a feature vector composed of the density of these perceptual features in the texture.

Partial methods focus on some specific aspect of texture properties at the expense of other aspects. Fractal methods explicitly measure how a texture varies with the scale it is measured at, but do not measure the structure of a texture at any given scale. Line methods measure properties of a texture along one-dimensional contours in a texture, and do not fully capture the two-dimensional structure of the texture.

Structural methods are characterized by their definition of texture as being composed of "texture elements" or primitives. In this case, the method of analysis usually depends upon the geometrical properties of these texture elements. Structural methods consider that the texture is produced by the placement of the primitives according to certain placement rules.

3.2.1 Parametric PDF Methods

This section reviews parametric PDF methods: these include auto-regressive methods, Gauss–Markov random field (GMRF) method, and uniform clique markov random field methods. Parametric PDF methods have the underlying assumption that textures are partially structured and partially stochastic. In practice, these methods also assume that the structure in the texture can be described locally.

Gauss–Markov Random Field. GMRF methods [13, 14, 54] model the intensity of a pixel as a stochastic function of its neighboring pixels' intensity. Specifically, GRMF methods use a Gaussian probability density function to model a pixel intensity. The mean of the Gaussian distribution is a linear function of the neighboring pixels' intensities. Typically, a least squares method is used to estimate the linear coefficients and the variance of the Gaussian distribution. As an alternative [18], a binomial distribution rather than a Gaussian distribution is used. However, in the parameter ranges used, the binomial distribution approximates a Gaussian distribution.

Chellappa and Chatterjee [13] give the following typical formulation:

$$\mathcal{I}(x,y) = \sum_{(\partial x, \partial y) \in \mathcal{N}_S} \theta_{(\partial x, \partial y)} (\mathcal{I}(x + \partial x, y + \partial y) + \mathcal{I}(x - \partial x, y - \partial y)) + e(x,y)$$

where $\mathcal{I}(x,y)$ is the intensity of a pixel at the location (x,y) in the image, $\mathcal{N}$ is the symmetric pixel neighborhood (excluding the pixel itself), $\mathcal{N}_S$ is a half of $\mathcal{N}$, $\theta_{(\partial x, \partial y)}$ are parameters estimated using a least squares method, and $e(x,y)$ is a zero mean stationary Gaussian noise sequence with the following properties:

$$E(e(x,y)\, e(\partial x, \partial y)) = \begin{cases} -\theta_{(x-\partial x, y-\partial y)}\nu & \text{if } (x - \partial x, y - \partial y) \in \mathcal{N}, \\ \nu & \text{if } (x,y) = (\partial x, \partial y), \\ 0 & \text{otherwise} \end{cases}$$

where ν is the mean square error of the least squares estimate.

Therrien [71] and de Souza [24] describe texture models which they term auto-regressive. The model used by Therrien is identical with the GMRF models. De Souza estimates the intensity of a pixel using a linear function of its neighbors but he does not use the variance of the distribution as one of his features, resuming only to use the mean of the distribution.

An important element of these methods is the neighborhood considered around a pixel. Cross and Jain [18] define the neighbors of a given pixel as those pixels which affect the pixel's intensity. By this definition the neighbors are not necessarily close. However, in practice these methods use a small neighborhood, typically a 5×5 window or smaller. Although long range interactions can be encoded in small neighborhoods, many researchers have found that these methods do not accurately model macrotextures. For example Cross and Jain [18] note: "This model seems to be adequate for duplicating the micro textures, but is incapable of handling strong regularity or cloud-like inhomogeneities."

GMRF models also assume that second order PDF models are sufficient to characterize a texture. Hall et al. [36] derived a test to determine whether a sample is likely to have been drawn from a 2D Gaussian distribution. They examined seven textures from the Brodatz album [8] and found that all fail the test.

Uniform Clique Markov Random Field. Uniform clique Markov random fields are described in Derin and Cole [22], and Derin and Elliott [23]. These methods are derived from the random field definition of texture and require several simplifying assumptions to make them computationally feasible.

Hassner and Slansky [39] propose MRFs as a model of texture. An MRF is a set of discrete values associated with the vertices of a discrete lattice. If we interpret the vertices of the lattice as the pixel locations, and the discrete values as the gray-level intensities at each pixel, we have the correspondence between an MRF and an image. Furthermore, by the definition of MRFs, the PDF of a value at a given lattice vertex is completely determined by the values at a set of neighboring vertices. The size of the neighborhood set is arbitrary, but fixed for any given lattice in a given MRF. The first order neighbors (n1) of a vertex are its four-connected neighbors and the second-order neighbors (n2) are its eight-connected neighbors. Within these neighborhoods the sets of neighbors which form cliques (single site, pairs, triples, and quadruples) (see Fig. 3.3) are usually used in the definition of the conditional probabilities. A clique type must be a subset of the neighborhood such that every pair of pixels in the clique are neighbors. Thus, for each pixel in a clique type, if that pixel is superimposed on the central pixel of a neighborhood, the entire clique must fall inside the neighborhood. In these conditions, the diagonally adjacent pairs of pixels are clique types within the n2 neighborhood system, but not within the n1 neighborhood system. Each different clique type has an associated potential function which maps from all combinations of values of its component pixels to a scalar potential.

We have the property that with respect to a neighboring system, there exists a unique Gibbs random field for every Markov random field and there

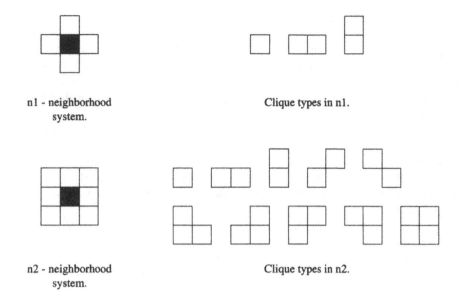

n1 - neighborhood
system.

Clique types in n1.

n2 - neighborhood
system.

Clique types in n2.

Fig. 3.3 n1 and n2 neighborhood systems and their associated clique types.

exists a unique Markov random field for every Gibbs random field. The consequence of this theorem is that one can model the texture either globally by specifying the total energy of the lattice or model it locally by specifying the local interactions of the neighboring pixels in terms of conditional probabilities. In the special case where the MRF is defined on a two-dimensional lattice with a homogeneous PDF, Gibbs random field theory states that the probability of the lattice having a particular state is:

$$P(I) = \frac{1}{Z} \exp\left\{ -\frac{U(I)}{T} \right\} \tag{3.1}$$

where I is the state of the lattice (or image), Z and T are normalizing constants, and $U(I)$ is an energy function given by:

$$U(I) = \sum_{C \in \mathcal{C}_\mathcal{N}} \left(\sum_{c \in \mathcal{C}_C(I)} V_C(c) \right) \tag{3.2}$$

where $\mathcal{N}$ is the neighborhood associated with the MRF, $\mathcal{C}_\mathcal{N}$ is the set of clique types generated by $\mathcal{N}$, $\mathcal{C}_C(I)$ is the set of instances of the clique type C in the image I, and $V_C(\cdot)$ is the potential function associated with clique type C.

In this way, given a suitably large neighborhood, the set of potential functions forms, by definition, a complete model of a texture. The parameters of this model are the potential values associated with each element of the domain of each of the clique types. However, real-world textures require a large neighborhood mask and a large number of gray-levels. The number of clique types grows rapidly with the size of the neighborhood and the domain of the potential function grows exponentially with the number of gray-levels. Consequently, the number of parameters required to model a real-world texture in this way is computationally infeasible.

To some extent, this difficulty is caused by a generality in the formulation of MRFs. The gray-levels in an image are discrete variables as required by the MRF formulation. However, there is considerable structure in the gray-levels, such as ordering, which is not exploited in the MRF formulation. Hassner and Slansky [39] advocate GMRF models, which do exploit the structure in gray-levels, as a "practical alternative."

Several simplifying assumptions which reduce the number of MRF texture model parameters were proposed. If the texture is assumed to be rotationally symmetric, the number of parameters is reduced considerably. Hassner and Slansky [39] also suggest "color indifference," where each potential function has only two values:

$$V_C(c) = \begin{cases} -\theta & \text{if all pixels in } c \text{ have the same intensity,} \\ +\theta & \text{otherwise} \end{cases} \tag{3.3}$$

where c is the clique instance and θ is a parameter associated with the clique type. This model is denoted as uniform clique MRF model [68]. There are clearly problems related to the use of this model. Firstly, it is important that a significant fraction of clique instances are uniform; otherwise most clique instances would have the potential θ and the model would have little discriminatory power. It follows that the intensity value in an image must be quantized to very few levels. This can be seen in Derin and Cole [22] and Derin and Elliott [23] where at most four levels of quantization are used. This degree of quantization is likely to discard some of the texture feature information in the original texture. Secondly, the potential function does not take into account whether distinct intensity levels are close in value or not. This may lead to the fact that for distinct textures the clique instances will have identical patterns of uniformity and non-uniformity.

In summary, uniform clique MRFs derive directly from the random field model of texture. However, practical considerations have forced unrealistic simplifications to be made.

Wold-Based Representations. Picard and Liu [60] developed a new model based on the Wold decomposition for regular stationary stochastic processes in 2D images. If an image is assumed to be a homogeneous 2D discrete random field, then the 2D Wold-like decomposition is a sum of three mutually orthogonal components: a harmonic field, a generalized evanescent field, and a purely indeterministic field. These three components are illustrated in Fig. 3.4 by three textures, each of which is dominated by one of these components. Qualitatively, these components appear as periodicity, directionality, and randomness, respectively.

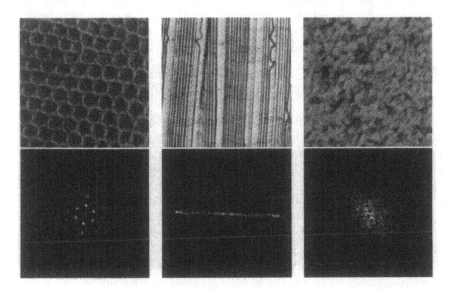

Fig. 3.4 The Wold decomposition transforms textures into three orthogonal components: harmonic, evanescent, and random. The upper three textures illustrate these components; below each texture is shown its discrete Fourier transform (DFT) magnitude.

The motivation for choosing a Wold-based model, in addition to its significance in random field theory, is its interesting relationship to independent psychophysical findings of perceptual similarity. Rao and Lhose [63] made a study where humans grouped patterns according to perceived similarity. The three most important similarity dimensions identified in this study were repetitiveness, directionality, and complexity. These dimensions might be considered the perceptual equivalents of the harmonic, evanescent, and indeterministic components, respectively, in the Wold decomposition.

Consider a homogeneous and regular random field $\{y(m,n)|(m,n) \in \mathcal{Z}^2\}$. The 2D Wold decomposition allows the field to be decomposed into two mutually orthogonal components [30]:

$$y(m,n) = v(m,n) + w(m,n) \tag{3.4}$$

where $v(m,n)$ is the deterministic component, and $w(m,n)$ is the indeterministic one. The deterministic component can be further decomposed into a mutually orthogonal harmonic component $h(m,n)$, and an evanescent component $g(m,n)$:

$$v(m, n) = h(m, n) + g(m, n). \tag{3.5}$$

In the frequency domain, the spectral distribution function (SDF) of $\{y(m, n)\}$ can be uniquely represented by the SDFs of its component fields:

$$F_y(\xi, \eta) = F_v(\xi, \eta) + F_w(\xi, \eta) \tag{3.6}$$

where $F_v(\xi, \eta) = F_h(\xi, \eta) + F_g(\xi, \eta)$, and functions $F_h(\xi, \eta)$ and $F_g(\xi, \eta)$ correspond to spectral singularities supported by point-like and line-like regions, respectively.

A Wold-based model can be built by decomposing an image into its Wold components and modeling each of the components separately. Two decomposition methods have been proposed in the literature. The first is a maximum likelihood direct parameter estimation procedure, which provides parametric descriptions of image Wold components [31]. The authors reported that the algorithm can be computationally expensive, especially when the number of spectral peaks is large or the energy in the spectral peaks is not very high compared to that in the neighboring Fourier frequencies. Unfortunately, these situations often arise in natural images. The second method is a spectral decomposition procedure [29] which applies a global threshold to the image periodogram, and selects the Fourier frequencies with magnitude values larger than the threshold to be the harmonic or the evanescent components. Although this method is computationally efficient, it is not robust enough for the large variety of natural texture patterns. The problem here is that the support region of an harmonic peak in a natural texture is usually not a point, but a small spread surrounding the central frequency. Therefore, two issues are essential for a decomposition scheme: locating the spectral peak central frequencies and determining the peak support regions. A new spectral decomposition-based approach which addresses these issues is presented in [50]. The algorithm decomposes an image by extracting its Fourier spectral peaks supported by point-like or line-like regions. After that, a spectral approach locates the peak central frequencies and estimates the peak support.

3.2.2 Non-Parametric PDF Methods

The distinction between parametric and non-parametric methods reflects the distinction made in statistics between parametric and non-parametric PDF modeling techniques. The methods described in this section also model the PDF of a pixel intensity as a function of the intensities of neighboring pixels. However, the methods described here use, in statistical parlance, non-parametric PDF models.

Gray-Level Co-occurrence Matrices. Spatial gray-level co-occurrence estimates image properties related to second order statistics. Haralick [37] suggested the use of gray-level co-occurrence matrices (GLCM) which have become one of the most well-known and widely used texture features. The

$G \times G$ gray-level co-occurrence matrix P_d for a displacement vector $\boldsymbol{d} = (dx, dy)$ is defined as follows. The entry (i, j) of P_d is the number of occurrences of the pair of gray-levels i and j which are a distance $\boldsymbol{d}$ apart. Formally, it is given as:

$$P_d(i, j) = |\{((r, s), (t, v)) : (t, v) = (r + dx, s + dy), \mathcal{I}(r, s) = i, \mathcal{I}(t, v) = j\}|$$

where $(r, s), (t, v) \in N \times N$ and $|\cdot|$ is the cardinality of a set.

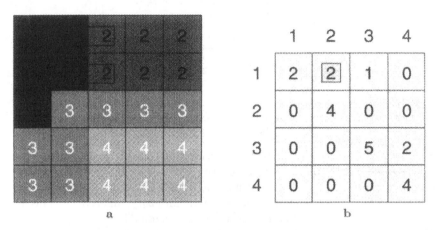

a   b

Fig. 3.5 Gray-level co-occurrence matrix computation; **a** an image is quantized to four intensity levels, **b** the corresponding GLC matrix is computed with the offset $(dx, dy) = (1, 0)$.

An example is given in Fig. 3.5. In Fig. 3.5(a) there are boxed two pairs of pixels which have $\mathcal{I}(x, y) = 1$ and $\mathcal{I}(x + dx, y + dy) = 2$. The corresponding bin in the GLC matrix is emphasized in Fig. 3.5(b).

Note that the co-occurrence matrix defined in this way is not symmetric. A symmetric variant can be computed using the formula: $P = P_d + P_{-d}$. The co-occurrence matrix reveals certain properties about the spatial distribution of the gray-levels in the texture image. For example, if most of the entries in the co-occurrence matrix are concentrated along the diagonals, then the texture is coarse with respect to the displacement vector $\boldsymbol{d}$. Haralick proposed a number of texture features that can be computed from the co-occurrence matrix. Some features are listed in Table 3.1. Here μ_x and μ_y are the means and σ_x and σ_y are the standard deviations of $P_d(x) = \sum_j P_d(x, j)$ and $P_d(y) = \sum_i P_d(i, y)$.

The co-occurrence matrix features suffer from a number of difficulties. There is no well established method of selecting the displacement vector $\boldsymbol{d}$ and computing co-occurrence matrices for a lot of different values of $\boldsymbol{d}$ is not feasible. Moreover, for a given $\boldsymbol{d}$, a large number of features can be computed. This means that some sort of feature selection method must be

Table 3.1 Some texture features extracted from gray-level co-occurrence matrices

Texture feature	Formula
Energy	$\sum_i \sum_j P_d^2(i, j)$
Entropy	$-\sum_i \sum_j P_d(i, j) \log P_d(i, j)$
Contrast	$\sum_i \sum_j (i - j)^2 P_d(i, j)$
Homogeneity	$\sum_i \sum_j \frac{P_d(i,j)}{1+|i-j|}$
Correlation	$\frac{\sum_i \sum_j (i-\mu_x)(j-\mu_y)P_d(i,j)}{\sigma_x \sigma_y}$

used to select the most relevant features. These methods have primarily been used in texture classification tasks and not in segmentation tasks.

Gray-Level Difference Methods. As noted by Weszka et al. [80], gray-level difference (GLD) methods strongly resemble GLC methods. The main difference is that, whereas the GLC method computes a matrix of intensity pairs, the GLD method computes a vector of intensity differences. This is equivalent to summing a GLC matrix along its diagonals.

Formally, for any given displacement $d = (dx, dy)$, let $\mathcal{I}_d(x, y) = |\mathcal{I}(x, y) - \mathcal{I}(x + dx, y + dy)|$. Let p_d be the probability density of $\mathcal{I}_d(x, y)$. If there are m gray-levels, this has the form of an m-dimensional vector whose ith component is the probability that $\mathcal{I}_d(x, y)$ will have value i. If the image $\mathcal{I}$ is discrete, it is easy to compute p_d by counting the number of times each value of $\mathcal{I}_d(x, y)$ occurs. Similar features as in Table 3.1 can be computed.

Texture Spectrum Methods. All methods described above model texture as a random field. They model the intensity of a pixel as a stochastic function of the intensities of neighboring pixels. However, the space of all possible intensity patterns in a neighborhood is huge. For example, if a 5×5 neighborhood is considered (excluding the central pixel), the PDF is a function in a 24-dimensional space. GMRF and GLC methods rely on assumptions that reduce the complexity of the PDF model. GRMF methods estimate the intensity as a function of all the neighboring pixels but assume that the distribution is Gaussian and is centered on a linear function of the neighboring intensities. The GLC method uses a histogram model; this requires the space of intensities to be partitioned into histogram bins. The partitioning is only sensitive to second order interactions but not to higher order interactions.

Texture spectrum methods use PDF models which are sensitive to high order interactions. Typically, these methods use a histogram model in which the partitioning of the intensity space is sensitive to high order interactions between pixels. This sensitivity is made feasible by quantizing the intensity values to a small number of levels, which considerably reduces the size of the space. The largest number of levels used is four but two levels, or thresholding, is more common.

Ojala et al. [56] proposed a texture unit represented by eight elements, each of which has two possible values $\{0, 1\}$ obtained from a neighborhood of 3×3 pixels. These texture units are called local binary patterns (LBP) and their occurrence of distribution over a region forms the texture spectrum. The LBP is computed by thresholding each of the noncenter pixels by the value of the center pixel, resulting in 256 binary patterns. The LBP method is a gray-scale invariant and can be easily combined with a simple contrast measure by computing for each neighborhood the difference of the average gray-level of those pixels which after thresholding have the value 1, and those which have the value 0, respectively. The algorithm is detailed below.

For each 3×3 neighborhood, consider P_i the intensities of the component pixels with P_0 the intensity of the center pixel. Then,

1. Threshold pixels P_i by the value of the center pixel: $P_i' = \begin{cases} 0 & \text{if } P_i < P_0, \\ 1 & \text{otherwise.} \end{cases}$

2. Count the number n of resulting non-zero pixels: $n = \sum\limits_{i=1}^{8} P_i'$.

3. Calculate the local binary pattern: $LBP = \sum\limits_{i=1}^{8} P_i' 2^{i-1}$.

4. Calculate the local contrast:

$$C = \begin{cases} 0 & \text{if } n = 0 \text{ or } n = 8, \\ \frac{1}{n} \sum\limits_{i=1}^{8} P_i' P_i - \frac{1}{8-n} \sum\limits_{i=1}^{8} (1 - P_i') P_i & \text{otherwise.} \end{cases}$$

A numerical example is given in Fig. 3.6.

Example			Thresholded			Weights		
6	5	2	1	0	0	1	0	0
7	6	1	1		0	8		0
9	3	7	1	0	1	32	0	128

LBP=1+8+32+128=169   C=(6+7+9+7)/4-(5+2+1+3)/4=4.5

Fig. 3.6 Computation of local binary pattern (LBP) and contrast measure (C).

The LBP method is similar to the methods described by Wang and He [78] and by Read et al. [64]. These methods generate more distinct texture units than LBP. Wang and He [78] quantize to three intensity levels, giving 3^8 or 6561 distinct texture units. Read et al. [64] quantize the image to four intensity levels, but use only a 3×2 neighborhood. This gives 4^6 or 4096 distinct texture units.

Another class of texture spectrum methods consists of N-tuple methods [2, 6]. Whereas the texture units in the methods described above use all the pixels in a small neighborhood, the N-tuple methods use a subset of pixels from a larger neighborhood. Typically, subsets of 6 to 10 pixels are used from a 6×6 to a 10×10 neighborhood. These subsets are selected randomly; a typical N-tuple memory algorithm might have 30 N-tuple units, each of which has a distinct random subset of pixels in the neighborhood. The texture unit histogram and texture class information are computed independently for each N-tuple unit. Texture class information from each N-tuple unit is combined to give the texture class information of the N-tuple memory.

In summary, texture spectrum methods are sensitive to high order interactions between pixels. This is made feasible by reducing the size of the space of intensities by quantization. Even in this reduced space, only a limited number of pixels, typically less than 10, contribute to the feature vector that characterizes the texture.

3.2.3 Harmonic Methods

The main methods in this category are the Fourier power spectrum and autocorrelation methods. These two methods are intimately related, since the autocorrelation function and the power spectral function are Fourier transforms of each other [38]. Harmonic methods model a texture as a summation of waveforms so they assume that intensity is a strongly periodic function of the spatial coordinates.

Autocorrelation Features. An important property of many textures is the repetitive nature of the placement of texture elements in the image. The autocorrelation function of an image can be used to assess the amount of regularity as well as the fineness/coarseness of the texture present in the image. Formally, the autocorrelation function ρ of an image $\mathcal{I}$ is defined as follows:

$$\rho(x,y) = \frac{\sum_{u=0}^{N} \sum_{v=0}^{N} \mathcal{I}(u,v)\mathcal{I}(u+x, v+y)}{\sum_{u=0}^{N} \sum_{v=0}^{N} \mathcal{I}^2(u,v)}. \tag{3.7}$$

Examples of autocorrelation functions for some Brodatz textures are shown in Fig. 3.7. The autocorrelation functions of non-periodic textures are dominated by a single peak. The breadth and elongation of the peak is determined by the coarseness and directionality of the texture. In a fine-grained texture, as in Fig. 3.7(a), the autocorrelation function will decrease rapidly, generating a sharp peak. On the other hand, in a coarse-grained texture, as in Fig. 3.7(b), the autocorrelation function will decrease more slowly, generating a broader peak. A directional texture (Fig. 3.7(c)) will generate an elongated peak. For regular textures (Fig. 3.7(d)) the autocorrelation function will exhibit peaks and valleys.

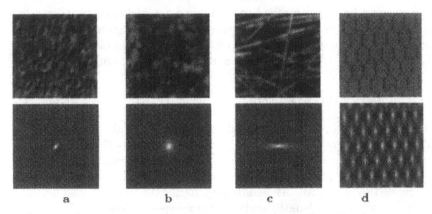

Fig. 3.7 Autocorrelation functions computed for four Brodatz textures.

The discriminatory power of the autocorrelation methods have been compared with other methods, empirically by Weszka et al. [80], and theoretically by Conners and Harlow [17]. Both studies find that the autocorrelation methods are less powerful discriminators than GLC methods. Weszka suggests that this is due to the inappropriateness of the texture model: "the textures ... may be more appropriately modeled statistically in the space domain (e.g., as random fields with specified autocorrelation), rather than as a sum of sinusoids."

Fourier Domain Features. The frequency analysis of the textured image is best done in the Fourier domain. As the psychophysical results indicated, the human visual system analyzes the textured images by decomposing the image into its frequency and orientation components [10]. The multiple channels tuned to different frequencies are also referred to as multi-resolution processing in the literature. Similar to the behavior of the autocorrelation function is the behavior of the distribution of energy in the power spectrum. Early approaches using spectral features divided the frequency domain into rings (for frequency content) and wedges (for orientation content). The frequency domain is thus divided into regions and the total energy in each of these regions is computed as a texture feature. For example the energy computed in a circular band is a feature indicating coarseness/fineness:

$$f_{r_1,r_2} = \int_0^{2\pi} \int_{r_1}^{r_2} |F(u,v)|^2 dr d\theta \tag{3.8}$$

and the energy computed in each wedge is a feature indicating directionality:

$$f_{\theta_1,\theta_2} = \int_{\theta_1}^{\theta_2} \int_0^\infty |F(u,v)|^2 dr d\theta \tag{3.9}$$

where $r = \sqrt{u^2 + v^2}$ and $\theta = \arctan(v/u)$.

In summary, harmonic methods are based on a model of texture consisted of sums of sinusoidal waveforms which extend over large regions. However, the modern literature, supported by empirical results, favors modeling texture as a local phenomenon, rather than a phenomenon with long range interactions.

3.2.4 Primitive Methods

This section reviews primitive methods including textural "edgeness," convolution, and mathematical morphology methods. Many of the texture primitives used in these methods have a specific scale and orientation. For example, lines and edges have a well-defined orientation, and the scale of a line is determined by its width. As seen before, harmonic methods also measure scale (frequency of wavelength) and orientation specific features. In particular, there is a strong relationship between primitive methods such as Gabor convolutions and Fourier transforms; essentially, a Gabor convolution is a spatially localized Fourier transform [20]. However, the distinction between these methods is straightforward: primitive methods measure spatially local features, whereas harmonic methods measure spatially dispersed features.

Primitive methods are also related to structural texture methods. Both methods model textures as being composed of primitives. However, structural methods tend to have one arbitrarily complex primitive, whereas primitive methods model texture as composed of many, simple primitives. Moreover, the relative placement of primitives is important in structural methods but plays no role in primitive methods.

Early Primitive Methods. Spatial domain filters are the most direct way to capture image texture properties. Earlier attempts at defining such methods concentrated on measuring the edge density per unit area. Fine textures tend to have a higher density of edges per unit area than coarser textures. The measurement of edgeness is usually computed by simple edge masks such as Robert's operator or the Laplacian operator. The edgeness measure can be computed over an image area by computing a magnitude from the response of Robert's mask or from the response of the Laplacian mask. Hsu [40] takes a different approach. He measures the pixel-wise intensity difference between the actual neighborhood and a uniform intensity; this distance is used as a measure of the density of edges.

Laws [48] describes a set of convolution filters which respond to edges, extrema, and segments of high frequency waveforms. A two-step procedure is proposed. In the first step, an image is convolved with a mask, and in the second the local variance is computed over a moving window. The proposed masks are small, separable, and simple and can be regarded as low cost, i.e., computationally inexpensive, Gabor functions.

Malik and Perona [53] proposed a spatial filtering to model the pre-attentive texture perception in the human visual system. They propose three stages: (1) convolution of the image with a bank of even-symmetric filters followed by a half-wave rectification, (2) inhibition of spurious responses in a localized area, and (3) detection of boundaries between the different textures. The even-symmetric filters used by them consist of differences of offset Gaussian functions. The half-wave rectification and inhibition are methods of introducing a non-linearity into the computation of texture features. This non-linearity is needed in order to discriminate texture pairs with identical mean brightness and identical second order statistics. The texture boundary detection is done by a straightforward edge detection method applied to the feature images obtained from stage (2). This method works on a variety of texture examples and is able to discriminate natural as well as synthetic textures with carefully controlled properties.

Gabor and Wavelet Models. The Fourier transform is an analysis of the global frequency content in the signal. Many applications require the analysis to be localized in the spatial domain. This is usually handled by introducing spatial dependency into the Fourier analysis. The classical way of doing this is through what is called the window Fourier transform. Considering a one-dimensional signal $f(x)$, the window Fourier transform is defined as:

$$F_w(u, \psi) = \int_{-\infty}^{\infty} f(x)w(x - \psi)e^{-2\pi i u x} dx. \tag{3.10}$$

When the window function $w(x)$ is Gaussian, the transform becomes a Gabor transform. The limits of the resolution in the time and frequency domain of the window Fourier transform are determined by the *time–bandwidth product* or the *Heisenberg uncertainty inequality* given by $\Delta t \Delta u \geq \frac{1}{4\pi}$. Once a window is chosen for the window Fourier transform, the time–frequency resolution is fixed over the entire time–frequency plane. To overcome the resolution limitation of the window Fourier transform, one lets the Δt and Δu vary in the time–frequency domain. Intuitively, the time resolution must increase as the central frequency of the analyzing filter is increased. That is, the relative bandwidth is kept constant in a logarithmic scale. This is accomplished by using a window whose width changes as the frequency changes. Recall that when a function $f(t)$ is scaled in time by a, which is expressed as $f(at)$, the function is contracted if $a > 1$ and it is expanded when $a < 1$. Using this fact, the wavelet transform can be written as:

$$W_{f,a}(u, \psi) = \frac{1}{\sqrt{a}} \int_{-\infty}^{\infty} f(t) h \left(\frac{t - \psi}{a} \right) dt. \tag{3.11}$$

Setting in Eq. 3.11,

$$h(t) = w(t) e^{-2\pi i u t} \tag{3.12}$$

we obtain the wavelet model for texture analysis. Usually the scaling factor will be based on the frequency of the filter.

Daugman [19] proposed the use of Gabor filters in the modeling of receptive fields of simple cells in the visual cortex of some mammals. The proposal to use the Gabor filters in texture analysis was made by Turner [74] and Clark and Bovik [16]. Gabor filters produce frequency decompositions that achieve the theoretical lower bound of the uncertainty principle [20]. They attain maximum joint resolution in space and frequency bounded by the relations $\Delta x \Delta u \geq \frac{1}{4\pi}$ and $\Delta y \Delta v \geq \frac{1}{4\pi}$, where $[\Delta x, \Delta y]$ gives the resolution in space and $[\Delta u, \Delta v]$ gives the resolution in frequency.

A two-dimensional Gabor function consists of a sinusoidal plane wave of a certain *frequency* and *orientation* modulated by a Gaussian envelope. It is given by:

$$g(x, y) = \exp \left(-\frac{1}{2} \left(\frac{x^2}{\sigma_x^2} + \frac{y^2}{\sigma_y^2} \right) \right) \cos(2\pi u_0 (x \cos\theta + y \sin\theta)) \tag{3.13}$$

where u_0 and θ are the frequency and phase of the sinusoidal wave, respectively. The values σ_x and σ_y are the sizes of the Gaussian envelope in the x and y directions, respectively. The Gabor function at an arbitrary orientation θ_0 can be obtained from Eq. 3.13 by a rigid rotation of the xy plane by θ_0.

The Gabor filter is a frequency and orientation selective filter. This can be seen from the Fourier domain analysis of the function. When the phase θ is 0, the Fourier transform of the resulting even-symmetric Gabor function $g(x, y)$ is given by:

$$G(u, v) = A \left(\exp \left(-\frac{1}{2} \left(\frac{(u - u_0)^2}{\sigma_u^2} + \frac{v^2}{\sigma_v^2} \right) \right) + \exp \left(-\frac{1}{2} \left(\frac{(u + u_0)^2}{\sigma_u^2} + \frac{v^2}{\sigma_v^2} \right) \right) \right)$$

where $\sigma_u = 1/(2\pi\sigma_x)$, $\sigma_v = 1/(2\pi\sigma_y)$ and $A = \pi\sigma_x\sigma_y$. This function is real-valued and has two lobes in the frequency domain, one centered around u_0, and another centered around $-u_0$. For a Gabor filter of a particular orientation, the lobes in the frequency domain are also appropriately rotated.

The Gabor filter masks can be considered as orientation and scale tunable edge and line detectors. The statistics of these microfeatures in a given region can be used to characterize the underlying texture information. A class of such self-similar functions referred to as Gabor wavelets is discussed by Ma and Manjunath [51]. This self-similar filter dictionary can be obtained by appropriate dilations and rotations of $g(x, y)$ through the generating function,

$$g_{mn}(x,y) = a^{-m}g(x',y'), \quad m = 0, 1, \ldots, S - 1, \tag{3.14}$$

$$x' = a^{-m}(x\cos\theta + y\sin\theta), \quad y' = a^{-m}(-x\sin\theta + y\cos\theta)$$

where $\theta = n\pi/K$, K the number of orientations, S the number of scales in the multiresolution decomposition, and $a = (U_h/U_l)^{-1/(S-1)}$ with U_l and U_h the lower and the upper center frequencies of interest, respectively.

An alternative to gain in the trade-off between space and frequency resolution without using Gabor functions is using a wavelet filter bank. The wavelet filter bank produces octave bandwidth segmentation in frequency. It allows simultaneously for high spatial resolutions at high frequencies and high frequency resolution at low frequencies. Furthermore, the wavelet tiling is supported by evidence that visual spatial-frequency receptors are spaced at octave distances [21]. A quadrature mirror filter (QMF) bank was used for texture classification by Kundu and Chen [47]. A two-band QMF bank utilizes orthogonal analysis filters to decompose data into low-pass and high-pass frequency bands. Applying the filters recursively to the lower frequency bands produces wavelet decomposition as illustrated in Fig. 3.8.

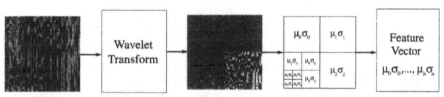

Fig. 3.8 Texture classifier for Brodatz textures samples using QMF-wavelet based features.

The wavelet transformation involves filtering and subsampling. A compact representation needs to be derived in the transform domain for classification and retrieval. The mean and the variance of the energy distribution of the transform coefficients for each subband at each decomposition level are used to construct the feature vector (Fig. 3.8) [67]. Let the image subband be $W_n(x,y)$, with n denoting the specific subband. Note that in the case of Gabor wavelet transform (GWT) there are two indexes m and n with m indicating a certain scale and n a certain orientation [51]. The resulting feature vector is $f = (\mu_n, \sigma_n)$ with,

$$\mu_n = \int |W_n(x,y)|dx\,dy, \tag{3.15}$$

$$\sigma_n = \sqrt{\int (|W_n(x,y)| - \mu_n)^2\,dx\,dy}. \tag{3.16}$$

Consider two image patterns i and j and let $f^{(i)}$ and $f^{(j)}$ represent the corresponding feature vectors. The distance between the two patterns in the features space is:

$$d(f^{(i)}, f^{(j)}) = \sum_n \left(\left| \frac{\mu_n^{(i)} - \mu_n^{(j)}}{\alpha(\mu_n)} \right| + \left| \frac{\sigma_n^{(i)} - \sigma_n^{(j)}}{\alpha(\sigma_n)} \right| \right) \tag{3.17}$$

where $\alpha(\mu_n)$ and $\alpha(\sigma_n)$ are the standard deviations of the respective features over the entire database.

Jain and Farrokhina [41] used a version of the Gabor transform in which window sizes for computing the Gabor filters are selected according to the central frequencies of the filters. They use a bank of Gabor filters at multiple scales and orientations to obtain filtered images. Each filtered image is passed through a sigmoidal nonlinearity. The texture feature for each pixel is computed as the absolute average deviation of the transform values of the filtered images from the mean with a window W of size $M \times M$. The filtered images have zero mean, therefore, the ith texture feature image $e_i(x,y)$ is given by the equation:

$$e_i(x,y) = \frac{1}{M^2} \sum_{(a,b) \in W} |\psi(r_i(a,b))| \tag{3.18}$$

where $r_i(x,y)$ is the filtered image for the ith filter and $\psi(t)$ is the nonlinearity having the form of $\tanh(\alpha t)$, where the choice of α is determined empirically.

Mathematical Morphology. Haralick [38] reviews mathematical morphology methods. These methods generate transformed images by erosion and dilation with *structural elements*. These structural elements correspond to texture primitives. For example, consider a binary image with pixel values *on* and *off*. A new image can be formed in which a pixel is *on* if the corresponding pixel and all pixels within a certain radius, in the original image are *on*. This transformation is an erosion with a circular structural element.

The number of *on* pixels in the transformed image will be a function of the texture in the original image and of the structural element. For example, consider a strongly directional texture, such as a striped image, and an elongated structural element. If the structural element is perpendicular to the stripes, there will be very few *on* pixels in the transformed image. However, if the structural element is elongated in the same direction as the stripes, most of the *on* pixels in the original image will be also *on* in the transformed image. Thus, the first order properties of the transformed image reflect the density, in the original image, of the texture primitive corresponding to structural element.

3.2.5 Fractal Methods

Fractal methods are distinguished from other methods by explicitly encoding the scaling behavior of the texture. Fractals are useful in modeling different properties of natural surfaces like roughness and self-similarity at different scales.

Self-similarity across scales in fractal geometry is a crucial concept. A deterministic fractal is defined using this concept of self-similarity as follows. Given a bounded set A in an Euclidean n-space, the set A is said to be self-similar when A is the union of N distinct (non-overlapping) copies of itself, each of which has been scaled down by a ratio of r. The fractal dimension D is related to the number N and the ratio r as follows:

$$D = \frac{\log N}{\log(1/r)}. \tag{3.19}$$

The fractal dimension gives a measure of the roughness of a surface. Intuitively, the larger the fractal dimension is, the rougher the texture is. Pentland [57] argued and gave evidence that images of most natural surfaces can be modeled as spatially isotropic fractals. However, most natural texture surfaces are not deterministic but have a statistical variation. This makes the computation of fractal dimension more difficult.

There are a number of methods proposed for estimating the fractal dimension. One method is the estimation of the box dimension [46]. Given a bounded set A in an Euclidean n-space, consider boxes of size $L_{\max}$ on a side which cover the set A. A scaled down version of the set A by ratio r will result in $N = 1/r^D$ similar sets. This new set can be covered by boxes of size $L = rL_{\max}$. The number of such boxes is related to the fractal dimension:

$$N(L) = \frac{1}{r^D} = \left(\frac{L_{\max}}{L}\right)^D. \tag{3.20}$$

The fractal dimension is estimated from Eq. 3.20 by the following procedure. For a given L, divide the n-space into a grid of boxes of size L and count the number of boxes covering A. Repeat this procedure for different values of L. The fractal dimension D can be estimated from the slope of the line:

$$\ln(N(L)) = -D\ln(L) + D\ln(L_{\max}). \tag{3.21}$$

Another method for estimating the fractal dimension was proposed by Voss [77]. Let $P(m, L)$ be the probability that there are m points within a box of side length L centered at an arbitrary point of a surface A. Let M be the total number of points in the image. When one overlays the image with boxes of side length L, then the quantity $(M/m)P(m, L)$ is the expected number of boxes with m points inside. The expected total number of boxes needed to cover the whole image is:

$$E[N(L)] = M \sum_{m=1}^{N} \frac{1}{m} P(m, L). \tag{3.22}$$

The expected value of $N(L)$ is proportional to L^{-D} and thus can be used to estimate the fractal dimension. Super and Bovik [69] proposed a method based on Gabor filters to estimate the fractal dimension in textured images.

The fractal dimension alone is not sufficient to completely characterize a texture. It has been shown [46] that there may be perceptually very different textures that have very similar fractal dimensions. Therefore, another measure called *lacunarity* [46, 77] has been suggested in order to capture the textural property that will distinguish between such textures. Lacunarity is defined as follows:

$$\Lambda = E\left[\left(\frac{M}{E[M]} - 1\right)^2\right] \tag{3.23}$$

where M is the mass of the fractal set and $E[M]$ is its expected value. Lacunarity measures the discrepancy between the actual mass and the expected value of the mass: lacunarity is small when the texture is fine and it is large when the texture is coarse. The mass of the fractal set is related to the length L by the power law: $M(L) = KL^D$.

Voss [77] computed the lacunarity from the probability distribution $P(m, L)$. Let $M(L) = \sum_{m=1}^{N} mP(m, L)$ and $M^2(L) = \sum_{m=1}^{N} m^2 P(m, L)$. Then the lacunarity Λ is given by:

$$\Lambda(L) = \frac{M^2(L) - (M(L))^2}{(M(L))^2}. \tag{3.24}$$

For fractal textures, the fractal dimension is a theoretically sound feature. However, many real-world textures are not fractal in the sense of obeying the self-similarity law. Any texture with sharp peaks in its power spectrum does not obey the self-similarity law.

3.2.6 Line Methods

Line methods are distinguished by measuring features from one-dimensional (though possibly curved) subsets of texture. These methods are computationally cheap. They were among the earliest texture methods being attractive in those times possibly because of their computational cheapness. These methods are not derived from the definition of texture as a locally structured two-dimensional homogeneous random field.

Barba and Ronsin [3], performed a *curvilinear integration* of the gray-level signal along some half scan lines starting from each pixel and in different orientations. The scan line ends when the integration reaches a preset value. The measure of texture is then given by the vector of displacements along the lines of different orientations.

Another line method is the *gray-level run length* method [33] which is based on computing the number of gray-level runs of various lengths. A gray-level run is a set of linearly adjacent image pixels having the same gray value. The length of the run is the number of pixels within the run. The element $r'(i, j|\theta)$ of the gray-level run length matrix specifies the number of times an image contains a run of length j for gray-level i in the angle θ direction. These matrices usually are calculated for several values of θ.

3.2.7 Structural Methods

The structural models of texture assume that textures are composed of texture primitives. They consider that the texture is produced by the placement of the primitives according to certain placement rules. Structural methods are related to primitive methods because both model textures as being composed of primitives. However, structural methods tend to have one arbitrarily complex primitive, whereas primitive methods model texture as composed of many, simple primitives. Moreover, the relative placement of primitives is important in structural methods but plays no role in primitive methods. Structural-based algorithms are in general limited in power unless one is dealing with very regular textures.

There are a number of ways to extract texture elements in images. Usually texture elements consist of regions in the image with uniform gray-levels (blobs or mosaics). Blob methods segment the image into small regions. In the mosaic algorithms the regions completely cover the image, forming a mosaic-like tessellation of the image [38, 72]. In some blob methods, the regions do not completely cover the image, forming a foreground of non-touching blobs. A blob or mosaic method models the image as composed of one simple but variable primitive; the measured features model the variations of the primitive found in the image.

Voorhees [76] and Chen et al. [15] describe methods in which features are extracted from non-contiguous blobs. Voorhees convolves the image with a center surround filter and thresholds the resulting convolution image slightly below zero. This gives an image consisting of many foreground blobs corresponding to dark patches in the original image. He characterizes a texture by the contrast, orientation, width, length, area, and area density of blobs within the texture. Chen et al. threshold the original image at several different intensities generating a separate binary image for each threshold. In these binary images contiguous pixels above the intensity threshold form bright blobs; likewise, contiguous pixels below the intensity threshold form dark blobs. The average area and irregularity of these blobs are measured for each intensity threshold, giving features which measure the granularity and elongation of the texture.

Tuceryan and Jain [72] proposed the extraction of texture tokens by using the properties of the Voronoi tessellation of the given image. Voronoi tessellation has been proposed because of its desirable properties in defining local spatial neighborhoods and because the local spatial distributions of tokens are reflected in the shapes of the Voronoi polygons. Firstly, texture tokens are extracted and then the tessellation is constructed. Tokens can be as simple as points of high gradient in the image or complex structures such as line segments or closed boundaries. Features of each Voronoi cell are extracted and tokens with similar features are grouped to construct uniform texture regions. Moments of the area of the Voronoi polygons serve as a useful set of features that reflect both the spatial distribution and shapes of the tokens in the textured image.

Zucker [81] has proposed a method in which he regards the observable textures (real textures) as distorted versions of ideal textures. The placement rule is defined for the ideal texture by a graph that is isomorphic to a regular or semiregular tessellation. These graphs are then transformed to generate the observable texture. Which of the regular tessellations is used as the placement rule is inferred from the observable texture. This is done by computing a two-dimensional histogram of the relative positions of the detected texture tokens.

Another approach to modeling texture by structural means is described by Fu [32]. Here the texture primitives can be as simple as a single pixel that can take a gray value, but is usually a collection of pixels. The placement rule is defined by a tree grammar. A texture is then viewed as a string in the language defined by the grammar whose terminal symbols are the texture primitives. An advantage of these methods is that they can be used for texture generation as well as texture analysis. The patterns generated by the tree grammars could also be regarded as ideal textures in Zucker's model [81].

3.3 Texture in Content-Based Retrieval

With content-based techniques, the important visual features of images are described mathematically using feature sets that are derived from the digital data. If chosen properly, the feature sets may coincide well with intuitive human notions of visual content while providing effective mathematical discrimination. Furthermore, the features may be extracted automatically without requiring human assistance. This approach allows the database to be indexed using the discriminating features that characterize visual content, and to be searched using visual keys. A content-based search of the database proceeds by finding the items most mathematically and visually similar to the search key.

Interest in visual texture was triggered by the phenomenon of texture discrimination which occurs when a shape is defined purely by its texture, with no associated change in color or brightness: color alone cannot distinguish between tigers and cheetahs! This phenomenon gives clear justification for texture features to be used in content-based retrieval together with color and shape. Several systems have been developed to search through image databases using texture, color, and shape attributes (QBIC [28], Photobook [58], Chabot [55], VisualSEEk [66], etc.). Although, in these systems texture features are used in combination with color and shape features, texture alone can also be used for content-based retrieval.

In practice, there are two different approaches in which texture is used as main feature for content-based retrieval. In the first approach, texture features are extracted from the images and then are used for finding similar images in the database [34, 52, 67]. Texture queries can be formulated in a similar manner to color queries, by selecting examples of desired textures from a palette, or by supplying an example query image. The system then retrieves images with texture measures most similar in value to the query. The systems using this approach may use already segmented textures as in the applications with Brodatz database [59], or they first have a segmentation stage after which the extracted features in different regions are used as queries [52]. The segmentation algorithm used in this case may be crucial for the content-based retrieval. In the second approach, texture is used for annotating the image [61]. Vision-based annotation assists the user in attaching descriptions to large sets of images and video. If a user labels a piece of an image as "water," a texture model can be used to propagate this label to other visually similar regions.

What distinguishes image search for database-related applications from traditional pattern classification methods is the fact that there is a human in the loop (the user), and in general there is a need to retrieve more than just the best match. In typical applications, a number of top matches with rank-ordered similarities to the query pattern will be retrieved. Comparison in the feature space should preserve visual similarities between patterns.

3.3.1 Texture Segmentation

As mentioned before, texture segmentation algorithms may be critical for the success of content-based retrieval. The segmentation process may be the first step in processing the user query. The segmentation is a difficult problem because one usually does not know a priori what types of textures exist in an image, how many different textures there are, and what regions have which texture. Generally speaking, one does not need to know which textures are present in the image in order to do segmentation. All that is needed is a way to tell that two textures, usually present in adjacent regions, are different. The performance of a segmentation method strongly depends on the quality of input features, i.e., the noise within homogeneous regions and the amplitude separation between them. In general, a segmentation method is judged according to its capacity to eliminate very small subregions within more substantial ones, in combination with its accuracy in locating region boundaries. These capabilities, which are strongly connected to an analysis of the local image content, distinguish segmentation from supervised pixel classification, where feature statistics are gathered over homogeneous training areas and pixels are classified without further using any local image information.

There are two general approaches to performing texture segmentation: region-based approaches and boundary-based approaches. In a region-based approach [52], one tries to identify regions in the image which have a uniform texture. Pixels or small local regions are merged based on the similarity of some texture property. The regions having different textures are then considered to be segmented regions. The advantage of this method is that the boundaries of regions are always closed and therefore, the regions with different textures are always well separated. However, in general, one has to specify the number of distinct textures present in the image in advance. Additionally, thresholds on similarity values are needed.

The boundary-based approaches [72, 76] are based upon the detection of differences in texture in adjacent regions. Here, one does not need to know the number of textured regions in the image in advance. However, the boundaries may not be closed and two regions with different textures are not identified as separate closed regions. Du Buf et al. [27] studied and compared the performance of various texture segmentation techniques and their ability to localize the boundaries.

Tuceryan and Jain [72] use the texture features computed from the Voronoi polygons in order to compare the textures in adjacent windows. The comparison is done using a Kolmogorov–Smirnoff test. A probabilistic relaxation labeling, which enforces border smoothness is used to remove isolated edge pixels and fill boundary gaps. Voorhees and Poggio [76] extract blobs and elongated structures from images (they suggest that these correspond to Julesz's textons [44]). The texture properties are based on blob characteristics such as their sizes, orientations, etc. They then decide whether the two sides of a pixel have the same texture using a statistical test called maximum frequency difference (MFD). The pixels where this statistic is sufficiently large are considered to be boundaries between different textures. A "blobworld" approach is used by Carson et al. [11]. In order to segment an image, the joint distribution of the color, texture, and position features of each pixel in the image is modeled. The expectation-maximization (EM) algorithm is used to fit a mixture of Gaussians to the data; the resulting pixel-cluster memberships provide the segmentation of the image. After the image is segmented into regions, a description of each region's color, texture, and spatial characteristics is produced.

Ma and Manjunath [52] propose a segmentation scheme which is appropriate for large images (large aerial photographs) and database retrieval applications. The proposed scheme utilizes Gabor texture features extracted from the image tiles and performs a coarse image segmentation based on local texture gradients. For such image retrieval applications accurate pixel level segmentation is often not necessary although the proposed segmentation method can achieve pixel-level accuracy. In the first stage, using the feature vectors, a local texture gradient is computed between each image tile and its surrounding eight neighbors. The dominant flow direction is identified in a competitive manner which is similar to a winner-takes-all representation. This is called texture edge flow as the gradient information is propagated to neighboring pixels (or tiles). The texture edge flow contains information about the (spatial) direction and energy of the local texture boundary. The next stage is the texture edge flow propagation. Following the orientation competition, the local texture edge flow is propagated to its neighbors, if they have the same directional preference. The flow continues until it encounters an opposite flow. This helps to localize the precise positions of the boundaries and concentrate the edge energies towards pixels where the image boundaries might exist. After the propagation reaches a stable state, the final texture edge flow energy is used for boundary detection. This is done by turning on the edge signals between two neighboring image tiles, if their final texture edge flow points in opposite directions. The texture energy is then defined to be the summation of texture edge flow energies in the two neighboring image tiles. This stage results in many discontinuous image boundaries. In the final stage, these boundaries are connected to form an initial set of image regions. At the end, a conservative region merging algorithm is used to group similar neighboring regions.

3.3.2 Texture Classification and Indexing

Texture classification involves deciding what texture category an observed image belongs to. In order to accomplish this, one needs to have a priori knowledge of the classes to be recognized. Once this knowledge is available and the texture features are extracted, the pattern classification techniques can be used in order to do the classification.

Examples where texture classification was applied as the appropriate texture processing method include the classification of regions in satellite images into categories of land use [37]. A recent extension of the technique is the texture thesaurus developed by Ma and Manjunath [52], which retrieves textured regions in images on the basis of similarity to automatically derived codewords (thesaurus) representing important classes of texture within the collection. This approach can be visualized as an image counterpart of the traditional thesaurus for text search. It creates the information links among the stored image data based on a collection of codewords and sample patterns obtained from the training set. Similar to parsing text documents using a dictionary or thesaurus, the information within images can be classified and indexed via the use of the texture thesaurus. The design of the thesaurus has two stages. The first stage uses a learning similarity algorithm to combine the human perceptual similarity with the low level feature vector information, and the second stage utilizes a hierarchical vector quantization technique to construct the codewords. The proposed texture thesaurus is domain dependent and can be designed to meet the particular need of a specific image data type by exploring the training data. It also provides an efficient indexing tree while maintaining or even improving the retrieval performance in terms of human perception. Furthermore, the visual codeword representation in the thesaurus can be used as information samples to help users browse through the database. A collection of visual thesauri for browsing large collections of geographic images is also proposed by Ramsey et al. [62]. Each texture region of an image is mapped to an output node in a self-organizing map (SOM) representing a unique cluster of similar textures. The output node number can be treated as an index for the image texture region and can be used to compute co-occurrence with other texture types (i.e., textures mapped to other output nodes). Then, users can browse the visual thesaurus and find a texture to start their query. A list of other classes of textures frequently co-occurring with the specified one is displayed in decreasing frequency order. From this list the user can refine the queries by selecting other textures they feel are relevant to their search.

3.3.3 Texture for Annotation

Traditionally, access to multimedia libraries has been in the form of text annotation – titles, authors, captions, and descriptive labels. Text provides a natural means of summarizing massive quantities of information. Text keywords consume little space and provide fast access into large amounts of data. When the data is text, it can be summarized using sophisticated computer programs based on natural language processing and artificial intelligence. When the data is not text but is for example image, then generating labels is considerably more difficult. Of course, some access to pictures is best achieved without text – for example, the user may wish to "find another image like this." In this case, a signal is compared to another signal and conversion to text is not required. However, annotation is both important for preparing multimedia digital libraries for query and retrieval, and useful for adding personal notes to the user's online collection. Tools that help annotate multimedia database need to be able to "see" as the human "sees" – so that if the human says "label this stuff grass" the computer will not only label that stuff but also find other grass that "looks the same" and label it too. Texture features, although low-level, play an important role in the high-level perception of visual scenes enabling the distinction of regions like "water" from "grass" or "buildings" from "sky." Such features alone do not solve the complete annotation problem but they are a key component of the solution.

Picard and Minka [61] proposed a system to help generate descriptions for annotating image and video. They noticed that a single texture model is not sufficient to reliably match human perception of similarity in pictures. Rather than using one model, the proposed system knows several texture models and is equipped with the ability to choose the one which "best explains" the regions selected by the user for annotating. If none of these models suffices, then the system creates new explanations by combining models. Determining which model gives the best features for measuring similarity is hampered by the fact that the users are fickle. People are nonlinear time-varying systems when it comes to predicting their behavior. A user may label the same scene differently, and expect different regions to be recognized as similar when his or her goals change. In the system, the user can browse through a database of scenes. He or she can select patches from one or more images as "positive examples" for a label. The system then propagates the label to new regions in the database that it thinks should also have the same label. The user can immediately view the results and can remove falsely labeled patches by selecting them to be "negative examples" for the label. The user can continue to add positive or negative examples for the same label or different labels. The system responds differently depending upon which patches are selected as positive or negative examples.

3.4 Summary

Texture is the term used to characterize the surface of a given object or phenomenon and it is undoubtedly one of the main features used in image processing and pattern recognition. In an image, texture is one of the visual characteristics that identifies a segment as belonging to a certain class. We recognize many parts of the image by texture rather than by shape, e.g., fur, hair, etc. If the texture belongs to a class that has a particular physical interpretation such as grass, hair, water, or sand, then it may be regarded as a "natural" texture. On the other hand, a texture may belong to a class identified by artificial visual characteristics that have a concise mathematical interpretation.

Despite its ubiquity, a formal definition of texture remains elusive. In the literature, various textural properties are often used to serve as the definition of texture, or more precisely to constrain the domain of problems. Some researchers formally define texture by describing it in terms of the human visual system, whereas others are completely driven in defining texture by the application in which this definition is used.

Identifying the perceived qualities of texture in an image is an important first step toward building mathematical models for texture. The intensity variations in an image which characterize texture are generally due to some underlying physical variations in the scene (e.g., waves in water). Modeling this physical variation is very difficult, so texture is usually characterized by the two-dimensional variations in the intensities present in the image. Various methods of texture representation were presented. The literature distinguishes between stochastic/statistical and structural methods for texture. The use of statistical features is one of the early methods proposed in the machine vision literature. These methods are based on the spatial distribution of gray values in the image. In a stochastic approach, a texture is assumed to be the realization of a stochastic process which is governed by some parameters. Analysis is performed by defining a model and estimating the parameters so that the stochastic process can be reproduced from the model and associated parameters. In a structural approach a texture is viewed as a two-dimensional pattern consisting of a set of primitives or subpatterns which are arranged according to certain placement rules.

Texture processing has been successfully applied in content-based retrieval. In practice, there are two different approaches in which texture is used as main feature for content-based retrieval. In the first approach, an image can be considered as a mosaic of different texture regions, and the image features associated with these regions can be used for search and retrieval. A typical query could be a region of interest provided by the user (e.g., a vegetation patch in a satellite image). This approach may involve an additional segmentation step which will identify the texture regions. In the second approach, texture is used for annotating the image. Here, vision-based annotation assists the user in attaching descriptions to large sets of images. The user is asked to label a piece of an image and a texture model can be used to propagate this label to other visually similar regions.

References

1. Ahuja, N and Rosenfeld, A, "Mosaic Models for Textures," IEEE Trans Patt Anal Mach Intell, 3, pp. 1–10, 1981.
2. Austin, J, "Grey Scale N-tuple Processing," International Conference on Pattern Recognition, pp. 110–120, 1988.
3. Barba, D and Ronsin, J, "Image Segmentation Using New Measure of Texture Feature," Digital Signal Processing, Elsevier-North-Holland, pp. 749–753, 1984.
4. Beck, J, Sutter, A, and Ivry, A, "Spatial Frequency Channels and Perceptual Grouping in Texture Segregation," Computer Vision Graphics Image Process, 37, pp. 299–325, 1987.
5. Bergen, JR and Adelson, EH, "Early Vision and Texture Perception," Nature, 333, pp. 363–364, 1988.
6. Bolt, G, Austin, J, and Morgan, G, "Uniform Tuple Storage in ADAM," Patt Recogn Lett, 13, pp. 339–344, 1992.
7. Bovik, A, Clarke, M, and Geisler, W, "Multichannel Texture Analysis Using Localized Spatial Filters," IEEE Trans on Patt Anal Mach Intell, 12, pp. 55–73, 1990.
8. Brodatz, P, "Textures: A Photographic Album for Artists and Designers," Dover Publications, 1966.
9. Caelli, T, Julesz, B, and Gilbert, E, "On Perceptual Analyzers Underlying Visual Texture Discrimination: Part II," Biol Cybern, 29(4), pp. 201–214, 1978.
10. Campbell, FW and Robson, JG, "Application of Fourier Analysis to the Visibility of Gratings," J Physiol, 197, pp. 551–566, 1968.
11. Carson, C, Thomas, M, Belongie, S, Hellerstein, J, and Malik, J, "Blobworld: A System for Region-Based Image Indexing and Retrieval," Conf. on Visual Information and Information Systems, pp. 509–516, 1999.
12. Chaudhuri, B, Sarkar, N, and Kundu, P, "Improved Fractal Geometry Based Texture Segmentation Technique," Proc IEE, 140, pp. 233–241, 1993.

13. Chellappa, R and Chatterjee, S, "Classification of Textures Using Gaussian Markov Random Fields," IEEE Trans Acoust Speech Sig Process, 33, pp. 959–963, 1985.
14. Chellappa, R, Chatterjee, S, and Bagdazian, R, "Texture Synthesis and Compression Using Gaussian Markov Random Fields Models," IEEE Trans Syst Man Cybern, 15, pp. 298–303, 1985.
15. Chen, Y, Nixon, M, and Thomas, D, "Statistical Geometrical Features for Texture Classification," Patt Recogn, 4, pp. 537–552, 1995.
16. Clark, M and Bovik, A, "Texture Segmentation Using Gabor Modulation/Demodulation," Patt Recogn Lett, 6, pp. 261–267, 1987.
17. Conners, R and Harlow, C, "A Theoretical Comparison of Texture Algorithms," IEEE Trans Patt Anal Mach Intell, 2, pp. 204–222, 1980.
18. Cross, G and Jain, AK, "Markov Random Field texture models," IEEE Trans Patt Anal Mach Intell, 5, pp. 25–39, 1983.
19. Daugman, J, "Two-Dimensional Spectral Analysis of Cortical Receptive Profiles," Vision Res, 20, pp. 847–856, 1980.
20. Daugman, J, "Uncertainty Relation for Resolution in Space, Spatial Frequency and Orientation Optimized by Two-Dimensional Visual Cortical Filters," J Opt Soc of Am, A 4, pp. 221–231, 1985.
21. Daugman, J, "Entropy Reduction and Decorrelation in Visual Coding by Oriented Neural Receptive Fields," IEEE Trans Biomed Eng, 36, 1989.
22. Derin, H and Cole, W, "Segmentation of Textured Images Using Gibbs Random Fields," Computer Vision, Graphics Image Process, 35, pp. 72–98, 1986.
23. Derin, H and Elliott, H, "Modeling and Segmentation of Noisy and Textured Images Using Gibbs Random Fields," IEEE Trans Patt Anal Mach Intell, 9, pp. 39–55, 1987.
24. De Souza, P, "Texture Recognition via Autoregression," Patt Recogn, 15, pp. 471–475, 1982.
25. De Valois, RL, Albrecht, DG, and Thorell, LG, "Spatial-Frequency Selectivity of Cells in Macaque Visual Cortex," Vision Res, 22, pp. 545–559, 1982.
26. Dubes, RC and Jain, AK, "Random Field Models in Image Analysis," J Appl Statist, 16(2), pp. 131–164, 1989.
27. Du Buf, J, Kardan, M, and Spann, M, "Texture Feature Performance for Image Segmentation," Patt Recogn, 23, pp. 291–309, 1990.
28. Flicker, M, Sawhney, H, Niblack, W, Ashley, J, Huang, Q, Dom, B, Gorkani, M, Hafner, J, Lee, D, Petkovic, D, Steele, D, and Yanker, P, "Query by Image and Video Content: The QBIC System," IEEE Computer, 28(9), pp. 23–32, 1995.
29. Francos, J, Meiri, A, and Porat, B, "A Unified Texture Model Based on a 2-D Wold-Like Decomposition," IEEE Trans Sig Process, 41, pp. 2665–2678, 1993.
30. Francos, J, "Orthogonal Decompositions of 2-D Random Fields and Their Applications in 2-D Spectral Estimation," Sign Process Applics, Handbook Statist, 41, pp. 207–227, 1993.
31. Francos, J, Narasimhan, A, and Woods, J, "Maximum Likelihood Parameter Estimation of Discrete Homogeneous Random Fields with Mixed Spectral Distributions," IEEE Trans Sig Process, 44(5), pp. 1242–1255, 1996.
32. Fu, K, Syntactic Pattern Recognition and Applications, Prentice-Hall, 1982.
33. Galloway, MM, "Texture Analysis Using Gray Level Run Lengths," Computer Vision Graphics Image Process, 4, pp. 172–179, 1975.
34. Gorkani, M and Picard, R, "Texture Orientation for Sorting Photos 'at a glance'," Int. Conf. on Patern Recognition, 1, pp. 459–464, 1994.
35. Haindl, M, "Texture Synthesis," CWI Quart, 4, pp. 305–331, 1991.

36. Hall, T and Gainnakis, G, "Texture Model Validation Using Higher-Order Statistics," Int Conf. on Acoustics, Speech and Signal Processing, pp. 2673–2676, 1991.
37. Haralick, RM, Shanmugam, K, and Dinstein, I, "Textural Features for Image Classification," IEEE Trans Syst Man Cybern, 3(6), pp. 610–621, 1973.
38. Haralick, RM, "Statistical and Structural Approaches to Texture," Proc IEEE, 67, pp. 786–804, 1979.
39. Hassner, M and Slansky, J, "The Use of Markov Random Fields as Models of Texture," Computer Vision Graphics Image Process, 12, pp. 357–360, 1980.
40. Hsu, S, "A texture-Tone Analysis for Automated Landuse Mapping with Panchromatic Images," Proc. of the American Society of Photogrammetry, pp. 203–215, 1977.
41. Jain, AK and Farrokhina, F, "Unsupervised Texture Segmentation Using Gabor Filters," Patt Recogn, 24, pp. 1167–1186, 1991.
42. Julesz, B, Gilbert, EN, Shepp, LA, and Frisch, HL, "Inability of Humans to Discriminate Between Visual Textures that Agree in Second-Order Statistics," Perception, 2, pp. 391–405, 1973.
43. Julesz, B, "Experiments in the Visual Perception of Texture," Scient Am, 232, pp. 34–43, 1975.
44. Julesz, B, "Textons, the Elements of Texture Perception and Their Interactions," Nature, 290, pp. 91–97, 1981.
45. Julesz, B, "A Theory of Preattentive Texture Discrimination Based on First-Order Statistics of Textons," Biol Cybern, 41(2), pp. 131–138, 1981.
46. Keller, J and Chen, S, "Texture Description and Segmentation Through Fractal Geometry," Computer Vision Graphics Image Process, 45, pp. 150–166, 1989.
47. Kundu, A and Chen, J-L, "Texture Classification Using QMF Bank-Based Subband Decomposition," CVGIP: Graph Models Image Process, 54, pp. 407–419, 1992.
48. Laws, KI, "Textured Image Segmentation," PhD thesis, University of Southern California, 1980.
49. Liu, F and Picard, R, "Periodicity, Directionality and Randomness: Wold Features for Image Modeling and Retrieval," IEEE Trans Patt Anal Mach Intell, 18, pp. 722–733, 1996.
50. Liu, F and Picard, R, "A Spectral 2-D Wold Decomposition Algorithm for Homogeneous Random Fields," IEEE Conf. on Acoustics, Speech and Signal Processing, 6, pp. 3501–3504, 1999.
51. Ma, WY and Manjunath, BS, "Texture Features and Learning Similarity," IEEE Conf. on Computer Vision and Pattern Recognition, pp. 425–430, 1996.
52. Ma, WY and Manjunath, BS, "A Texture Thesaurus for Browsing Large Aerial Photographs," J Am Soc Inform Sci, 49(7), pp. 633–648, 1998.
53. Malik, J and Perona, P, "Preattentive Texture Discrimination with Early Vision Mechanisms," J Opt Soc Am, A 7, pp. 923–932, 1990.
54. Manjunath, BS, Simchony, T, and Chellappa, R, "Stochastic and Deterministic Networks for Texture Segmentation," IEEE Trans Acoust Speech Sig Process, 38, pp. 1039–1049, 1990.
55. Ogle, V and Stonebracker, M, "Chabot: Retrieval from a Relational Database of Images," Computer, 28(9), pp. 40–48, 1995.
56. Ojala, T, Pietikainen, M, and Harwood, D, "A Comparative Study of Texture Measures with Classification Based on Feature Distribution," Patt Recogn, 29, pp. 51–59, 1996.
57. Pentland, A, "Fractal-Based Description of Natural Scenes," IEEE Trans Patt Anal Mach Intell, 6, pp. 661–672, 1984.

58. Pentland, A, Picard, R, and Sclaroff, S, "Photobook: Content-Based Manipulation of Image Databases," Int J Computer Vision, 18, pp. 233–254, 1996.
59. Picard, R, Kabir, T, and Liu, F, "Real-Time Recognition with the Entire Brodatz Texture Database," IEEE Conf. on Computer Vision and Pattern Recognition, pp. 638–639, 1993.
60. Picard, R and Liu, F, "A New Wold Ordering for Image Similarity," IEEE Conf. on Acoustics, Speech and Signal Processing, 5, pp. 129–132, 1994.
61. Picard, R and Minka, T, "Vision Texture for Annotation," Multimed Syst, 3(1), pp. 3–14, 1995.
62. Ramsey, M, Chen, H, Zhu, B, and Schatz, B, "A Collection of Visual Thesauri for Browsing Large Collections of Geographic Images," J Am Soc Informa Sci, 50(9), pp. 826–834, 1999.
63. Rao, A and Lohse, G, "Towards a Texture Naming System: Identifying Relevant Dimensions of Texture," IEEE Conf. on Visualization, 1993.
64. Read, J and Jayaramamurthy, S, "Automatic Generation of Texture Feature Detectors," IEEE Trans Computers, C-21, pp. 803–812, 1972.
65. Richards, W and Polit, A, "Texture Matching," Kybernetic, 16, pp. 155–162, 1974.
66. Smith, JR and Chang, SF, "VisualSEEk: A Fully Automated Content-Based Image Query System," ACM Multimedia, pp. 87–98, 1996.
67. Smith, JR and Chang, SF, "Transform Features for Texture Classification and Discrimination in Large Image Databases," Int. Conf. Image Processing, pp. 407–411, 1996.
68. Smith, G, "Image Texture Analysis Using Zero Crossing Information," PhD Thesis, University of Queensland, 1998.
69. Super, B and Bovik, A, "Localized Measurement of Image Fractal Dimension Using Gabor Filters," J Visual Commun Image Represent, 2, pp. 114–128, 1991.
70. Tamura, H, Mori, S, and Yamawaki, Y, "Textural Features Corresponding to Visual Perception," IEEE Trans Syst Man Cybern, 8, pp. 460–473, 1978.
71. Therrien, C, "An Estimation-Theoretic Approach to Terrain Image Segmentation," Computer Vision Graphics Image Process, 22, pp. 313–326, 1983.
72. Tuceryan, M and Jain, AK, "Texture Segmentation Using Voronoi Polygons," IEEE Trans Patt Anal Mach Intell, 12, pp. 211–216, 1990.
73. Tuceryan, M and Jain, AK, "Texture Analysis," Handbook of Pattern Recognition and Computer Vision, pp. 235–276, 1993.
74. Turner, M, "Texture Discrimination by Gabor Functions," Biol Cybern, 55, pp. 71–82, 1986.
75. van Gool, L, Dewael, P, and Oosterlinck, A, "Texture Analysis Anno 1983," Computer Vision Graphics Image Process, 29, pp. 336–357, 1985.
76. Voorhees, H and Poggio, T, "Computing Texture Boundaries in Images," Nature, 333, pp. 364–367, 1988.
77. Voss, R, "Random Fractals: Characterization and Measurement," in Scaling Phenomena in Disordered Systems, Plenum, New York, 1986.
78. Wang L and He, D, "Texture Classification Using Texture Spectrum," Patt Recogn, 23, pp. 905–910, 1991.
79. Wechsler, H, "Texture Analysis – A Survey," Sig Process, 2, pp. 271–282, 1980.
80. Weszka, J, Dyer, C, and Rosenfeld, A, "A Comparative Study of Texture Measures for Terrain Classification," IEEE Trans Syst Man Cybern, 6, pp. 269–265, 1976.
81. Zucker, S, "Toward a Model of Texture," Computer Graphics Image Process, 5, pp. 190–202, 1976.

4. State of the Art in Shape Matching

Remco C. Veltkamp and Michiel Hagedoorn

4.1 Introduction

Large image databases are used in an extraordinary number of multimedia applications in fields such as entertainment, business, art, engineering, and science. Retrieving images by their content, as opposed to external features, has become an important operation. A fundamental ingredient for content-based image retrieval is the technique used for comparing images. There are two general methods for image comparison: intensity based (color and texture) and geometry based (shape). A recent user survey about cognition aspects of image retrieval shows that users are more interested in retrieval by shape than by color and texture [62]. However, retrieval by shape is still considered one of the most difficult aspects of content-based search. Indeed, systems such as IBM's Query By Image Content, QBIC [57], perhaps one of the most advanced image retrieval systems to date, is relatively successful in retrieving by color and texture, but performs poorly when searching on shape. A similar behavior is exhibited in the new Alta Vista photo finder [10].

Shape matching is a central problem in visual information systems, computer vision, pattern recognition, and robotics. Applications of shape matching include industrial inspection, fingerprint matching, and content-based image retrieval. Figs 4.1, 4.2, and 4.3 illustrate a few typical problems which need to be solved:

Fig. 4.1 Shape matching in fruit inspection.

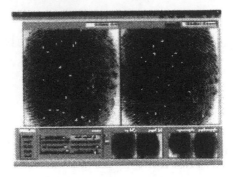

Fig. 4.2 Fingerprint matching.

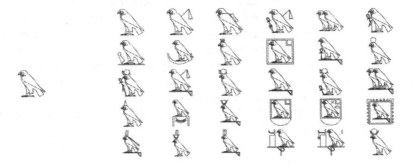

Fig. 4.3 Query hieroglyph (left), and hieroglyphs retrieved from database [77].

1. Fig. 4.1 illustrates an application in agricultural inspection. A typical problem here is to find a matching transformation. Based on shape characteristics, we can find the transformation that matches one piece of fruit with another.
2. Fig. 4.2 shows a point set matching application in fingerprint identification. After extraction of featuring points, two point sets must be matched. The difficulty here is that there is typically no one-to-one correspondence between the two point sets. The matching technique should be robust against noise and occlusion.
3. Fig. 4.3 shows an application in multimedia retrieval. Given the query shape at the left, the task is to find all pictures that contain similar shapes. The typical problem is that only pieces of the query shape appear in only parts of some of the database pictures.

This chapter deals with the matching of geometric shapes, with an emphasis on techniques from computational geometry. We are concerned with geometric patterns such as finite point sets, curves, and regions. For an overview of more general shape analysis, see Loncaric [48].

Matching deals with transforming a pattern, and measuring the resemblance with another pattern using some dissimilarity measure. Pattern matching and shape matching are commonly used interchangeably. However, more formally, the shape of a pattern is the pattern under all transformations in a transformation group. The matching problem is studied in various forms. Given two patterns and a dissimilarity measure:

Computation problem. Compute the dissimilarity between the two patterns.

Decision problem. For a given threshold, decide whether the dissimilarity between two patterns is smaller than the threshold.

Decision problem. For a given threshold, decide whether there exists a transformation such that the dissimilarity between the transformed pattern and the other pattern is smaller than the threshold.

Optimization problem. Find the transformation that minimizes the dissimilarity between the transformed pattern and the other pattern.

Sometimes the time complexities in solving these problems are rather great, so that it makes sense to devise approximation algorithms that find an approximation:

– Given two patterns, find a transformation that gives a dissimilarity between the two patterns that is within a specified factor from the minimum dissimilarity.

There are several variations on these problems. A pattern can be compared to a single pattern or to many other patterns, in which case an indexing structure is needed to speed up the comparisons. Another variation is to take artifacts such as noise into account, or to perform partial matching, i.e., finding a finding within a larger pattern.

There are various ways to approach the shape matching problem. In this chapter we focus on methods from computational geometry. Computational geometry is the subarea of algorithms design that deals with the design and analysis of algorithms for geometric problems involving objects like points, lines, polygons, and polyhedra. The standard approach taken in computational geometry is the development of exact, provably correct and efficient solutions to geometric problems. Aspects that play a crucial role in the algorithmic solutions to matching are the representation of patterns, the transformation group, and the dissimilarity measure.

4.2 Approaches

Matching has been approached in a number of ways, including tree pruning [71], the generalized Hough transform or pose clustering [14, 68], geometric hashing [80], the alignment method [37], statistics [66], deformable templates [64], relaxation labeling [58], Fourier descriptors [48], wavelet transform [43], curvature scale space [49], and neural networks [31]. Without being complete, in the following subsections we will describe and group a number of these methods together.

4.2.1 Global Image Transforms

There are a number of techniques that transform the image from color information in the spatial domain to color variation information in the frequency domain. Although such approaches do not explicitly encode shape for matching and retrieval, they represent color or intensity transitions in the image, which typically occurs at object boundaries.

A specific class of image transformations are wavelet-based transforms. Wavelets are functions that decompose signals (here two-dimensional color signals) into different frequency components. Each component is then analyzed at a resolution corresponding to its scale. Because the original image can be represented as a linear combination of wavelet functions, similar to the Fourier transform, we can process the images by the wavelet coefficients. By truncating the coefficients below a threshold, image data can be sparsely represented, at the cost of loss of detail. A set of such coefficients can be used as a feature vector for image matching.

The wavelet transform can be done with different basis functions. The Haar basis functions, used by Jacobs et al. [43], do not perform well when the query image consists of a small translation of the target image. This problem is less visible in the approach of Wang et al. [78] using Daubechies basis functions.

For the purpose of shape matching, a drawback of global image transforms is that shape information is not explicitly represented, and that the whole image is encoded, including color and texture information that need not indicate object transitions. As a result, it is not possible to measure how much two different images are similar in terms of shape. Also, due to the global nature, it is not possible to match a query shape with only a part of an image.

4.2.2 Global Object Methods

Below, we mention a few methods that work on an object as a whole, i.e., a complete object area or contour. An important drawback of all these methods is that complete objects in images must be clearly segmented, which is in itself an ill-posed problem. Typically the result of a segmentation process is a partitioning into regions that need not correspond to whole objects. However, the global object methods work only for whole objects. In general, such methods are not robust against noise and occlusions.

Moments. When a complete object in an image has been identified, it can be described by a set of moments $m_{p,q}$. The (p,q) moment of an object $O \subseteq \mathbb{R}^2$ is given by

$$m_{p,q} = \int_{(x,y)\in O} x^p y^q \, dx \, dy \tag{4.1}$$

or, in terms of pixels in a binary $[1,n] \times [1,m]$ image f:

$$\sum_{x=1}^{n} \sum_{y=1}^{m} x^p y^q f(x,y) \tag{4.2}$$

where the background pixels have value zero, and the object pixels have value one. The infinite sequence of moments, $p, q = 0, 1, \ldots$, uniquely determines the shape, and vice versa. Variations are described in [20] and [44].

Based on such moments, a number of functions, moment invariants, can be defined that are invariant under certain transformations such as translation, scaling, and rotation. Using only a limited number of low order moment invariants, the less critical and noisy high order moments are discarded. A number of such moment invariants can be put into a feature vector, which can be used for matching. Global object features such as area, circularity, eccentricity, compactness, major axis orientation, Euler number, concavity tree, shape numbers, and algebraic moments can all be used for shape description [15, 56]. A number of such features are for example used by the QBIC system [52].

Modal Matching. Rather than working with the area of an object, the boundary can be used instead. Samples of the boundary can be described with Fourier descriptors, the coefficients of the discrete Fourier transform [73].

Another form of shape decomposition is the decomposition into an ordered set of eigenvectors, also called principal components. Again, the noisy high order components can be discarded, using only the most robust components. The idea is to consider n points on the boundary of an object, and to define a matrix D such that element D_{ij} determines how boundary points i and j of the object interact, typically involving the distance between points i and j.

The eigenvectors e_i of D, satisfying $D\,e_i = \lambda e_i$, $i = 1, \ldots, n$, are the *modes* of D, also called eigenshapes. To match two shapes, take the eigenvectors e_i of the query object, and the eigenvectors e'_j of the target object, and compute a mismatch value $m(e_i, e'_j)$. For simplicity, let us assume that the eigenvectors have the same length. For a fixed $i = i_0$, determine the value j_0 of j for which $m(e_{i_0}, e'_j)$ is minimal. If the value of i for which $m(e_i, e'_{j_0})$ is minimal is equal to i_0, then point i of the query and point j of the target match each other. See for example [33] and [65] for variations on this basic technique of modal matching.

Curvature Scale Space. Another approach is the use of a scale space representation of the curvature of the contour of objects. Let the contour C be parameterized by arclength s: $C(s) = (x(s), y(s))$. The coordinate functions of C are convolved with a Gaussian kernel ϕ_σ of width σ:

$$x_\sigma(s) = \int x(s)\phi_\sigma(t-s)\, dt, \qquad \phi_\sigma(t) = \frac{1}{\sqrt{2\pi\sigma^2}} e^{-\frac{t^2}{2\sigma^2}}, \qquad (4.3)$$

and the same for $y(s)$. With increasing value of σ, the resulting contour gets smoother, see Fig. 4.4, and the number of zero crossings of the curvature along it decreases, until finally the contour is convex and the curvature is positive.

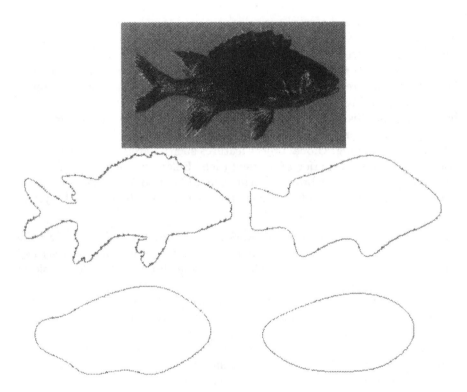

Fig. 4.4 Contour evolution reducing curvature changes, see http://www.ee.surrey.ac.uk/Research/VSSP/imagedb/demo.html.

For continuously increasing σ, the positions of the curvature zero crossings continuously move along the contour, until two such positions meet and annihilate each other. Matching of two objects can be done by matching points of annihilation in the s, σ plane [49].

Another way of reducing curvature changes is based on the turning angle function (see Section 4.4.1), or tangent space representation [47].

4.2.3 Voting Schemes

The voting schemes discussed here generally work on so-called interest points. For the purpose of visual information systems, such points are for example corner points detected in images.

Geometric hashing [46, 80] is a method that determines if there is a transformed subset of the query point set that matches a subset of a target point set. The method first constructs a single hash table for all target point sets together. Each point is represented as $e_0 + \kappa(e_1 - e_0) + \lambda(e_2 - e_0)$, for some fixed choice of points e_0, e_1, e_2, and the (κ, λ)-plane is quantized into a two-dimensional table, mapping each real coordinate pair (κ, λ) to an integer index pair (k, ℓ).

Let there be N target point sets B_i. For every target point set, the following is done. For every three non-collinear points e_0, e_1, e_2 from the point set, express the other points as $e_0 + \kappa(e_1 - e_0) + \lambda(e_2 - e_0)$, and append the tuple (i, e_0, e_1, e_2) to entry (k, ℓ). If there are $\mathcal{O}(m)$ points in each target point set, the construction of the hash table is of complexity $\mathcal{O}(Nm^4)$.

Now, given a query point set A, choose three non-collinear points e'_0, e'_1, e'_2 from the point set, and express each other point as $e'_0 + \kappa(e'_1 - e'_0) + \lambda(e'_2 - e'_0)$, and tally a vote for each tuple (i, e_0, e_1, e_2) in entry (k, ℓ) of the table. The tuple (i, e_0, e_1, e_2) that receives most votes indicates the target point set T_i containing the query point set. The affine transformation that maps (e'_0, e'_1, e'_2) to the winner (e_0, e_1, e_2) is assumed to be the transformation between the query and the target. The complexity of matching a single query set of n points is $\mathcal{O}(n)$. There are several variations of this basic method, such as balancing the hashing table, or avoiding taking all possible $\mathcal{O}(n^3)$ 3-tuples.

The generalized Hough transform [14], or pose clustering [68], is also a voting scheme. Here, affine transformations are represented by six coefficients. The quantized transformation space is represented as a six-dimensional table. Now for each triplet of points in the query set, and each triplet of points from the target set, compute the transformation between the two triples, and tally a vote in the corresponding entry of the table. This must be done for all target point sets. The entry with the highest score is assumed to be the transformation between the query and the target. The complexity of matching a single query set is $\mathcal{O}(Nm^3n^3)$.

In the alignment method [37, 70], for each triplet of points from the query set, and each triplet from the target set, we compute the transformation between them. With each such transformation, all the other points from the target set are transformed. If they match with query points, the transformation receives a vote, and if the number of votes is above a chosen threshold, the transformation is assumed to be the matching transformation between the query and the target. The complexity of matching a single query set is $\mathcal{O}(Nm^4n^3)$.

Variations of these methods also work for geometric features other than points, and for other transformations than affine transformations. A comparison between geometric hashing, pose clustering, and the alignment method is made by Wolfson [79]. Other voting schemes exist, for example taking a probabilistic approach [53].

4.2.4 Computational Geometry

Computational geometry is the subarea of algorithms design that deals with geometric problems involving operations on objects like points, lines, polygons, and polyhedra. Over the past twenty years the area has grown into a mainstream worldwide research activity. The success of the field as a research discipline can be explained by the beauty of the problems and their solutions, and by the many applications in which geometric problems and algorithms play a fundamental role. The standard approach taken in computational geometry is the development of exact, provably correct and efficient solutions to geometric problems. See for example the textbooks [16, 27, 51, 55], and the handbook of Ref. [32].

The impact of computational geometry on application domains was minor up to a few years ago. On one hand, the research community has been developing more interest in application problems and real-world conditions, and develops more software implementations of the most efficient algorithms available. On the other hand, there is more interest from the application domains in computational geometry techniques, and companies even start to specifically require computational geometry expertise.

Aspects that play an important role in the algorithmic solutions to matching are the representation, decomposition, approximation, and deformation of shapes, the transformation of one shape to another, the measurement of shape similarity, and the organization of shapes into search structures. In the following we give an overview of the state of the art in geometric shape matching from the computational geometry point of view. It should be noted though that the boundary of the field of computational geometry is not sharp, and considering a method a computational geometry method or not is somewhat arbitrary.

First we consider properties of dissimilarity measures, then we list a number of problems in shape matching, together with the best known result to solve them. We are primarily concerned with patterns defined by finite point sets, curves, and regions. Unless otherwise stated, patterns are a subset of $\mathbb{R}^2$, and the underlying distances are Euclidean.

Dissimilarity Measures. Many pattern matching and recognition techniques are based on a similarity measure between patterns. A similarity measure is a function defined on pairs of patterns indicating the degree of resemblance of the patterns. It is desirable that such a similarity measure is a metric. Furthermore, a similarity measure should be invariant for the geometrical transformation group that corresponds to the matching problem. Below, we discuss a number of properties of metrics, such as invariance for transformation groups.

Let S be any set of objects. A *metric* on S is a function $d : S \times S \to \mathbb{R}$ satisfying the following three conditions for all $x, y, z \in S$ [25]:

(*i*) $d(x, x) = 0$
(*ii*) $d(x, y) = 0$ implies $x = y$
(*iii*) (strong triangle inequality) $d(x, y) + d(x, z) \geq d(y, z)$

If a function satisfies only (*i*) and (*iii*), then it is called a *pseudometric*, or sometimes semimetric. Symmetry follows from (*i*) and (*iii*): $d(y, z) \leq d(z, y) + d(z, z) = d(z, y)$, and $d(z, y) \leq d(y, z) + d(y, y) = d(y, z)$, so $d(y, z) = d(z, y)$. An alternative triangle inequality is the following:

(*iii'*)   (triangle inequality) $d(x, y) + d(y, z) \geq d(x, z)$

but (*i*) and (*iii'*) do not imply symmetry:

(*iv'*) $d(x, y) = d(y, x)$

So d is a pseudometric only if it satisfies (*iv'*) in addition to (*i*) and (*iii'*). Any (pseudo)metric is non-negative: $d(x, y) + d(y, x) \geq d(x, x)$, so $d(x, y) \geq 0$.

A set S with a fixed metric d is called a metric space. Given two elements x and y of S, the value $d(x, y)$ is called the *distance* between x and y. By identifying elements of S with zero distance, any pseudometric induces a metric on the resulting partition.

A set of bijections G in S is a *transformation group* if $g^{-1}h \in G$ for all $g, h \in G$. A (pseudo)metric d on a set S is said to be *invariant* for the transformation group G acting on S if $d(g(x), g(y)) = d(x, y)$ for all $g \in G$ and $x, y \in S$.

The *orbit* of G passing through $x \in S$ is the set of images of x under G:

$$G(x) = \{g(x) \mid g \in G\}$$

The orbits form a partition of S. The collection of all orbits is called the *orbit set*, denoted by S/G.

The following theorem shows that a pseudometric invariant under a transformation group results in a natural pseudometric on the orbit set. Rucklidge [60] used this principle to define a shape distance based on the Hausdorff distance.

Theorem 4.1. *Let G be a transformation group for a set S; let d be a pseudometric on S invariant for G. Then $\tilde{d} : S/G \times S/G \to \mathbb{R}$ defined by*

$$\tilde{d}(G(x), G(y)) = \inf\{d(g(x), y) \mid g \in G\}$$

is a pseudometric.

Let $\mathcal{P}$ be a fixed collection of subsets of $\mathbb{R}^2$. Any element of $\mathcal{P}$ is called a *pattern*. We call the collection $\mathcal{P}$ with a fixed metric d a *metric pattern*

Fig. 4.5 Affine invariance: $d(A, B) = d(g(A), g(B))$.

space. A collection of patterns $\mathcal{P}$ and a transformation group G determine a family of shapes $\mathcal{P}/G$. For a pattern $A \in \mathcal{P}$, the corresponding *shape* equals the orbit

$$G(A) = \{g(A) \mid g \in G\}.$$

The collection of all these orbits forms a *shape space*. If d is invariant for G, then Theorem 4.1 gives a pseudometric $\tilde{d}$ on the shape space $\mathcal{P}/G$.

Shape matching often involves computing the similarity between two patterns, independent of transformation. This is exactly what the shape metric $\tilde{d}$ is good for. Given two patterns A and B, it determines the greatest lower bound of all $d(g(A), B)$ under transformations $g \in G$, resulting in a transformation-independent distance between the corresponding shapes $G(A)$ and $G(B)$.

A collection of patterns $\mathcal{P}$ uniquely determines a maximal subgroup T of the homeomorphisms under which $\mathcal{P}$ is closed. (Homeomorphisms are continuous, bijective functions having a continuous inverse.) The subgroup T consists of all homeomorphism t such that both the image $t(A)$ and the inverse image $t^{-1}(A)$ are members of $\mathcal{P}$ for all patterns $A \in \mathcal{P}$.

The metric pattern space $(X, \mathcal{P}, d)$ is invariant for a transformation $g \in T$ if $d(g(A), g(B))$ equals $d(A, B)$ for all A and B in $\mathcal{P}$. The invariance group G of a metric pattern space consists of all transformations in T for which it is invariant. Affine invariance is often desired in many pattern matching and shape recognition tasks. Fig. 4.5 shows patterns A and B in the Euclidean plane, and image patterns $g(A)$ and $g(B)$ under an affine transformation g. Invariance for affine transformations makes the distance between two patterns independent of the choice of coordinate system.

Finding an affine invariant metric for patterns is not so difficult. Indeed, a metric that is invariant not only for affine transformations, but for general homeomorphisms is the discrete metric:

$$d(A, B) = \begin{cases} 0 & \text{if } A \text{ equals } B, \\ 1 & \text{otherwise.} \end{cases} \tag{4.4}$$

However, this metric lacks useful properties. For example, if a pattern A is only slightly distorted to form a pattern A', the discrete distance $d(A, A')$ is already maximal.

Therefore it makes sense to devise metrics with specific properties. A frequently used dissimilarity measure is the Hausdorff distance, which is defined for arbitrary non-empty bounded and closed sets A and B as the infimum of the distance of the points in A to B and the points in B to A. This can be formulated as follows:

$$d(A, B) = \inf\{\epsilon > 0 \mid A \subseteq B^\epsilon \text{ and } B \subseteq A^\epsilon\} \tag{4.5}$$

where A^ϵ denotes the union of all disks with radius ϵ centered at a point in A. The Hausdorff distance is a metric. The invariance group for the Hausdorff distance consists of isomorphisms (rigid motions and reflections). The Hausdorff distance is robust against small deformations, but it is sensitive to noise: a single outlier, a far away noise point, drastically increases the Hausdorff distance, see Fig. 4.6.

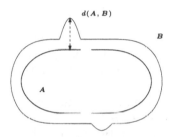

Fig. 4.6 Hausdorff distance.

In the next few sections, we give an overview of dissimilarity measures for more restricted patterns: finite point sets, curves, and regions. Then, in Section 4.6 we will list a number of robustness properties for these measures.

4.3 Finite Point Sets

Let A and B be point sets of sizes n and m, respectively. Matching the point sets means finding a correspondence between points of A and points of B. An optimal matching minimizes some dissimilarity measure between the point sets. The correspondence can be many-to-many, but also one-to-one, both have their applications. Matching has been studied extensively in a graph theory setting, where the problem is to find a matching in a graph (V, E) with vertices $V = A \cup B$, and given edges E with weights. Exploiting the geometric nature if the vertices are points and the weights are distances between points, results in more efficient algorithms, see Vaidya [72] for example.

For the purpose of multimedia retrieval, exact point set matching and congruence matching are less important. This topic is not treated here; the reader is referred to Alt, et al.[4] and Brass and Knauer [19].

4.3.1 Bottleneck Matching

Let A and B be two point sets of size n, and $d(a, b)$ a distance between two points. The bottleneck distance is the minimum over all one-to-one correspondences f between A and B of the maximum distance $d(a, f(a))$. The results on bottleneck distance mentioned in this section are due to Efrat and Itai [28].

If $d(a, b)$ is the Euclidean distance, the bottleneck distance between A and B can be computed in time $\mathcal{O}(n^{1.5} \log n)$. It is computed using a technique called parametric search. This is usually considered an impractical method, although it has been implemented for other problems [63].

An alternative is to compute an approximation d to the bottleneck distance d^*. An approximate matching between A and B with d the furthest matched pair, such that $d^* < d < (1 + \epsilon)d^*$, can be computed in time $\mathcal{O}(n^{1.5} \log n)$. This algorithm makes use of an optimal approximate nearest neighbor algorithm [13].

So far we have considered only the computation problem, computing the distance between two point sets. The decision problem for translations, deciding whether there exists a translation ℓ such that $d(A + \ell, B) < \epsilon$, can be done in time $\mathcal{O}(n^5 \log n)$.

Because of the high degree in the complexity, it is interesting to look at approximations with a factor ϵ: $d(A + \ell, B) < (1 + \epsilon)d(A + \ell^*, B)$. Finding such a translation can be done in time $\mathcal{O}(n^{2.5})$ [61].

The optimization problem considers the computation of the minimum distance under a group of transformations. It finds the optimal transformation f^* such that $d(f(A), B)$ is minimized. For rigid motions (translations plus rotations, sometimes called congruences), this can be found in time $\mathcal{O}(n^6 \log n)$ [4]. For translations only, it can be computed in time $\mathcal{O}(n^5 \log^2 n)$ [28].

An approximation translation ℓ within factor two, $d(A + \ell, B) \leq 2d(A + \ell^*, B)$, can be obtained by translating A such that the lower left corner of the axis parallel bounding box (called reference point) coincides with the one of B. An approximation with factor $1 + \epsilon < 2$ can be obtained in time $\mathcal{O}(C(\epsilon, d)n^{1.5} \log n)$, with $C(\epsilon, d)$ a constant depending on ϵ and dimension d: $C(\epsilon, d) = (\frac{1+\epsilon}{\epsilon^2})^d \log(1/\epsilon)$.

Some variations on computing the bottleneck distance between point sets are the following. If A is a set of points, and B a set of segments, computing the bottleneck distance can be done in $\mathcal{O}(n^{1.5+\epsilon})$ time. When the points are in $\mathbb{R}^d$ and the distance is L_∞, it can be computed in time $\mathcal{O}(n^{1.5} \log^{d-1} n)$.

Let A and B be two point sets of size m and n, and k a number not larger than m and n. The problem of finding the smallest bottleneck distance over all one-to-one matchings between k points in A and k points in B can be computed in time $\mathcal{O}(m \log m + n^{1.5} \log m)$. Typical application of this result is in situations where we search a query pattern A in a larger target pattern B and have to deal with noise points.

4.3.2 Minimum Weight Matching

The minimum total distance (weight) is the minimum over all one-to-one correspondences f between A and B of the sum of the distances $d(a, f(a))$. It can be computed in time $\mathcal{O}(n^{2+\epsilon})$ [2]. Here, the constant ϵ stands for a positive constant which can be chosen arbitrarily small with an appropriate choice of other constants of the algorithm. For the L_∞ distance, it can be computed in time $\mathcal{O}(n^2 \log^3 n)$ [72].

4.3.3 Uniform Matching

The "most uniform" distance is the minimum over all one-to-one correspondences f between A and B of the difference between the maximum and the minimum $d(a, f(a))$. The most uniform matching is also called balanced or fair matching. The distance can be computed in time $O(n^{10/3} \log n)$ [29]. It is based on batched range searching, where the ranges are congruent annuli.

The problem of finding the smallest uniform distance over all one-to-one matchings beteen k points in A and k points in B can be computed with the same time complexity.

4.3.4 Minimum Deviation Matching

The minimum deviation distance is the minimum over all one-to-one correspondences f between A and B of the difference between the maximum and average distance $d(a, f(a))$. This can be computed in time $\mathcal{O}(n^{10/3+\epsilon})$ [29].

4.3.5 Hausdorff Distance

In many applications, for example stereo matching, not all points from A need to have a corresponding point in B, due to occlusion and noise. Typically, the two point sets are of different size, so that no one-to-one correspondence exists between all points. In that case, a dissimilarity measure that is often used is the Hausdorff distance. The Hausdorff distance was defined in Section 4.2.4 for general sets. For finite point sets, it can equivalently be defined as follows.

The *directed* Hausdorff distance $\vec{d}(A, B)$ is defined as the maximum over all points in A of the distances to a point from B. The Hausdorff distance $d(A, B)$ is the maximum of $\vec{d}(A, B)$ and $\vec{d}(A, B)$:

$$d(A, B) = \max\{\vec{d}(A, B), \vec{d}(B, A)\}, \qquad \vec{d}(A, B) = \max_{a \in A} \min_{b \in B} d(a, b)$$

with $d(a, b)$ the underlying (Euclidean, say) distance.

It can be computed using Voronoi diagrams in time $\mathcal{O}((m+n) \log(m+n))$ [7]. The use of Voronoi diagrams for computing the Hausdorff distance is explained in Section 4.5.3 for matching polygons.

Given two point sets A and B, the translation ℓ^* that minimizes the Hausdorff distance $d(A + \ell, B)$ can be determined in time $\mathcal{O}(mn(\log mn)^2)$ when the underlying metric is L_1 or L_∞ [21]. This is done using a search structure called segments tree. For other L_p metrics, $p = 2, 3, \ldots$ it can be computed in time $\mathcal{O}(mn(m + n)\alpha(mn)\log(m + n))$ [41]. ($\alpha(n)$ is the inverse Ackermann function, a very slowly increasing function.) This is done using the upper envelopes of Voronoi surfaces.

Computing the optimal rigid motion r (translation plus rotation), minimizing $H(r(A), B)$ can be done in time $\mathcal{O}((m + n)^6 \log(mn))$ [39]. This is done using dynamic Voronoi diagrams. Given a real value ϵ, deciding if there is a rigid motion such that $H(r(A), B) < \epsilon$ can be done in time $\mathcal{O}((m + n)m^2n^2 \log mn)$ [22].

Given the high complexities of these problems, it makes sense to look at approximations. Computing an approximate optimal Hausdorff distance under translation and rigid motion can be done in time $\mathcal{O}((m+n)\log(m+n))$ [3].

4.3.6 Transformation Space Subdivision

Matching of finite points, from images, under homotheties (translation and scaling) is done by subdividing the transformation space [40]. Rather than the Hausdorff distance itself, the partial Hausdorff distance is used, which is the maximum of the two directed partial Hausdorff distances $\vec{d}_k(A, B)$ and $\vec{d}_k(B, A)$:

$$d_k(A, B) = \max\{\vec{d}_k(A, B), \vec{d}_k(B, A)\}, \qquad \vec{d}_k(A, B) = \underset{a \,\in\, A}{\overset{k\text{th}}{}} \min_{b \in B} d(a, b).$$

The partial Hausdorff distance is not a metric since it fails the triangle inequality. The running time depends on the depth of subdivision of transformation space.

The subdivision of transformation space is generalized to a general framework by Hagedoorn and Veltkamp [35]. Here the matching can be done with respect to other transformations as well, for example, similarity (translation, rotation, and scaling), or affine transformation (translation, rotation, scaling, and shear). The method works for many dissimilarity measures, but we used a technique for constructing metrics using functions $\mathbf{f}_A, \mathbf{f}_B : \mathbb{R}^2 \to \mathbb{R}$ defined on patterns A and B, and the affine invariant metric defined by integrating the absolute difference of $\mathbf{f}_A$ and $\mathbf{f}_B$. Fig. 4.7 illustrates matching with this metric, compared to the partial Hausdorff distance.

4.4 Curves

The most direct way of representing curves is by their position function, defining all the positions of the curve. A parametric curve A is defined in

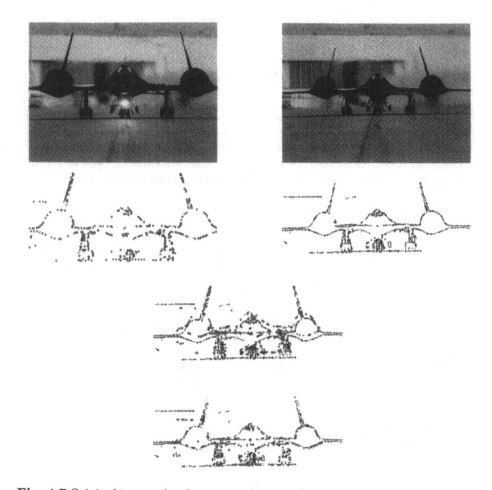

Fig. 4.7 Original images (top), extracted points (second row), matching with partial Hausdorff distance (third row), and matching with the affine invariant metric from Ref. [35] (bottom).

terms of a parameter: $A(t) = (x(t), y(t))$. In general, many parameterizations result in the same shape of the curve, but have different derivative vectors along the curve [74]. A standard parameterization is by arc length along the curve; the arc length is usually denoted by s. Polygonal curves (polylines) are usually represented by their sequence of vertices. An implicit definition of the curve, $A : f(x, y) = 0$, is less often used in matching.

Polylines from real-world applications often contain many spurious vertices, which can be removed by approximating the polygon. There are many heuristics for approximating polygonal curves, see, e.g., Rosin [59] for a comparison. Two methods of optimal approximation are the following:

– Given a polyline A and a number k, construct an approximation polyline A_k of k vertices, minimizing the approximation error, or dissimilarity, $d(A, A_k)$.
– Given a polyline and an error bound ϵ, construct an approximation polyline A_ϵ with dissimilarity $d(A, A_\epsilon) < \epsilon$, minimizing the number of vertices.

Both approximations can be computed in time $\mathcal{O}(n^2 \log n)$ for various error measures [42]. However, these optimal approximations are not suitable for constructing a hierarchy of approximations, in the sense that each segment at one level may be refined at the next level of approximation. Approximating polygons at various levels allows the hierarchical processing of curves [75].

4.4.1 Turning Function

Representations other than the position function are also useful in matching. From the position function, other representations can be derived, such as the tangent, acceleration, tangent angle, cumulative angle, periodic cumulative angle, and the curvature functions [73].

The cumulative angle function, or turning function, $\Theta_A(s)$ of a polygon A gives the angle between the counterclockwise tangent and the x-axis as a function of the arc length s. $\Theta_A(s)$ keeps track of the turning that takes place, increasing with left hand turns, and decreasing with right hand turns. Clearly, this function is invariant under translation of the polyline. Rotating a polyline over an angle θ results in a vertical shift of the function with an amount θ.

For polylines, the turning function is a piecewise constant function, increasing or decreasing at the vertices, and constant between two consecutive vertices, see Fig. 4.8.

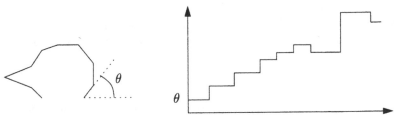

Fig. 4.8 Polygonal curve and turning function.

Matching polylines based on the turning functions can be done as follows. For simplicity, first assume that the two curves have the same length. The L_p metric on function spaces, applied to Θ_A and Θ_B, gives a dissimilarity measure on A and B:

$$d_{A,B} = \left(\int |\Theta_A(s) - \Theta_B(s)|^p \, ds \right)^{1/p}.$$

Minimizing this dissimilarity under rotation θ amounts to minimizing $d(A, B)$ $= \int |\Theta_A(s) - \Theta_B(s) + \theta|^p \, ds$. For $p = 2$, the minimum is obtained for $\theta = \int \Theta_B(s) \, ds - \int \Theta_A(s) \, ds$.

In Vleugels and Veltkamp [77], for the purpose of retrieving hieroglyphic shapes, the polygonal curves do not have the same length, so that partial matching can be performed. In that case we can move the starting point of the shorter one along the longer one, and consider only the turning function where the arc lengths overlap. This is a variation of the algorithms for matching closed polygons with respect to the turning function, which can be done in time $\mathcal{O}(mn \log(mn))$ [12], see Section 4.5.

Partial matching under scaling, in addition to translation and rotation, is more involved. This can be done in time $\mathcal{O}(m^2 n^2)$ [24]. The dissimilarity balances the length of a match against the squared error. Given two matches with the same squared error, the match involving the longer part of the polylines has a better dissimilarity. The dissimilarity measure is a function of the scale, rotation, and the shift of one polyline along the other. An analytic formula of the dissimilarity in terms of scale and shift yields a search problem in scale-shift plane. This space is divided into regions. A minimum of the dissimilarity is found by a line sweep over the plane.

4.4.2 Signature Function

A less discriminative function is the so-called signature function. At every point along the curve, the signature function σ value is the arc length of the curve to the left or on the tangent line at that point, see Fig. 4.9. It is invariant under similarity: combinations of translation, rotation, and scaling. For convex curves, the signature function is one everywhere, because at every point, the whole curve lies to the left of the tangent. For a single polyline curve, the signature function can be computed in time $\mathcal{O}(n^2)$ [54].

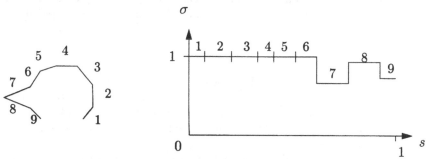

Fig. 4.9 A curve and its signature function.

For polylines, dissimilarity measures can be used that are based on "time warps" of sequences of elements (vertices or segments), pairing elements of A

to elements of B. The pairing need not be one-to-one: the pairing of element i of A to element j of B, may be followed by a pairing of i to $j+1$, $i+1$ to j, or $i+1$ to $j+1$. Using dynamic programming, this takes time $\mathcal{O}(nm)$ [45].

4.4.3 Affine Arclength

Instead of turning functions, affine invariant representations of curves may be used as a basis for shape matching. An example of such a representation is *affine arclength*. While turning functions are invariant only under similarity transformations the normalized affine arclength is invariant for all affine transformations. Huttenlocher and Kedem [38] use the one-dimensional Hausdorff distance to compare affine arclength descriptions of curves.

Let $A : [0,1] \to \mathbb{R}$ be a twice continuously differentiable curve, and let A' and A'' denote the first and second order derivates, respectively. The affine arclength is the function $\sigma : \mathbb{R} \to \mathbb{R}$ given by

$$\sigma(t) = \int_0^t |\det(A'(x), A''(x))|^{\frac{1}{3}} \, dx. \tag{4.6}$$

The *normalised arclength* is defined as follows:

$$\sigma^*(t) = \frac{\sigma(t)}{\sigma(1)}. \tag{4.7}$$

Instead of these definitions, Huttenlocher and Kedem use a discretized version of affine arclength to represent the boundary of a simple polygon. This discretized representation is a finite set of numbers between 0 and 1, one number for each boundary vertex. Two simple polygons are equal if the respective discretized arclengths are equal up to translation modulo 1. This problem can be solved in a perturbation-robust manner by minimizing the Hausdorff distance between the two representations (seen as one-dimensional finite point sets). The latter problem can be solved in time $\mathcal{O}(mn \log(mn))$.

4.4.4 Reflection Metric

Affine arclength can be used to define affine invariant similarity measures on curves. However, there is no straightforward generalization of it to patterns that consist of more than one connected component. The *reflection metric* [34] is an affine-invariant metric that is defined on finite unions of curves in the plane.

The reflection metric is defined as follows. First, unions of curves are converted into real-valued functions on the plane. Then, these functions are compared using integration, resulting in a similarity measure for the corresponding patterns.

The functions are formed as follows, for each finite union of curves A. For each $x \in \mathbb{R}^n$, the *visibility star* V_A^x is defined as the union of open line segments connecting points of A that are visible from x:

$$V_A^x = \bigcup \{ \overline{xa} \mid a \in A \text{ and } A \cap \overline{xa} = \varnothing \}. \tag{4.8}$$

The *reflection star* R_A^x is defined by intersecting V_A^x with its reflection in x:

$$R_A^x = \{ x + v \in \mathbb{R}^2 \mid x - v \in V_A^x \text{ and } x + v \in V_A^x \}. \tag{4.9}$$

The function $\rho_A : \mathbb{R}^2 \to \mathbb{R}$ is the area of the reflection star in each point:

$$\rho_A(x) = \text{area}(R_A^x). \tag{4.10}$$

Observe that for points x outside the convex hull of A, this area is always zero. The reflection metric between patterns A and B defines a normalised difference of the corresponding functions ρ_A and ρ_B:

$$d(A, B) = \frac{\int_{\mathbb{R}^2} |\rho_A(x) - \rho_B(x)| \, dx}{\int_{\mathbb{R}^2} \max(\rho_A(x), \rho_B(x)) \, dx}. \tag{4.11}$$

From the definition it follows that the reflection metric is invariant under all affine transformations. In contrast with single-curve patterns, this metric is defined also for patterns consisting of multiple curves. In addition, the reflection metric is deformation, blur, crack, and noise robust.

Here, we focus at the computation of the reflection metric for finite unions of line segments in the plane. First, compute partitions of the plane in which the combinatorial structure of the reflection star is constant. Using the latter partition, the reflection distance can be computed in time $\mathcal{O}(rI(m + n))$ for two separate collections of segments with m and n segments, where r is the complexity of the overlay of two partitions, and $I(k)$ denotes the time needed to integrate the absolute value of quotients of polynomials with at most degree k over a triangle. Assuming $I(k)$ is linear in k, the overall complexity amounts to $\mathcal{O}(r(m + n))$. The complexity of the overlay, r, is $\mathcal{O}(m^4 + n^4)$.

The reflection metric can be generalized to finite unions of $(d - 1)$-dimensional hyper-surfaces in d dimensions. The generalization consists of replacing the two-dimensional area by the d-dimensional volume.

4.4.5 Hausdorff Distance

The Hausdorff distance is not only defined for finite point sets, but for any two compact sets. Special cases are sets of polylines. The results for polylines are the same as for polygons, see Section 4.5.3.

4.4.6 Fréchet Distance

The Hausdorff distance is often not appropriate to measure the dissimilarity between curves. For all points on A, the distance to the closest point on B may be small, but if we walk forward along curves A and B simultaneously, and measure the distance between corresponding points, the maximum of these distances may be larger, see Fig. 4.10. This is what is called the Fréchet distance. More formally, let A and B be two parameterized curves $A(\alpha(t))$ and $B(\beta(t))$, and let their parameterizations α and β be continuous functions of the same parameter $t \in [0,1]$, such that $\alpha(0) = \beta(0) = 0$, and $\alpha(1) = \beta(1) = 1$. The Fréchet distance is the minimum over all monotone increasing parameterizations $\alpha(t)$ and $\beta(t)$ of the maximal distance $d(A(\alpha(t)), B(\beta(t)))$, $t \in [0,1]$.

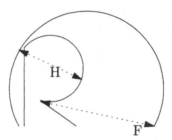

Fig. 4.10 Hausdorff (H) and Fréchet (F) distance between two curves.

Alt and Godeau [8] consider the computation of the Fréchet distance for the special case of polylines. Deciding whether the Fréchet distance is smaller than a given constant, can be done in time $\mathcal{O}(mn)$. Based on this result, and the "parametric search" technique, it is derived that the computation of the Fréchet distance can be done in time $\mathcal{O}(mn \log(mn))$. Although the algorithm has low asymptotic complexity, it is not really practical. The parametric search technique used here makes use of a sorting network with very high constants in the running time. A simpler sorting algorithm leads to an asymptotic running time of $\mathcal{O}(mn (\log mn)^3)$. Still, the parametric search is not easy to implement. A simpler algorithm, which runs in time $\mathcal{O}(mn(m+n) \log(mn))$ is given in Godau [30].

A variation of the Fréchet distance is obtained by dropping the monoticity condition of the parameterization. The resulting Fréchet distance $d(A, B)$ is a pseudometric: zero distance need not mean that the objects are the same, see Section 4.2.4. For this the decision problem, deciding whether $d(A, B) < \epsilon$ for a given ϵ, can be decided in time $\mathcal{O}(mn)$. The actual distance can be computed in time $\mathcal{O}(mn \log(mn))$.

Another variation is to consider partial matching: finding the part of one curve to which the other has the smallest Fréchet distance. The corresponding decision problem can be solved in time $\mathcal{O}(mn\log(mn))$, the computation problem in time $\mathcal{O}(mn(\log(mn))^2)$.

4.4.7 Size Function

Relatively new are so-called size functions [76]. Size functions can be defined for arbitrary planar graphs and a "measuring function" D. An example of such a measuring function is the distance from each pattern point to the center of mass. The size function $s_D(x, y)$ is then defined as the number of connected components of the set of points with $D \leq y$ that have at least one point with $D \leq x$. Size functions do not uniquely represent a shape, but classes of shapes, depending on the measuring function.

4.4.8 Pixel Chains

Given two sets of pixel chains, the root mean square of the distances from one set of pixels to the other can be computed with the relatively efficient hierarchical chamfer matching algorithm, which works on the basis of the distance transform and the chamfer distance [18].

4.5 Regions

As mentioned in Section 4.2.2, normalization of regions, filled contours, is often done using algebraic moments. For the special case of polygons, this can be done in time linear in the number of vertices [67].

A representation that has proven to be relevant in human vision is the medial axis, producing a skeleton and a width value at each point on the skeleton (the so-called quench function). For polygonal contours, the medial axis and the quench function can be computed in time linear in the number of vertices [23]. For pixel chain contours, this can be computed using the distance transform [17].

The dissimilarity of contours can be based on sample points along the contour curve, the whole contour curve, or the enclosed area. For example, Fourier descriptors are based on samples of the contour. A number of methods based on the contour curve and the area are mentioned below.

4.5.1 Turning Function

As already mentioned in Section 4.4.1, the turning function is also applicable for matching regions, and was used by Arkin et al. [12] for matching polygons under translation, rotation, and scaling. For the special case of polygons, matching based on turning functions can be done as follows. First rescale both polygons so that the perimeter has length one. The L_p metric on function spaces, applied to Θ_A and Θ_B, gives a dissimilarity measure on A and B:

$$d_{A,B} = \left(\int |\Theta_A(s) - \Theta_B(s)|^p \, ds \right)^{1/p}.$$

If the starting point of the arc length parameter of $\Theta_A(s)$ is shifted by an amount t, the new function is $\Theta_A(s+t)$. If the polygon is rotated by an angle θ, the new function is $\Theta_A(s) + \theta$. Making the dissimilarity invariant for the starting point of the arc length parameter, and minimizing under rotation θ, amounts to minimizing

$$d_{A,B}(t,\theta) = \left(\int |\Theta_A(s+t) - \Theta_B(s) + \theta|^p \, ds \right)^{1/p}$$

for t and θ.

For any fixed t and $p = 2$, $d_{A,B}(t,\theta)$ is minimal for $\theta = \int \Theta_B(s) \, ds - \int \Theta_A(s) \, ds - 2\pi t$. For polygons, the turning functions are piecewise constant step functions. Therefore $d_{A,B}(t,\theta)$ can be evaluated as the sum of $\mathcal{O}(m+n)$ terms corresponding to the areas between the dotted lines, see Fig. 4.11. The minimum $d_{A,B}(t,\theta)$ is obtained when two steps of the step functions coincide, of which are $\mathcal{O}(mn)$ possible solutions. This leads to a straightforward $\mathcal{O}(mn(m+n))$ algorithm. This can be sped up by incremental evaluation of $d_{A,B}(t,\theta)$ for all the $\mathcal{O}(mn)$ possible solutions, giving an algorithm of time complexity $\mathcal{O}(mn\log(mn))$ [12].

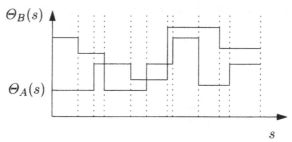

Fig. 4.11 Rectangles enclosed by $\Theta_A(s)$, $\Theta_B(s)$, and dotted lines are used for evaluation of dissimilarity.

It should be noted that nonuniform noise in the form of perturbation of vertices unevenly spread along the polygon is problematic for this distance function.

4.5.2 Fréchet Distance

Parameterized contours are curves where the starting point and ending point are the same. However, the starting and ending point could as well lie somewhere else on the contour, without changing the shape of the contour curve.

Deciding whether the Fréchet distance of two contours is smaller than ϵ, irrespective the starting point, can done in time $\mathcal{O}(mn \log(mn))$. The corresponding computation problem, computing the Fréchet distance, can be solved in time $\mathcal{O}(mn(\log(mn))^2)$ [8].

For convex contours curves, the Fréchet distance is equal to the Hausdorff distance, which can be computed in time $\mathcal{O}(mn \log(mn))$ [5].

4.5.3 Hausdorff Distance

Given two polygons A and B, the directed Hausdorff distance from A to B can be computed using the Voronoi diagram of B, which assigns to each vertex and edge of A a region of points that lie closer to that vertex or edge than to any other, as shown in Fig. 4.12. If the edges in the Voronoi diagram separate regions of two edges (e.g., $l(e_1, e_2)$), or two vertices (e.g., $l(v_1, v_3)$), or the regions of an edge and its endpoint vertex (e.g., $l(v_1, e_1)$), then they are line segments. The Voronoi edge is a parabolic segment if it separates regions of a polygon edge and a vertex that not its endpoint (e.g., $p(v_1, e_2)$). The Voronoi diagram of B has $\mathcal{O}(n)$ edges, and it can be computed in time $\mathcal{O}(n \log n)$.

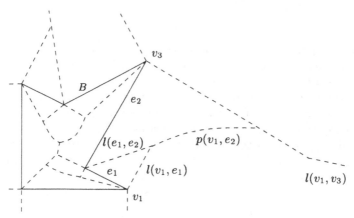

Fig. 4.12 Polygon and its Voronoi diagram.

To compute the directed Hausdorff distance from A to B, let us consider the part of B that falls within a single region of the Voronoi diagram of A, for example the thick line segments in Fig. 4.13. Moving along the thick polyline, the distance to B first decreases, than increases, so the maximal distance is obtained at the intersection of the thick segments with the Voronoi diagram. In general, the maximal distance is obtained at a vertex of A or at an intersection point of A with the Voronoi diagram. Note that there can be multiple intersection points on an edge of the Voronoi diagram, and the largest distance is obtained at the intersection with the largest or the smallest coordinates; there are $\mathcal{O}(m + n)$ of these points. At those points of A where the maximal distance can occur, we have to actually compute to distance to B, and take the maximum. This can be done in time $\mathcal{O}((m + n) \log(m + n))$ by a plane sweep algorithm, see Alt et al. [7] for details.

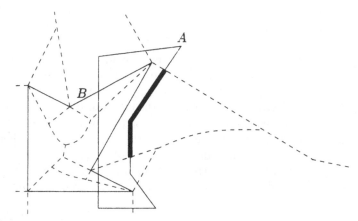

Fig. 4.13 Overlay of polygon A with the Voronoi diagram of B.

Given two polygons, the minimal Hausdorff distance under translation can be computed in time $\mathcal{O}((mn)^2(\log(m + n))^3)$ using parametric search [1], or simpler in time $\mathcal{O}((mn)^3(m + n) \log(m + n))$ [6].

Given the high complexities, it makes sense to implement approximation algorithms to find a transformation that gives a Hausdorff distance that is at most a constant times the minimum distance. For matching under translations, this can be done the following way. Let ℓ_A be the lower left corner of the axis parallel bounding box of A, i.e., it has the smallest x-coordinate of all points in A, and also, independently, the smallest y-coordinate of all points in A. Suppose that the optimal translation of A would be f, so that the Hausdorff distance $d_H = d(f(A), B)$ is minimal. Then the distance between ℓ_A and ℓ_B cannot be larger than $d_H\sqrt{2}$. So if g is the translation that maps ℓ_A onto ℓ_B, then the Hausdorff distance $d(g(A), B)$ is at most a factor $(1 + \sqrt{2})$ times the optimal d_H [7]. Determining g can obviously be done in time $\mathcal{O}(m + n)$, but computing the resulting distance still takes $\mathcal{O}(m + n) \log(m + n)$, as above.

The minimal Hausdorff distance under rigid motions (not only translations, but also rotations) can be computed in time $\mathcal{O}((mn)^4(m+n) \log(m+n))$ [6]. So again, an approximation algorithm is interesting. Let k_A be the centroid of the edges of the convex hull of A. Suppose that the optimal rigid motion of A would be f, so that the Hausdorff distance $d_H = d(f(A), B)$ is minimal. There are many rigid motions of A that map k_A onto k_B. If

g is the one that gives the smallest Hausdorff distance, then the Hausdorff distance $d(g(A), B)$ is at most a factor $(4\pi + 3)$ times the optimal d_H. For details about how to determine g, see Alt et al. [7]. The time complexity is $\mathcal{O}((mn) \log(mn) \log^*(mn))$. (The notation $\mathcal{O}(\log^* n)$ means $\inf\{k| \log\log^{(k \ \mathrm{times})} \log n \leq 1\}$). In words, it is the number of times that log has be applied to get down from n to below one. For example $\log^* 2^{4294967296}$ is only 6.)

4.5.4 Area of Overlap and Symmetric Difference

Two dissimilarity measures that are based on the area enclosed by the polygons rather than the boundaries, are the area of overlap and the area of symmetric difference. For two compact sets A and B, the area of overlap is defined as $area(A \cap B)$, and the area of symmetric difference is defined as $area((A - B) \cup (B - A))$, see Fig. 4.14. The area of symmetric difference is a metric, but the area of overlap is not. The invariance group is the class of diffeomorphisms with unit Jacobi determinant. For translations, the transformation that maximizes the area overlap also minimizes the area of symmetric difference.

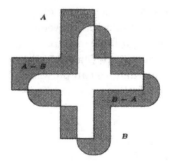

Fig. 4.14 Area of overlap and symmetric difference.

Given two polygons, computing the area of overlap can be done by computing the arrangement of two simple polygons, the combinatorial structure of point, edges, and facets resulting from overlaying the two polygons. This can be done in time $\mathcal{O}(n \log^* n + C)$, with C the complexity of the arrangement (number of vertices, edges, and facets). After preprocessing, taking time $\mathcal{O}((mn)^2)$, the area of overlap can be computed more efficiently, even for any translation of one polygon with respect to the other, in time $\mathcal{O}(\log(m + n))$ [50].

If the polygons are convex, computing the smallest area of overlap under translations can be done in time $O((m + n) \log(m + n))$ [26]. It turns out that translating the polygons so that their centroids coincide gives an overlap of at least 9/25 of the optimal solution [26].

Translating convex polygons so that their centroids coincide also gives an approximate solution for the symmetric difference, which is at most 11/3 of the optimal solution under translations [9]. This also holds for a set of transformations F other than translations, if the following holds: the centroid of A, $c(A)$, is equivariant under the transformations, i.e., $c(f(A)) = f(c(A))$ for all f in F, and F is closed under composition with translation.

The computation of the centroids can be done in linear time by triangulating each polygon, determining the centroids and areas of the triangles, and then determining the total centroids as the weighted sum of the triangle centroids. This takes time linear in the number of vertices.

Normalizing the area of overlap and symmetric difference by the area of the union of the two polygons makes these measures invariant under a larger transformation group, namely the group of all diffeomorphisms $f(x)$ with a Jacobi determinant that is constant over all points $x \in \mathbb{R}^2$ [34].

4.6 Robustness

We have already seen in Section 4.2.4 that the Hausdorff distance is not robust against noise. There are other types of distortions that can also have its effect on the measure of dissimilarity between two patterns. Fig. 4.15 shows the effect that discretization can have on a pattern, such as deformation and blurring, as well as the formation of cracks and noise. If we have a robust, invariant metric on patterns, then we can perform shape matching in a robust manner by using the shape metric.

Fig. 4.15 Discretization effects: deformation, blur, cracks, and noise.

Below, we formalize four types of robustness. We introduce four axioms expressing robustness for what we call "deformation", "blur", "cracks" and "noise." Deformation robustness says that each point in a pattern may be moved a little bit without seriously affecting the value of the metric. Blur robustness says that new points may be added close to the original pattern.

Crack robustness says that components of patterns may be broken up as long as the cracks are relatively thin. Noise robustness says that new small parts may be added to a pattern.

Let $\mathcal{P}$ be a collection of patterns in $\mathbb{R}^2$, and let T be the maximal group of homeomorphisms under which $\mathcal{P}$ is closed. A metric d on $\mathcal{P}$ is called *deformation robust* if it satisfies the following axiom:

Axiom 4.1. *For each $A \in \mathcal{P}$ and $\epsilon > 0$, there is a $\delta > 0$ such that $\|x - t(x)\| < \delta$ for all $x \in bd(A)$ implies $d(A, t(A)) < \epsilon$ for all $t \in T$.*

Deformation robustness is equivalent to saying that for each pattern $A \in \mathcal{P}$, the map $t \mapsto t(A)$ with domain T and range $\mathcal{P}$ is continuous. Fig. 4.16 shows the image of A under a transformation with a small δ in the sense of Axiom 4.1.

In the following, the boundary of a pattern is denoted with $bd(A)$. We call a metric pattern space *blur robust* if the following holds:

Axiom 4.2. *For each $A \in \mathcal{P}$ and $\epsilon > 0$, an open neighborhood U of $bd(A)$ exists, such that $d(A, B) < \epsilon$ for all $B \in \mathcal{P}$ satisfying $B - U = A - U$ and $bd(A) \subseteq bd(B)$.*

The axiom says that additions close to the boundary of A do not cause discontinuities. Fig. 4.17 shows a neighborhood U of A in which parts of B occur that are not in A.

Fig. 4.16 Deformation robust.

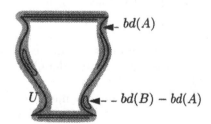

Fig. 4.17 Blur robust.

We say that a metric pattern space is *crack robust* if the next axiom holds:

Axiom 4.3. *For each $A \in \mathcal{P}$, each "crack" $x \in bd(A)$, and $\epsilon > 0$, an open neighborhood U of x exists such that $A - U = B - U$ implies $d(A, B) < \epsilon$ for all $B \in \mathcal{P}$.*

The axiom says that applying changes to A within a small enough neighborhood of a boundary point of A results in a pattern B close to A in pattern space as in Fig. 4.18 . Whether the connectedness is preserved does not matter.

If the following axiom is satisfied, we call a metric pattern space *noise robust*:

Axiom 4.4. *For each $A \in \mathcal{P}$, $x \in \mathbb{R}^2 - bd(A)$, and $\epsilon > 0$, an open neighborhood U of x exists such that $B - U = A - U$ implies $d(A, B) < \epsilon$ for all $B \in \mathcal{P}$.*

This axiom says that changes in patterns do not cause discontinuities in pattern distance, provided the changes happen within small regions. By means of the triangle inequality, we obtain an equivalent axiom when neighborhoods of finite point sets instead of singletons are considered.

Fig. 4.19 shows a pattern A and a point x. Addition of noise $B - A$ within a neighborhood U of x results in a new pattern B. Axiom 4.4 says that the distance between A and B can be made smaller by making U smaller.

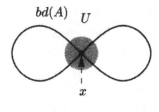

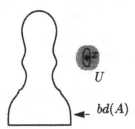

Fig. 4.18 Crack robust.   **Fig. 4.19** Noise robust.

All these robustness axioms can also be formulated for patterns in higher dimensions. For a more detailed description, see Hagedoorn and Veltkamp [36].

For the dissimilarity measures treated in the previous sections, Table 4.1 lists the invariance group, and which robustness axioms are satisfied. For a more detailed treatment, see Hagedoorn and Veltkamp [34]. The distance measure that is most suitable for any particular application totally depends on the application at hand.

4.7 Software

Most of these results are so recent that almost no implementations are available. Code for matching point sets under the Hausdorff distance is made available via `http://www3.cs.cornell.edu/dph/docs/`. Code for polygon similarity testing using turning angles is available via the Stony Brook Algorithm Repository, see `http://www.cs.sunysb.edu/~algorith/files/shape-similarity.shtml`. Software for size functions and matching with size functions is available via `http://www.dm.unibo.it/~ferri/vismath/sizefcts/sizehom2.htm`. No matching software is available via Netlib

Table 4.1 Patterns, metrics, invariance group, and robustness. "Iso" means the group of isomorphisms, "sim" means similarities, "diff-cj" means diffeomorphisms with constant Jacobian determinant. "N.a." formally means that the axiom is satisfied, but that this is meaningless for the pattern and distance (not applicable)

Distance	Pattern	Inv. group	Deform	Blur	Crack	Noise
Bottleneck	Finite point sets	Iso	Yes	N.a.	N.a.	N.a.
Minimum weight	Finite point sets	Iso	Yes	N.a.	N.a.	N.a.
Most uniform	Finite point sets	Iso	Yes	N.a.	N.a.	N.a.
Minimum deviation	Finite point sets	Iso	Yes	N.a.	N.a.	N.a.
Fréchet	Curves	Iso	Yes	N.a.	N.a.	N.a.
Turning func + L_p	Curves	Sim	Yes	N.a.	N.a.	N.a.
Signature func. + warp	Curves	Sim	Yes	N.a.	N.a.	N.a.
Norm. aff. arc length + Hausd.	Curves	Aff	no	N.a.	N.a.	N.a.
Reflection	Sets of curves	Aff	Yes	Yes	Yes	Yes
Norm. area symm. diff.	Regions	Diff-cj	Yes	Yes	Yes	Yes
Hausdorff	Non-empty compact sets	Iso	Yes	Yes	Yes	No
Discrete	Point sets	Hom	No	No	No	No

(http://netlib.bell-labs.com/netlib/index.html), and no matching software at all is mentioned in the overview of computational geometry software of Amenta [11].

For implementing geometric algorithms, CGAL, the Computational Geometry Algorithms Library is available via http://www.cs.uu.nl/CGAL/. The library provides geometric primitives such as points, lines, and triangles, basic operations such as distance and intersection calculations, as well as higher level data structures and algorithms such as triangulation, convex hull, planar map, etc.

References

1. Agarwal, PK, Sharir, M, and Toledo, S, "Applications of Parametric Searching in Geometric Optimization," J Algorith, 17, pp. 292–318, 1994.
2. Agarwal, PK, Efrat, A, and Sharir, M, "Vertical Decomposition of Shallow Levels in 3-Dimensional Arrangements and its Applications," 11th Annual ACM Symposium on Computational Geometry, pp. 39–50, 1995.

3. Aichholzer, O, Alt, H, and Rote, G, "Matching Shapes with a Reference Point," Int J Comput Geometry and Applic, 7, pp. 349–363, August, 1997.
4. Alt, H, Mehlhorn, K, Wagener, H, and Welzl, E, "Congruence, Similarity, and Symmetries of Geometric Objects," Discrete Comput Geometry, 3, pp. 237–256, 1988.
5. Alt, H, Blömer, J, Godau, G, and Wagener, H, "Approximation of Convex Polygons," 17th International Colloquium on Automata, Languages, and Programming (ICALP), Lecture Notes in Computer Science 443, pp. 703–716, Springer, 1990.
6. Alt, H, Behrends, B, and Blömer, J, "Approximate Matching of Polygonal Shapes," 7th Annual ACM Symposium on Computational Geometry, pp. 186–193, 1992.
7. Alt, H, Behrends, B, and Blömer, J, "Approximate Matching of Polygonal Shapes," Ann Math Artif Intell, pp. 251–265, 1995.
8. Alt, H and Godeau, M, "Computing the Fréchet Distance Between Two Polygonal Curves," Int J Comput Geometry & Applic, pp. 75–91, 1995.
9. Alt, H, Fuchs, U, Rote, G, and Weber, G, "Matching Convex Shapes with Respect to the Symmetric Difference," Algorithms ESA '96, 4th Annual European Symposium on Algorithms, Barcelona, Spain, pp. 320–333, September 1996. Lecture Notes in Computer Science 1136, Springer, 1996.
10. Alta Vista Photo Finder, http://image.altavista.com/cgi-bin/avncgi.
11. Amenta, N, "Computational Geometry Software," In Goodman and O'Rourke [32], chapter 52, pp. 951–960, 1997.
12. Arkin, E, Chew, P, Huttenlocher, D, Kedem, K, and Mitchel, J, "An Efficiently Computable Metric for Comparing Polygonal Shapes," IEEE Trans Patt Anal Mach Intell, 13(3), pp. 209–215, 1991.
13. Arya, S, Mount, DM, Netanyahu, NS, Silverman, R, and Wu, A, "An Optimal Algorithm for Approximate Nearest Neighbor Searching," 5th Annual ACM-SIAM Symposium on Discrete Algorithms, pp. 573–582, 1994.
14. Ballard, DH, "Generalized Hough Transform to Detect Arbitrary Patterns," IEEE Trans Patt Anal Mach Intell, 13(2), pp. 111–122, 1981.
15. Ballard, DH and Brown, CM, Computer Vision, Prentice Hall, 1982.
16. Boissonnat, JD and Yvinec, M, Algorithmic Geometry, Cambridge University Press, 1998.
17. Borgefors, G, "Distance Transforms in Digital Images," Computer Vision Graphics Image Process, 34, pp. 344-371, 1986.
18. Borgefors, G, "Hierarchical Chamfer Matching: a Parametric Edge Matching Algorithm," IEEE Trans Patt Anal Mach Intell, 10(6), pp. 849–865, November 1988.
19. Brass, P and Knauer, C, "Testing the Congruence of d-Dimensional Point Sets," 16th Annual Symposium on Computational Geometry, 2000.
20. Chen, CC, "Improved Moment Invariants for Shape Discrimination," Patt Recogn, 26(5), pp. 683–686, 1993.
21. Chew, P and Kedem, K, "Improvements on Approximate Pattern Matching," 3rd Scandinavian Workshop on Algorithm Theory, Lecture Notes in Computer Science 621, pp. 318–325. Springer, 1992.
22. Chew, LP, Goodrich, MT, Huttenlocher, DP, Kedem, K, Kleinberg, JM, and Kravet, D, "Geometric Pattern Matching Under Euclidean Motion," Comput Geometry Theory and Applic, 7, pp. 113–124, 1997.
23. Chin, F, Snoeyink, J, and Wang, CA, "Finding the Medial Axis of a Simple Polygon in Linear Time," 6th Annual International Symposium on Algorithms Computation (ISAAC 95), Lecture Notes in Computer Science 1004, pp. 382–391. Springer, 1995.

24. Cohen, SD and Guibas, LJ, "Partial Matching of Planar Polylines Under Similarity Transformations," 8th Annual Symposium on Discrete Algorithms, pp. 777–786, 1997.
25. Copson, ET, Metric Spaces, Cambridge University Press, 1968.
26. de Berg, M, Devillers, O, van Kreveld, M, Schwarzkopf, O, and Teillaud, M, "Computing the Maximum Overlap of Two Convex Polygons Under Translation," 7th Annu. Internat. Sympos. Algorithms Comput, 1996.
27. de Berg, M, van Kreveld, M, Overmars, M, and Schwarzkopf, O, Computational Geometry: Algorithms and Applications, Springer-Verlag, 1997.
28. Efrat, A and Itai, A, "Improvements on Bottleneck Matching and Related Problems Using Geometry," 12th Symposium on Computational Geometry, pp. 301–310, 1996.
29. Efrat, A and Katz, MJ, "Computing Fair and Bottleneck Matchings in Geometric Graphs," 7th International Symposium on Algorithms and Computation, pp. 115–125, 1996.
30. Godau, M, "A Natural Metric for Curves - Computing the Distance for Polygonal Chains and Approximation Algorithms," Symposium on Theoretical Aspects of Computer Science (STACS), Lecture Notes in Computer Science 480, pp. 127–136, Springer, 1991.
31. Gold, S, Matching and Learning Structural and Spatial Representations with Neural Networks, PhD thesis, Yale University, 1995.
32. Goodman, JE and O'Rourke, J, editors, Handbook of Discrete and Computational Geometry, CRC Press, 1997.
33. Günsel, B and Tekalp, M, "Shape Similarity Matching for Query-by-Example," Patt Recogn, 31(7), pp. 931–944, 1998.
34. Hagedoorn, M and Veltkamp, RC, "Metric Pattern Spaces," Technical Report UU-CS-1999-03, Utrecht University, 1999.
35. Hagedoorn, M and Veltkamp, RC, "Reliable and Efficient Pattern Matching Using an Affine Invariant Metric," Int J Computer Vision, 31(2/3), pp. 203–225, 1999.
36. Hagedoorn, M and Veltkamp, RC, "A Robust Affine Invariant Metric on Boundary Patterns," Int J Pattern Recogn Patt Anal, 13(8), pp. 1151–1164, December, 1999.
37. Huttenlocher, D and Ullman, S, "Object Recognition Using Alignment," International Conference on Computer Vision, London, pp. 102–111, 1987.
38. Huttenlocher, DP and Kedem, K, "Computing the Minimum Hausdorff Distance for Point Sets Under Translation," Proc. 6th Annual ACM Symp. Computational Geometry, pp. 340–349, 1990.
39. Huttenlocher, DP, Kedem, K, and Kleinberg, JM, "On Dynamic Voronoi Diagrams and the Minimum Hausdorff Distance for Point Sets Under Euclidean Motion in the Plane," 8th Annuual ACM Symposium on Computational Geometry, pp. 110–120, 1992.
40. Huttenlocher, DP, Klanderman, GA, and Rucklidge, WJ, "Comparing Images Using the Hausdorff Distance," IEEE Trans Patt Anal Mach Intell, 15, pp. 850–863, 1993.
41. Huttenlocher, DP, Kedem, K, and Sharir, M, "The Upper Envelope of Voronoi Surfaces and its Applications," Discrete Comput Geometry, 9, pp. 267–291, 1993.
42. Imai, H and Iri, M, "Polygonal Approximation of a Curve - Formulations and Algorithms," in Toussaint [69], pp. 71–86, 1988.
43. Jacobs, C, Finkelstein, A, and Salesin, D, "Fast Multiresolution Image Querying," Computer Graphics Proceedings SIGGRAPH, pp. 277–286, 1995.

44. Khotanzad, A and Hong, YH, "Invariant Image Recognition by Zernike Moments," IEEE Trans Patt Anal Mach Intell, 12(5), pp. 489–497, 1990.
45. Kruskal, JB, "An Overview of Sequence Comparison: Time Warps, String Edits, and Macromolecules," SIAM Rev, 25, pp. 201–237, 1983.
46. Lamdan, Y and Wolfson, H J, "Geometric Hashing: a General and Efficient Model-Based Recognition Scheme," 2nd International Conference on Computer Vision, pp. 238–249, 1988.
47. Latecki, LJ and Lakämper, R, "Convexity Rule for Shape Decomposition Based on Discrete Contour Evolution," Computer Vision Image Understand, 73(3), pp. 441–454, 1999.
48. Loncaric, S, "A Survey of Shape Analysis Techniques," Patt Recogn, 31(8), pp. 983–1001, 1998.
49. Mokhtarian, F, Abbasi, S, and Kittler, J, "Efficient and Robust Retrieval by Shape Content Through Curvature Scale Space" Image Databases and Multi-Media Search, First International Workshop IDB-MMS'96, Amsterdam, The Netherlands, pp. 35–42, 1996.
50. Mount, DM and Wu, AY, "On the Area of Overlap of Translated Polygons," Comput Vision Image Understand, 64, pp. 53–61, July, 1996.
51. Mulmuley, K, Computational Geometry: An Introduction Through Randomized Algorithms, Prentice Hall, 1993.
52. Niblack, W, Barber, R, Equitz, W, Flickner, M, Glasman, E, Petkovic, D, Yanker, P, Faloutsos, C, and Taubin, G, "The QBIC project: Querying Image by Content Using Color, Texture, and Shape," Electronic Imaging: Storage and Retrieval for Image and Video Databases, Proceedings SPIE, 1908, pp. 173–187, 1993.
53. Olson, CF, "Efficient Pose Clustering Using a Randomized Algorithm," Int J Computer Vision, 23(2), pp. 131–147, 1997.
54. O'Rourke, J, "Curve Similarity via Signatures," in Toussaint [69], pp. 295–317, 1985.
55. O'Rourke, J, Computational Geometry in C, Cambridge University Press, 1994.
56. Prokop, RJ and Reeves, AP, "A Survey of Moment-Based Techniques for Unoccluded Object Representation and Recognition," CVGIP: Graphics Models Image Process, 54(5), pp. 438–460, 1992.
57. QBIC project, http://wwwqbic.almaden.ibm.com/.
58. Ranade, S and Rosenfeld, A, "Point Pattern Matching by Relaxation," Patt Recogn, 12, pp. 269–275, 1980.
59. Rosin, PL, "Techniques for Assessing Polygonal Approximations of Curves," IEEE Trans Patt Recogn Mach Intell, 19(5), pp. 659–666, 1997.
60. Rucklidge, W, Efficient Visual Recognition Using the Hausdorff Distance, Lecture Notes in Computer Science, Springer, 1996.
61. Schirra, S, "Approximate Decision Algorithms for Approximate Congruence," Inform Process Lett, 43, pp. 29–34, 1992.
62. Schomaker, L, de Leau, E, and Vuurpijl, L, "Using Pen-Based Outlines for Object-Based Annotation and Image-Based Queries," Visual Information and Information Systems – Third International Conference VISUAL'99, Amsterdam, The Netherlands, Lecture Notes in Computer Science 1614, pp. 585–592, June, 1999.
63. Schwerdt, J, Smid, M, and Schirra, S, "Computing the Minimum Diameter for Moving Points: An Exact Implementation Using Parametric Search," 13th Annual ACM Symposium on Computational Geometry, pp. 466–468, 1997.
64. Sclaroff, S and Pentland, AP, "Modal Matching for Correspondence and Recognition," IEEE Trans Patt Anal Mach Intell, 17(6), pp. 545–561, 1995.

65. Sclaroff, S, "Deformable Prototypes for Encoding Shape Categories in Image Databases," Patt Recogn, 30(4), pp. 627–641, 1997.

66. Small, CG, The Statistical Theory of Shapes, Springer Series in Statistics, Springer, 1996.

67. Steger, C, "On the Calculation of Arbitrary Moments of Polygons," Technical Report, Dept. Computer Science, University of München, October, 1996.

68. Stockman, G, "Object Recognition and Localization via Pose Clustering," Computer Vision Graphics Image Process, 40(3), pp. 361–387, 1987.

69. Toussaint, GT editor, Computational Morphology, North-Holland, 1988.

70. Ullman, S, High-Level Vision, MIT Press, 1996.

71. Umeyama, S, "Parameterized Point Pattern Matching and its Application to Recognition of Object Families," IEEE Trans Patt Anal and Mach Intell, 15(1), pp. 136–144, 1993.

72. Vaidya, PM, "Geometry Helps in Matching," SIAM J Comput, 18(6), pp. 1201–1224, 1989.

73. van Otterloo, PJ, A Contour-Oriented Approach to Shape Analysis, Hemel Hampstead, Prentice Hall, 1992.

74. Veltkamp, RC, "Survey of Continuities of Curves and Surfaces," Computer Graphics Forum, 11(2), pp. 93–112, 1992.

75. Veltkamp, RC, "Hierarchical Approximation and Localization," Visual Computer, 14(10), pp. 471–487, 1998.

76. Verri, A and Uras, C, "Metric-Topological Approach to Shape Recognition and Representation," Image Vision Comput, 14, pp. 189–207, 1996.

77. Vleugels, J and Veltkamp, RC, "Efficient Image Retrieval Through Vantage Objects," Visual Information and Information Systems – Third International Conference VISUAL'99, Amsterdam, The Netherlands, Lecture Notes in Computer Science 1614, pp. 575–584, June 1999.

78. Wang, JZ, Wiederhold, G, Firschein, O, and Wei, SX, "Wavelet-Based Image Indexing Techniques With Partial Sketch Retrieval Capability," Fourth Forum on Research and Technology Advances in Digital Libraries, IEEE, 1997.

79. Wolfson, HJ, "Model-Based Object Recognition by Geometric Hashing," 1st European Conference on Computer Vision, Lecture Notes in Computer Science 427, pp. 526–536, Springer, 1990.

80. Wolfson, HJ and Rigoutsos, I, "Geometric Hashing: An Overview," IEEE Comput Sci Eng, pp. 10–21, October-December, 1997.

5. Feature Similarity

Jean-Michel Jolion

5.1 Introduction

In the field of image retrieval, one is often interested in achieving a compact representation of the image content using concise but effective features. One of the most popular approaches to retrieval from a large image or video database is iconic request, e.g., find some images similar to the one given as an example. Some well-known products are available (for instance QBIC [8]). However, their effectiveness always depends on the fact that the notion of similarity between two images is subjective and thus not uniquely defined [1, 12]. All of the existing states of the art of pre-attentive search techniques emphasize the need of descriptions of images and powerful similarity measures or metrics to compare these descriptions. In this chapter, we will address the choice of a metric.

5.2 From Images to Similarity

5.2.1 Pre-attentive vs. Attentive Features and Similarities

A distinction can be made between pre-attentive and attentive similarities which can be expressed in a Bayesian framework. Let Q be the query proposed by the user U, and I an image of the database. We can estimate the similarity between Q and I by the conditional probability $P(I|Q,U)$. Using Bayes' theorem, we have:

$$\begin{aligned} P(I,Q,U) &= P(I|Q,U)P(Q,U) = P(I|Q,U)P(Q|U)P(U), \\ P(I,Q,U) &= P(Q|I,U)P(I,U) = P(Q|I,U)P(I|U)P(U). \end{aligned} \quad (5.1)$$

So,

$$P(I|Q,U) = \frac{P(Q|I,U)P(I|U)}{P(Q|U)}. \quad (5.2)$$

We look for the maximum values of this probability over all the images I so the term $P(Q|U)$ does not interact in this search and what we have to maximize is the product

$$P(Q|I,U)P(I|U). \quad (5.3)$$

It is always difficult to maximize a product. The term $P(Q|I,U)$ refers to the a posteriori validity of I and $P(I|U)$ is the a priori validity. This a priori stands for the constraints the user puts on the database and is clearly related to the concepts the user will try to summarize by its request. This leads to *attentive* similarity or semantic constraints. The a posteriori term $P(Q|I,U)$ also induces an attentive mechanism assuming that the stimuli to be compared have previously been interpreted (i.e., they belong to the same class and can be characterized by some high level features like keywords). It is, however, possible to consider this term as being made of two parts, e.g., $P(Q|U)$ and the *pre-attentive* part, e.g., $P(Q|I)$. However, there is no way, in this probabilistic framework, to separate these two points of view on the request. This mainly explains the inherent drawback of this approach.

In order to go further, we have to make a strong assumption about the independence of these two parts yielding a new expression (again $P(Q|U)$ is neglected as we want to maximize the product over I):

$$P(Q|I)P(I|U). \tag{5.4}$$

As Santini and Jain [23] did, we assume that "attentive similarity is based on features specific of the recognition process, while pre-attentive similarity is based on image features only." So, the database we are working with will not be organized in any sense, nor be related to any specific application field.

Basically, a classic approach is thus first, to extract these features from the images of the database, and summarize these features in a reduced set of k indexes. Then, given an image example, the retrieval process consists in extracting features from this image, projecting them onto the indexes space and looking for the nearest neighbors based on some particular distance.

An overview of a generic architecture of such system is proposed in Fig. 5.1. Every image is characterized by a set of k features. The query Q is processed the same way as the images of the database. Firstly, one computes the k similarities between the features. Then, these similarities are combined in a global scalar similarity $S(I,Q)$. The system selects the N images having the greatest overall similarities to the query Q. A feedback process allows a better combination of the individual similarities without any extra computation of these similarities.

5.2.2 Distance vs. Similarity

Given a feature and its representation associated with each image, we need a metric to compare an image I of the database and the query Q. A basic way is to use a distance d. However, do we really need a distance? A distance is defined as follows:

$$d : \Im \times \Im \rightarrow \Re^+$$

where $\Im$ is the set of images and $\Re^+$ the set of positive real numbers. d must satisfy the following properties for all the images I, J and K in $\Im$.

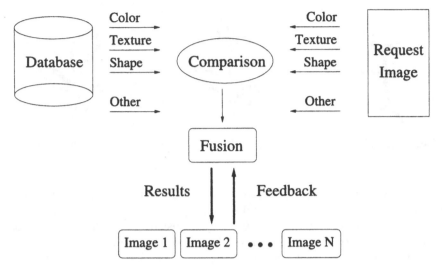

Fig. 5.1 Global architecture of our system.

$P_1 : d(I, I) = d(J, J)$ self-similarity
$P_2 : d(I, J) \geq d(I, I)$ minimality
$P_3 : d(I, J) = d(J, I)$ symmetry
$P_4 : d(I, K) + d(K, J) \geq d(I, J)$ triangular inequality

Any application satisfying P_1, P_2, and P_4 is a *metric*. Any application satisfying P_1, P_2, and P_3 is a *(di)similarity*. Many works, starting in the early 1950s with Attneave [2], have investigated the Euclidean nature of perception. A very nice survey is Santini and Jain [24]. Many similarity experiments like, Tversky [31], showed that the properties P_1 to P_3 are not validated by human perception. For instance, we are searching a database using a particular query Q so we cannot assume the symmetry as obvious because searching the database with Q is not equivalent to searching the same database with an other image. Furthermore, Tversky [31] showed that in general, the less salient stimulus is more similar to the more salient than the more salient is to the less salient. The triangular inequality is more complex and there is no proof that a similar property exists for human perception. For instance, in Fig. 5.2, the distance between the drawing on the left a and the drawing in the middle b is small, as is the distance from middle drawing to the right one c. However, the distance from the left drawing to the right one is large, and therefore, $d(a, b) + d(b, c) > d(a, c)$ does not hold.

The self-similarity and minimality properties are often preserved at least for normalization purposes and constitute the minimal requirements for the similarity. Of course, if one refutes all the properties of a metric, he must introduce new constraints. Tversky and Gati [32] proposed three ordinal properties, *dominance*, *consistency*, and *transitivity* which are more relevant to

Fig. 5.2 Under partial matching, the triangular inequality does not hold.

human perception. For instance, in a k-dimensional feature space, the dominance sets that the k-dimensional dissimilarity exceeds all the k'-dimensional projections of that distance (for $k' < k$). Consistency implies that the ordinal relation between dissimilarities along one dimension is independent of the other coordinates. The transitivity ensures that the "in between" relation behaves as in the metric model.

One of the most famous non-metric models is the feature contrast model proposed by Tversky [31]. An image is characterized as a set of binary features, i.e., I is characterized by the set A_I of features that I possesses. The similarity function $s(I, J)$ between two images I and J must satisfy three properties.

Matching. $s(I, J) = F(A_I \cap A_J, A_I - A_J, A_J - A_I)$.
Monoticity. $s(I, J) > s(I, K)$ if $A_I \cap A_K \subseteq A_I \cap A_J$, $A_I - A_J \subseteq A_I - A_K$,
   $A_J - A_J \subseteq A_K - A_I$.
Independence [24]. If the pairs (I, J) and (K, L), as well as the pairs (I', J')
   and (K', L') agree on the same two features, while the pairs (I, J) and
   (I', J') as well as (K, L) and (K', L') agree on a remaining (third) component, then $s(I, J) \geq s(I', J') \Leftrightarrow s(K, L) \geq s(K', L')$.

Tversky showed that if a similarity function s verifies these three properties, then there exist a similarity function S, a non-negative function f, and two positive constants $\alpha, \beta \geq 0$ such that $\forall (I, J, K, L) \in \Im^4$

$$S(I, J) \geq S(K, L) \Leftrightarrow s(I, J) \geq s(K, L),$$
$$S(I, J) = f(A_I \cap A_J) - \alpha f(A_I - A_J) - \beta f(A_J - A_I).$$

This representation is called the *contrast model* and shows that any similarity satisfying matching, monoticity, and independence can be obtained using a linear combination of a function of the common features and the distinctive features. This model accounts for violation of the properties P_1 to P_4 depending on the values of the constants α and β (for instance if $\alpha \neq \beta$, P_3 is not validated).

However, the main disadvantage of this model is that it requires binary features, thus it is more appropriate for shape comparisons than for texture or color analysis. On this particular topic of shape similarity measures, a complete survey can be found in Veltkamp and Hagedoorn [33].

5.3 From Features to Similarity

The similarity measurements are based on comparisons between images features which thus have to be previously extracted. These features will mainly characterize:

- Gray levels, for instance the so-called differential invariants [25]
- Colors with histograms [29, 30] or moments [26]
- Texture with the coefficients in the Fourier or wavelets domains or the responses of a set of Gabor filters characterizing the orientations and scales [17, 36]
- Shape for special geometric features like lines or curves [10]
- Structure [19]

The previous chapters of this book present a complete description of the extraction processes associated with these features. Many works have already proposed specific feature extraction processes and it is not easy to compare them [5]. However, the way the similarity between two images, and consequently their representation, is computed highly depends on the way these representations have been built. In this section, we will introduce the main criteria that will set up the constraints on the similarity computation.

5.3.1 Complete vs. Partial Feature

One can simply take the overall distribution of a given feature in the input image or video as an index. This leads to SNR- or correlation-like tools (see Section 5.5). This is applied for close comparisons between very similar images in order to evaluate a distortion effect as in compression techniques and is thus not relevant for the image indexing domain where the similarity is not so strict. Moreover, we are often interested in more compact and informative descriptions.

Carson et al. [4] proposed to first segment the image as a set of regions. This approach, however, is complex because it requires a segmentation step (and we know that segmentation is never optimal). Moreover, it assumes that the regions extracted by the segmentation process are close to the objects in the scene. This is only valid for objects of homogeneous color and/or texture and thus is very domain dependent.

Another approach is based on key points. It argues that two signals are similar if they have particular feature values spatially located in a consistent order. The locations of these particular values are called the *interest points* or *key points* of the signal [3]. This approach has first been used in the image indexing field by Schmid and Mohr [25]. A feature consists of a list of values or a histogram. A very large set of interest point detectors has been already proposed in the image indexing literature [25]. This wide variety of detectors is mainly due to a lack of definition of the concept of interest points. However, most of these works refer to the same literature and basically assume that key points are equivalent to corners or more generally to image points characterized by a significant gradient amount in more than one direction. Obviously, a differential framework results from these definitions [6].

Studies on visual attention, more related to the human vision, propose very different models. The basic information is still the variation in the stimuli but it is no longer taken into account in a differential way but mainly from an energy point of view [13]. We proposed a detector which uses the contrast.* As variations and thus contrasts exist at almost any scale in an image, we must use a multiresolution framework. The contrast pyramid has been defined in Ref. [15].

Referring to this framework, we say that a pixel is an interest point if its contrast is significantly higher than those of its neighbors. When extracted at a low level in the pyramid, this pixel is related to a local contrast; when extracted at a higher level in the pyramid, this pixel is related to a region contrast. These contrasts are then combined in a top-down scheme in a pyramid yielding a final representation where both local and global information are taken into account. More details on this technique are presented by Bres and Jolion [3]. Figure 5.3 shows an example of interest point extraction.

5.3.2 Global vs. Local Feature

A feature value can be global, e.g., it can take into account all the pixels in the image. However, we are dealing with so many different images that it is not obviously possible to summarize the behavior of an image regarding a particular feature, for instance color, by a scalar or even by a set of scalars characterizing the entire image.

It is sometime difficult to separate the global vs. local choice from the complete vs. partial. Let us explain it for the example of gray-level index. For a given image, one can compute the mean gray-level and use it as a global index or select the histogram of gray-level values yielding a local feature-based index. Local means that a value is computed based on a subset of the image usually in a neighborhood of a given point. On the other hand, these computations can be performed using the entire set of image pixels, i.e., yielding a complete index, or only on a subset of the image pixels, yielding a partial index (see Table 5.1).

---

* From a mathematical point of view, the contrast is less constrained than the differential model of variation.

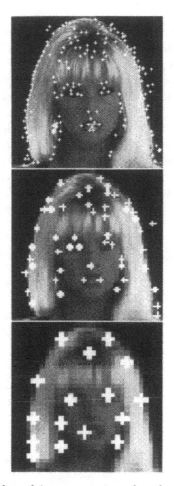

Fig. 5.3 Contrast-based interest points for three levels of a portrait (the significance level has been set to 0.2).

Table 5.1 Example of complete/partial vs. local/global indexes

	Complete	Partial
Local	Histogram of $N \times N$ values	Histogram of $k \times k$ values
Global	Mean of $N \times N$ values	Mean of $k \times k$ values

5.3.3 About Histograms

A classic representation is a set of local values for each image and each feature. For instance, let an image be summarized by its key points.* The neighborhood of each of these points can be texture characterized by the maximum

---

* This holds for every set of image pixels, and so for the entire image.

responses of a set of Gabor filters (typically with eight orientations and three scales). The information is stored in an ordered set of 24 histograms, one for each filter. Such histograms are a very efficient representation as they are invariant to translation. However, they do not keep any information about the 2D spatial distribution of the values as does a spectral representation like the Fourier domain.

The 2D pairwise histograms introduced by Huet [10] can be used as an alternative. Let A and B be two (interest) points in the input image and let v_A and v_B be their associated local value for the feature currently being studied. The pair (v_A, v_B) is an entry in a 2D histogram (Fig. 5.4). Scanning all the pairs leads to an accumulation process defined by $H(v_A, v_B) = H(v_A, v_B) + w(A, B)$ where w can be one of the following:

- $\forall (A, B)$: $w(A, B) = 1$
- $\forall (A, B)$: $w(A, B) = \frac{1}{1+d(A,B)}$ where d stands for an arbitrary spatial distance function
- $\forall (A, B)$ such that A belongs to the k nearest neighbors of B, $w(A, B) = 1$, otherwise $w(A, B) = 0$. Using $k = 6$ leads to a pseudo Voronoï tessellation of the spatial domain. A study on the choice of k can be found in Huet and Hancock [10].

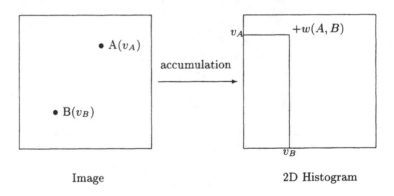

Fig. 5.4 The 2D histogram updating process for a pair of interest points.

Figure 5.5 shows an example image and two 2D texture histograms out of its 24-histogram description. The image contains frequencies mainly in the horizontal orientation (orientation 0). Therefore, the histogram for orientation 0 shows a distribution of strong responses, i.e., strong bins from indexes 4 to 6. The histogram for orientation index 2, which corresponds to structures in orientations around 45°, shows only one strong bin at index $(0,0)$, i.e., almost no response.

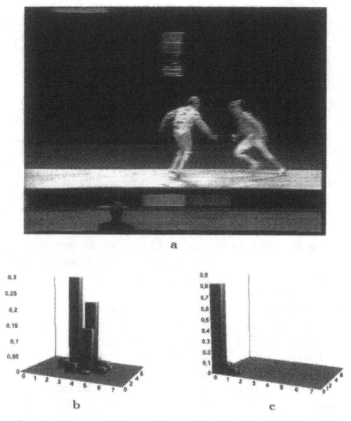

Fig. 5.5 Image representation using ordered sets of histograms containing amplitudes: Example picture **a** and two histograms of the ordered set with orientation $0 = 0^o$ **b** and orientation $2 = 45^o$ **c**, both for scale index 0.

5.3.4 Global vs. Individual Comparison of Features

One of the assumptions of the system described in Fig 5.1 is that the features are compared individually and then combined. This is a strong hypothesis whose main advantages are to allow a parallel processing of all the features (and so hopefully some speed-up) and to simplify the comparison operators by reducing the dimensionality of the comparison to be carried out.

Let k be the number of features related to features of different nature like color, texture, etc. It is difficult to build a global distance or similarity measure in such a space because color is not texture and also because the way the indexes have been computed may induce different behavior. It is not because the range of two indexes is $[0, 1]$ that we can easily do some linear combination. Indeed this kind of computation also assumes that both distributions of values are of the same nature, which is most of the time not true. That is why one often assumes that the overall similarity is based on a combination of individual similarities, e.g., one for each feature yielding a point in $\Re^k$. However, this could lead to some misunderstanding of the actual nature of the relation between two images, if this one is purely multidimensional and cannot be analyses through this set of projections.

As a final step, the k-dimensional vector is summarized into a scalar in order to sort the images of the database and then output the N more similar images.

The third advantage of separately computing k similarities is related to the pre-attentive approach. Indeed, the request has been selected by a user and he should be a part of the global decision process. If we do not assume any attentive information, he cannot specify some keywords in order to reduce the search. He can only interact with the system by giving some weights for the overall similarity computation as in now classic systems like QBIC [8] or Jacobs [14]. This is obviously not the right way as the user is most of the time not a specialist so is not able to write down significant values for these weights, e.g., he is not able to translate what he means by "similar" in terms of color, texture, shape, etc.

This is why we argue that the process must be iterative. The first step is a classic one and the user interaction with the system is limited to some global choices. At the end of this step, the system returns a set of answers which covers different points of view of what "similar" could be. The user then specifies which images he considers as good answers and which ones as wrong answers. The system automatically recomputes all its internal weights between similarities and proposes a new set of answers. This will be detailed in Section 5.5.

5.4 Similarity Between Two Sets of Points

In this section, we assume that the image index is made of a set of feature vectors collected from part of the image.

5.4.1 Multidimensional Voting

We can use the Euclidean distance to compare two vectors in the feature space: $d_E(\mu, \nu) = \sqrt{\sum_i (\mu_i - \nu_i)^2}$ where μ and ν denote two vectors.

A similarity measure between two images can now be built. Schmid and Mohr [25] used a voting algorithm: each vector V_i of the query image is compared to all vectors V_j in the database, which are linked to their images M_k. If the distance between V_i and V_j is below a threshold t, then the respective image M_k gets a vote. The images having maximum votes are returned to the user. As an alternative, a vote can be associated with the image corresponding to the nearest neighbor of a vector of the query image. On one hand, the nearest neighbor strategy is too strict but on the other, the threshold strategy is sensitive to the choice of t. Both approaches suffer from the fact that the local geometric constraints are not taken into account.

We changed this method by explicitly searching for corresponding points in both images [36]. The means to qualify two points as being a pair is the minimum distance in feature space. To do this we build a matrix which stores the distances of all possible feature pairs, the rows i denoting the key points of the query image, the columns j the key points of the compared image, and the elements $E_{i,j}$ the distance between point i of the query image and point j of the compared image. Then we search the minimum element of the matrix. The column and row number denote the first pair of corresponding interest points. Both column and row are deleted from the matrix, since these two points are not available to other pairs. Then we again search the minimum element of the remaining matrix. This algorithm is continued until the matrix is eliminated (the points of at least one image are exhausted) or the minimum distance does not exceed a given threshold (there are no more points having a corresponding partner). The similarity between the two images is calculated using the number of corresponding points found:

$$s(A, B) = \frac{2 \times \text{Number of corresponding points}}{N(A) + N(B)} \tag{5.5}$$

where $N(A)$ and $N(B)$ denote the number of interest points of the image A and b, respectively. The cost of the algorithm is $O(N^2)$ for calculating the distance matrix plus the cost for the search of the corresponding pairs, which is dependent on the similarity of the two images. The higher the similarity the higher the cost, since the search for the minimum distance in the matrix has to be done more often. Still the overall cost of the search algorithm is $O(N^3)$.

A post-processing step can modify the matching between points (Fig. 5.6) in order to validate a set of predefined geometric constraints like "if A is above B, then match(A) is above match(B)." This kind of strategy is very useful to prune the set of match candidates during the distance based search which can then be less sensitive to the choice of the distance threshold. This approach takes into account the spatial structure of the interest points in an image. This kind of information can also be represented as a graph as presented in next section. It can also gives rise to shape comparison for which an extended survey is Veltkamp and Hagedoorn [33].

5.4.2 Graph-Based Matching

We still assume that an image is represented, for retrieval purposes, by a set of points, each associated with a feature vector. This information can be abstracted as an attributed relational graph, the vertices of which are the points, the vertice attributes are the vector values, and the edges can be established by computing the n-nearest neighbor graph. An example of such

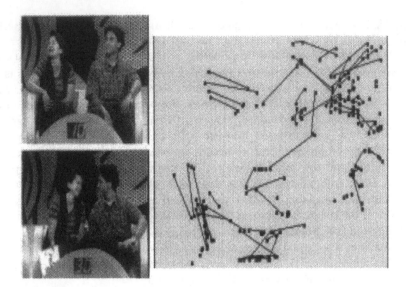

Fig. 5.6 Two images and their corresponding points.

a model has been proposed by Huet et al. [11] for image retrieval based on line patterns.

Computing a similarity between two images is thus equivalent to graph matching. Many different algorithms have been proposed in the past for this kind of problem. However, their high complexity is generally a strong limitation as an image gives rise to graphs having a large number of vertices. Huet et al. [11] proposed a tractable graph matching technique which first matches the nodes of the query with their best neighbor in the database (so very much resembles the previous technique) and then iteratively modifies the set of correspondences to optimize the *a posteriori* probability of the database image given the query image. Retrieval is realized by identifying the image of largest matching probability.

This approach can be extended to vertices having geometric shapes, e.g., from points to objects. However, the classic "is neighbor of" relation has to be extended to the B-strings relationship categories. Matching is thus divided into three types called $type_0$, $type_1$, and $type_2$ with increasing strictness of match. For instance, $type_0$ matching includes disjoint, meets, overlaps, contains, and inside; $type_1$ matching is made more strict by ensuring that two vertices having a $type_0$ match also display the same orthogonal relationships (above, below, to the right of, to the left of); $type_2$ matching requires the same actual relationship between the objects along each axis.

Computing similarity between such graphs is equivalent to the search for the largest common subgraph matching. Classic graph algorithms performing such tasks (for instance the maximal clique detection graph algorithm proposed by Levi [16]) has complexity of $O(L(mn)^n)$ in the worst case for a request graph of n vertices and a database of L models of size m vertices. This performance shows that classic graph matching algorithms are often unsuitable as the response time is unacceptable. Shearer et al. [27] proposed an efficient graph matching algorithm with complexity independent of the number of models (L) and their size (m) and only polynomial in the size of the query (n). Of course, one has to pay for this efficiency by a pre-processing step of exponential complexity in order to build a decision tree.

It is obvious that the image retrieval domain will benefit from research in graph matching as the huge amount of data to be processed sets up a new challenge for more and more powerful algorithms.

5.5 Similarity Between Histograms

One of the most classical representations for a set of values is the histogram, mainly because it is very easy to compute and there exists a large number of distances and similarities. We will survey them in this section.

5.5.1 Classic Distances

We now consider the computation of a distance or similarity between two distributions. We assume that these two distributions are two histograms H_0 and H_1 having n bins each. Distances between histograms are of two different natures: vector distances and global distances.

Vector distances are fast to compute with complexity $O(n)$. A very well-known example of this first category, is the Minkowski form distance

$$d_{L_p}(H_0, H_1) = \left[\sum_{i=1}^{n} |H_0(i) - H_1(i)|^p \right]^{1/p}. \tag{5.6}$$

For $p = 2$, this yields the Euclidean distance. For $p = 1$, we get the Manhattan or city block distance. The max metric is the limit for $p \to \infty$. This type of distance is very sensitive to shifts, even if these shifts are small. This comes from the fact that no other column but the exactly matching one is seen by the method. When dealing with images, this is a great disadvantage. For instance, variation in illumination shifts the color histogram. Histograms of orientations (in the case of texture or shape analysis) must also be processed in a particular way, e.g., some shifting must be introduced to compensate for rotation [37].* So the distance between two histograms is actually the minimum over a set of distances.

---

* In classical image indexation applications like video indexing, we are not interested in a complete rotation invariance but we must take into account small variations of orientations (typically $\pm 22°$).

Moreover, as these distances are mainly based on the mean operator, they are not robust to noise. Robustness against noise (and especially impulse noise) can be achieved with the Jeffrey distance, which is based on the Kullback–Leibler distance [20]:

$$d_J(H_0, H_1) = \sum_{i=1}^{i} \left[H_0(i) \log \left(\frac{H_0(i)}{m_i} \right) + H_1(i) \log \left(\frac{H_1(i)}{m_i} \right) \right] \qquad (5.7)$$

with $m_i = \frac{H_0(i)+H_1(i)}{2}$.

However, this distance still does not behave well when a histogram is a shifted version of another histogram. Another distance is the Bhattacharyya distance. This choice seems to be competitive based on the comparative study of Huet and Hancock [10]. A special case of the Chernoff bound of the error probability in binary classification, and very close to a classic correlation, Bhattacharyya's distance is defined as:

$$d(H_1, H_2) = -\log \sum_{\alpha} \sqrt{H_1(\alpha)H_2(\alpha)} \qquad (5.8)$$

where the histograms are normalized, e.g., $\sum_{\alpha} H_i(\alpha) = 1$.

It has the symmetric property but the triangular property is only satisfied for specific configurations.

One can also use the *signal-to-noise ratio* which evaluates the difference between two distributions by the ratio of the number of differences relative to the number of variations in one distribution:

$$d_{snr}(H_0, H_1) = 10 \log_{10} \left(\frac{Var[H_0]}{Var[H_0 - H_1]} \right) \qquad (5.9)$$

where $Var[\cdot]$ stands for the variance operator.

This measure is non-symmetric but this property is not required in our application. This measure is often considered as part of the second class, e.g., global distances, as it uses the global parameters of the histograms.

On the other hand, there exists a large set of globally defined distances which are more sensitive to the shape of the histogram than to the exact values. So, they are obviously more robust to noise, small shifts, and outliers (up to a certain amount of course). An example of this kind of distance is the correlation coefficient defined as:

$$\rho(H_0, H_1) = \frac{E[H_0 H_1] - E[H_0]E[H_1]}{\sqrt{Var[H_0]Var[H_1]}} \qquad (5.10)$$

where $E[\cdot]$ stands for the mean operator.

However, this distance is only looking for linear comparisons between distributions and mainly assumes an unimodal model of these distributions.

Global distances try to account for shift and noise, yielding more complex algorithms. We will now focus on two particular global distances.

5.5.2 The Unfolded Distance

The unfolded distance has been designed in the field of image processing and was first proposed by Shen and Wong [28]. It has then been generalized by Werman et al. [35] to multidimensional histogram comparisons. It tries to overcome the problem of many vector distances which can be summarized in the following example [35]. Consider three histograms having only one non-null bin in location 0, 1, and 6, respectively (Fig. 5.7).

$$H_1 \; : \; 5 \ldots \ldots$$
$$H_2 \; : \; . \; 5 \ldots \ldots$$
$$H_3 \; : \; \ldots \ldots 5 \; .$$

Fig. 5.7 Three histograms of 8 bins each. The symbol "." stands for a null value of the histogram.

These three histograms are very similar, based on vector distances, because they are based on pairwise differences without taking into account the locations of the bins. Indeed, these distances are (mainly) invariant to any permutations of the bins as long as that the permutation is applied the same way on the compared histograms. We would prefer to have a distance which better discriminates the third histogram from the other ones.

The main idea is first to unfold the histograms (we thus have nothing to do if we already get a sorted list of values). The distance between the two resulting vectors is computed using the city-block distance (sum of absolute differences between pairwise values). One constraint is thus to have histograms of the same size, and a normalization has to be performed first if this constraint is not satisfied. The complexity is $O(m)$ where m is the size of the images or the number of key points extracted from the images.

This distance gave better results than vector distances when used for the comparison of texture features [34].

5.5.3 The Earth Mover's Distance

The earth mover's distance (EMD) was first proposed by Rubner et al. [21] in 1998. Let us consider the columns of one histogram as earth hills and the columns of the others as holes. The sizes of the hills and the holes are proportional to the values of the bins in the histograms. The EMD is calculated as the minimum cost it takes to fill all the holes with the earth coming from the hills. Figure 5.8 shows a short example of the way this distance works.

A partial cost is proportional to the distance between a hole and a hill from which you decide to take the earth. It can thus be $c_{ij} = |i - j|$.

This problem has been known for many years as the transportation problem [22]. The algorithm which solves this problem is $O(n^2)$ with n being the number of bins of the histograms. The EMD is thus defined as:

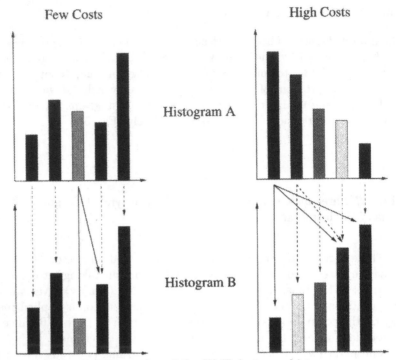

Fig. 5.8 Examples of the EMD between histograms.

$$EMD(H_0, H_1) = \frac{minimum\ cost}{\min(area(H_0), area(H_1))}. \qquad (5.11)$$

It seems that this distance has many properties which are needed in image analysis. It is robust against noise, scaling, and shift because it mainly compares the shapes of the histograms. So histograms where most of the "earth" is concentrated on different locations, which means they have high values far apart, will be judged as very different. Histograms of similar shapes but with noise everywhere will be judged as similar because the earth can be moved around to account for this noise. The main constraint when using this distance is to have similar areas for the histograms.

This distance has proved to have nice properties for image indexing [21]. The optimized algorithm is as follows:

Let H_0 and H_1 be two histograms with values $H_0(i)$ and $H_1(i)$, for $i = 1, \cdots, n$. Let $c_{ij} = |i - j|$. We want to minimize $\sum_i \sum_j c_{ij} x_{ij}$, where x_{ij} is the amount we shift from location i in H_0 to location j in H_1.

Step 1. Let $w_i = max_j c_{ij}$ and $y_j = max_i c_{ij}$.
Step 2. Find (i, j) such that $w_i + y_j - c_{ij} = max_{(k,l)}[(w_k + y_l - c_{kl}) > 0]$.
Step 3. Set x_{ij} to $min(H_0(i), H_1(j))$.

Step 4. Subtract x_{ij} from $H_0(i)$ and $H_1(j)$. Eliminate from the problem the row and column which results in a zero availability ($H_0(i) = 0$) or requirement ($H_1(j) = 0$). Stop if all the values are 0; otherwise return to Step 1.

Step 5. Calculate

$$EMD(H_0, H_1) = \frac{\sum_{i=1}^{n} \sum_{j=1}^{n} |i - j| x_{ij}}{\min \left[\sum_{i=1}^{n} H_0(i), \sum_{j=1}^{n} H_1(j) \right]}$$

The extension to multidimensional histograms is straightforward but very expensive, especially if many histograms have to be compared.

5.6 Merging Similarities

5.6.1 An Information Retrieval Approach

Any image I in the database is now characterized by a vector $\Phi^I = (\phi_1^I, \ldots, \phi_k^I)$ of k values corresponding to the k individual similarities obtained for the k features used by the system.

The simplest way to combine these similarities yielding a scalar is of course to compute the weighted sum:

$$S(I, Q) = \sum_{i=1}^{k} \alpha_i \phi_i^I. \tag{5.12}$$

However, this will produce a mean similarity which is not always what we are looking for. This step is a crucial one. Indeed, the features the system uses are not of the same importance. Moreover they do not have the same meaning, especially for a non-specialist user. So it is undesirable to let the user choose the weights the system will use to combine the individual similarities as most of the systems do.

The choice of the weights has been widely studied in the field of information retrieval. The model of Fagin and Wimmers [7] is a solution to combine the relevance of several features (it is the unique solution under some classic constraints like local linearity). Let $x_1, \ldots, x_k$ be an ordered set of keys/features, $\theta_1, \ldots, \theta_k$ their respective weights (assuming that $\theta_1 \geq \ldots \geq \theta_k$), and $f(x_1, \ldots, x_i)$ the relevance of a document based on the keys x_1 to x_i. The overall relevance of a document is computed as:

$$\begin{aligned} f_{\theta_1, \ldots, \theta_k}(x_1, \ldots, x_k) = {} & (\theta_1 - \theta_2) f(x_1) + 2(\theta_2 - \theta_3) f(x_1, x_2) \\ & + 3(\theta_3 - \theta_4) f(x_1, x_2, x_3) \\ & + \ldots + k \theta_k f(x_1, \ldots, x_m) \end{aligned} \tag{5.13}$$

The main point is that the overall relevance only depends on k terms $f(x_1), f(x_1, x_2), \ldots, f(x_1, x_2, \ldots, x_k)$ and not on terms such as $f(x_2)$, $f(x_1, x_3)$ and so on. An example of such a overall relevance function will be given in the next section.

5.6.2 A Probabilistic Model

One possible way to determine the weights θ_i is to look for a particular behavior. In a pre-attentive search, we must allow that an image only matches the request for some but not all the features, e.g., partial matches. This induces that we have to emphasize the highest values of Φ^I. A non-linear merging process is thus required such as the α-trimmed mean which consists in taking into account (in a classic weighted mean) only the α-percent highest values among the k similarities. The α value is thus related to the amount of partial match we decide to allow and has to be *a priori* determined with the so-called threshold effect. Moreover, when k is not large, talking about percentage is somehow tricky. A probabilistic model can be used in order to overcome this threshold effect.

Since the distance function used to compare a pair of images - using one feature at a time - returns values between 0 and 1, it can be assumed to represent the probability that the two images match regarding this feature. The probability that two images match depends on the number of features that match and their relative match values. Let us assume that the similarities ϕ_j^I are sorted in decreasing order yielding $\tilde{\phi}_j^I$.

The probability that j features match is given by:

$$P_j^I = \frac{1}{j} \sum_{i=1}^{j} \tilde{\phi}_i^I. \tag{5.14}$$

The probability that the query and an image I match is given by:

$$S_{Q,I} = \frac{1}{k} \sum_{j=1}^{k} P_j^I. \tag{5.15}$$

We propose $S_{Q,I}$ as the global similarity value. With this function, the probability that an image matches the query is not only based on all the features but it also takes into account the partial matches of 1, 2, 3, $\ldots$, k features between the two images. It is important to stress that for the first computation of the similarity value, the distances ϕ_j^I are sorted before computing the probabilities of the matchings. This is done, so that greater weight is assigned to features that match better than those that result in a poor match. We get from Eqs 5.14 and 5.15:

$$\tilde{S}_{(Q,I)} = \frac{1}{k} \sum_{j=1}^{k} \frac{1}{j} \sum_{i=1}^{j} \tilde{\phi}_i^I = \sum_{i=1}^{k} \tilde{\alpha}_i \tilde{\phi}_i^I. \tag{5.16}$$

The weights $\tilde{\alpha}_i$ only depend on k. For $k = 5$, we get:

$$\tilde{\alpha} = (0.4566, 0.2567, 0.1567, 0.090, 0.040)^t \qquad (5.17)$$

Note that this function will not be affected by small variations in the distance values, nor the values themselves, because an average value is used to characterize the matching probability each time.

The proposed similarity measure is a scalar value in the interval $[0, 1]$ and is related to the probability that the two compared images are the same. When all the features match, the returned value will be 1, and when there is no match, the function will return zero. In all the other cases, the returned value will be an overall assessment of the match taking into account how many features match between the two images (this is similar to the Tversky's assumption). The function has the following properties: $S_{I,I} = 1$ and $S_{I,Q} = S_{Q,I}$. However, the symmetry property $S_{Q,I} = S_{I,Q}$ will only hold if the similarity function between the features is symmetric too.

This function validates the Fagin and Wimmers [7] model for:

$$f(x_1, \ldots, x_i) = \sum_{j=1}^{i} \phi_j^I. \qquad (5.18)$$

Our weights, $\tilde{\alpha}$ can be related to the weights of the Fagin and Wimmers by:

$$
\begin{aligned}
\tilde{\alpha}_1 &= \theta_1 + \theta_2 + \ldots + \theta_k, \\
\tilde{\alpha}_2 &= 2\theta_2 + \theta_3 + \ldots + \theta_k, \\
\tilde{\alpha}_3 &= 3\theta_3 + \theta_4 + \ldots + \theta_k, \\
&\ldots \\
\tilde{\alpha}_{k-1} &= (k-1)\theta_{k-1} + \theta_k, \\
\tilde{\alpha}_k &= k\theta_k.
\end{aligned}
\qquad (5.19)
$$

This function emphasizes strong matches between features. Therefore, it is expected that images with partial strong matches will appear earlier in the resulting set than images with equivalent uniform global matchings. This property widens the range of the selected images and lets the user specify which images he/she is interested in as a next step. It is also important that the returned value does not depend on any assumption but only on the number of features used for the comparison.

5.7 Similarity Improvement

At first sight, it seems that an average of the partial matches would suffice to describe the similarity between the images. However, two images can be similar in some of the features, but not highly ranked by the system because of the low similarities for the other features. What is important is that the system must return all images that gave a strong match in most of the features (if not all of them) and let the user define the features that are more crucial in the search.

The advantage of using a linear combination of the similarities is that the weights can be modified so as to achieve special kinds of matches. After the initial retrieval, the system may feedback automatically or manually some new weights for the similarity function. This relates to the field of relevance feedback, which is fully presented in Chapter 9. We will present here some trends based on our probabilistic model.

Given a user selection, we want to modify the weights $\{\alpha_j, j \in [1, k]\}$, so as to retrieve new images closer to the query. So, after the weights are modified, the new similarity measures between the selected images and the query should be higher than the new similarity measures between the remaining images and the query. We will investigate a navigation-like process.

If we assume that the overall image similarity is a weighted sum of the individual distances between the features and the query, this overall similarity for each image can be written as: $A^T \Phi^i$, where $A^T = (\alpha_1 \alpha_2 \ldots \alpha_k)$ and $(\Phi^i)^T = (\phi_1^i \phi_2^i \ldots \phi_k^i)$.

So, if the new weights after user selection are represented by: $A^* = \{\alpha_j^*, j \in [1, k]\}$, then the following constraints must be satisfied:

$$(A^*)^T \Phi^{k_1} \leq (A^*)^T \Phi^{k_2} \leq \ldots \leq (A^*)^T \Phi^{l_1} \leq (A^*)^T \Phi^{l_2} \ldots \qquad (5.20)$$

where $k_1 \ldots$, and $l_2, \ldots$ stand for the user-based order of the images.

This system of inequalities can be solved with an optimization algorithm such as Simplex [18], to obtain the new set of weights. However, this requires a sufficiently large set of constraints.

Given a set of responses, we know that some of them will not be appropriate from the user's point of view. This is due to our choice limited to a pre-attentive similarity. In order to evolve towards a better result, we can introduce a simple relevance feedback process.

First, if we assume that most of the images of the first set of N answers returned by the system belong to the right class, it is possible to use a maximum a posteriori process which does not need to interact with the user, i.e., non supervised relevance feedback. Indeed, the probability of a given feature C_i, given the query Q is defined as:

$$P(C_i | Q) \propto \sum_{j=1}^{N} S(J_j, Q) \phi_i^{J_j} \qquad (5.21)$$

where $J_1, \ldots, J_N$ are the N answers returned by the system during the first step. The next set of weights at iteration $q + 1$ is thus given by:

$$\alpha_i^{q+1} = \sum_{j=1}^{N} S^q(J_j^q, Q) \phi_i^{q, J_j}. \qquad (5.22)$$

This non-supervised relevance feedback assumes some kind of Gaussian distribution of the images which of course is not robust and does not obviously fit the user requirements. This must be included in supervised relevance feedback in order to evolve from our pre-attentive computations to a better estimate of the probabilistic attentive model we adopted at the beginning of this chapter.

For instance, we show that the a priori $P(I|U)$ term (see Eq. 5.3) cannot be explicitly expressed by the user for any image. However, this can be done for a small set of images, e.g., the set of system responses based on the pre-attentive similarities. The user can select, through an interactive interface, the best and the worst responses. From these two subsets, different strategies have been proposed to determine a new set of weights based on probabilistic or rank criteria [9].

5.8 Conclusions

We surveyed in this chapter the panel of theories, models, and techniques relevant to the computation of similarities between features extracted from images. There is no optimality in such a field. Firstly, the choice of a technique depends on the nature of the features to be compared. The best strategy is of course to get a learning set and to compare on this set the different distance measures. As they have different properties, they will not consider similarity in the same way. This application-dependent strategy is the only one that can efficiently work up to now. Next, the quality of an image retrieval system is not just dependent on the similarity but mainly on the features used to represent the images. Finally, as we pointed out, we are too often working in the pre-attentive domain and there is still no solution to fill the gap between this domain and the attentive domain which is what the user actually wants. This field will clearly benefit from on-going research.

References

1. Aigrain, P, Zhang, H, and Petkovic, DP, "Content-Based Representation and Retrieval of Visual Media: a State-of-the-Art Review," Multimed Tools Applic, 3(3), pp. 179–202, 1996.
2. Attneave, F, "Dimensions of Similarity," Am J Psychol, 63, pp. 516–556, 1950.
3. Bres, S and Jolion, JM, "Detection of Interest Points for Image Indexation," in Ref. [12], pp. 427–434, 1999.
4. Carson, C, Thomas, M, Belongie, S, Hellerstein, JM, and Malik, J, Blobworld: A System for Region-Based Image Indexing and Retrieval, in Ref. [12], pp. 509–516, 1999.
5. Del Bimbo, A, Visual Information Retrieval, Morgan Kauffman, 1999.
6. Deriche, R and Giraudon, G, "A Computational Approach for Corner and Vertex Detection," Int J Computer Vision, 10(2), pp. 101–124, 1993.
7. Fagin, R and Wimmers, EL, "Incorporating User Preferences in Multimedia Queries," Int. Conf. on Database Theory, pp. 247–261, 1997.

8. Flickner, M, Sawney, H, Niblack, W, Ashley, J, Huang, Q, Dom, B, Gorkani, M, Hafner, J, Lee, D, and Petkovic, DP, "Query by Image and Video Content: the QBIC System," IEEE Computer, special issue on content-based picture retrieval systems, 28(9), pp. 23–32, 1995.
9. Heinrichs, A, Koubaroulis, D, Levienaise-Obadia, B, Rovida, P and Jolion, JM, "Robust Image Retrieval in a Statistical Framework," 6th Int. Conf. on Content-Based Multimedia Information Access, Paris, France, April 12–14, 2000, pp. 1616–1631, 1999.
10. Huet, B and Hancock, ER, "Relational Histograms for Shape Indexing," Int. Conf. on Computer Vision (ICCV98), pp. 563–569, 1998.
11. Huet, B, Cross, A, and Hancock, ER, "Sensitivity Analysis for Graph Matching from Large Structural Libraries," 2nd Int. IAPR-TC-15 Workshop on Graph-Based Representations (GbR'99), Austria, May 10–12, 1999, Osterreichische Computer Gesellschaft, band 126, W Kropatsch and JM Jolion, eds, 1999.
12. Huijsmans, DP and Smeulders, A, Visual Information and Information Systems: Third International Conference, VISUAL'99, Springer, Lecture Notes in Computer Science, 1614, 1999.
13. Itti, L, Koch, C, and Niebur, E, "A Model of Saliency-Based Visual Attention for Rapid Scene Analysis," IEEE Trans Patt Anal Mach Intell, 20(11), pp. 1254–1259, 1998.
14. LaCascia, M and Ardizzone, E, "JACOBS: Just A COntentBased Query System for Video Databases," Proc. of ICASSP'96, Atlanta, 1996.
15. Jolion, JM, "Analyse multirésolution du contraste dans les images numériques," Traitement Sig, 11(3), pp. 245–255, 1994.
16. Levi, G, "A Note on the Derivation of Maximal Common Subgraphs of Two Directed or Undirected Graphs," Calcolo, 9, pp. 341–354, 1972.
17. Manjunath, BS and Ma, WY, "Texture Features for Browsing and Retrieval of Image Data," IEEE Trans Patt Anal Mach Intell, 18(8), pp. 837–842, 1996.
18. Press, H, Flannery, BP, Teukolsky, SA, and Vetterling, WT, Numerical Recipes in C: the Art of Scientific Computing, Cambridge University Press, Chapter 10, section 8, 1987.
19. Popescu, O, "Utilisation des points d'intérêt pour l'indexation d'images," Technical Report RR-99-07, Laboratoire Reconnaissance de Formes et Vision, INSA Lyon, 1999.
20. Puzicha, J, Hofmann, T, and Buhman, JM, "Non-Parametric Similarity Measures for Unsupervised Texture Segmentation and Image Retrieval," IEEE Conf. on Computer Vision and Pattern Recognition (CVPR), pp. 267–272, 1997.
21. Rubner, Y, Tomasi, C, and Guibas, LJ, "A Metric for Distributions with Applications to Image Databases," IEEE Int. Conf. on Computer Vision (ICCV'98), pp. 207–214, 1998.
22. Russell, EJ, "Extension of Dantzig's Algorithm to Finding an Initial Near Optimal-Basis for the Transportation Problem," Operat Res, 17, pp. 187–191, 1969.
23. Santini, S and Jain, R, "The Graphical Specification of Similarity Queries," J Visual Lang Comput, 1997.
24. Santini, S and Jain, R, "Similarity Measures," IEEE Trans Patt Anal Mach Intell, 21(9), pp. 871–883, 1999.
25. Schmid, C and Mohr, R, "Local Grayvalue Invariants for Image Retrieval," IEEE Trans Patt Anal Mach Intell, 19(5), pp. 530–535, 1997.

26. Sebe, N, Tian, Q, Loupias, E, Lew, MS, and Huang, TS, "Color Indexing Using Wavelet-Based Salient Points," IEEE Workshop on Content-Based Access of Image and Video Libraries (CBAIVL-2000), June, 2000.
27. Shearer, K, Bunke, H, Venkatesh, S, and Kieronska, D, "Graph Matching for Video Indexing," Computing (JM Jolion and W Kropatsch, eds), 12, pp. 53–62, 1998.
28. Shen, HC and Wong, AKC, "Generalised texture representation and metric," Computer Vision Graphics Image Process, 23, pp. 187–206, 1983.
29. Stricker, M and Dimai, A, "Color Indexing with Weak Spatial Constraints," Storage and Retrieval for Still Image and Video Databases IV, SPIE 2670, 1996.
30. Swain, M and Ballard, D, "Color Indexing," Int J Computer Vision, 7(1), pp. 11–32, 1991.
31. Tversky, A, "Features of Similarity," Psychol Rev, 84(4), pp. 327–352, 1977.
32. Tversky, A and Gati, I, "Similarity, Separability, and the Triangle Inequality," Psychol Rev, 89, pp. 123–154, 1982.
33. Veltkamp, RC and Hagedoorn, M, "Shape Similarity Measures, Properties and Constructions," Visual Information and Information Systems: Fourth International Conference, VISUAL'2000, Springer, Lecture Notes in Computer Science, 2000.
34. Werman, M and Peleg, S, "Multiresolution Texture Signatures Using Min-Max Operators," 7th Int. Conf. on Pattern Recognition, pp. 97–99, 1984.
35. Werman, M, Peleg, S and Rosenfeld, A, "A Distance Metric for Multidimensional Histograms," Computer Vision Graphics Image Process, 32(3), pp. 328–336, 1985.
36. Wolf, C, Jolion, JM, Kropatsch, W, and Bischof, H, "Content based Image retrieval using Interest Points and Texture Features," 15th Int. Conf. on Pattern Recognition, Barcelona, Spain, pp. 234–237, September, 2000.
37. Wolf, C, Jolion, JM, and Bischof, H, "Histograms for Texture Based Image Retrieval," 24th Workshop of the Autrian Assoc. for Pattern Recognition (OEAGM2000), Villach, Austria, May 25–26, 2000.

6. Feature Selection and Visual Learning

Michael S. Lew

6.1 Introduction

In many content-based retrieval systems, the user is asked to understand how the computer sees the world. An emerging trend is to try to have the computer understand how people see the world. However, understanding the world is a fundamental computer vision problem which has withstood decades of research. The critical aspect to these emerging methods is that they have modest ambitions. Petkovic [26] has called this finding "simple semantics." From recent literature, this generally means finding computable image features which are correlated with visual concepts. The key distinction is that we are not trying to fully understand how human intelligence works. This would imply creating a general model for understanding all visual concepts. Instead, we are satisfied to find features which describe some small, but useful domains of visual concepts.

6.2 Simple Semantics

In this chapter we discuss learning simple semantics, or in another way, visual learning of concepts. This brings into mind the question, "What is visual learning?" Such a general term could refer to anything having to do with artificial or human intelligence in all its sophistication and complexity. Rather than a vague description, we seek to define it clearly at least within the boundaries of this chapter as either (1) feature tuning; (2) feature selection; or (3) feature extraction/construction. Feature tuning refers to determining the parameters which optimize the use of the feature. This is often called parameter estimation. Feature selection means choosing one or more features from a given initial set of candidate features. The chosen features typically optimize the discriminatory power regarding the ground truth which consists of positive and negative examples. Feature construction is defined as creating new features from a base set of atomic features and integration rules. In this chapter, our focus is on feature selection-based methods.

What is a visual concept? For our purposes, a visual concept is anything which we can apply a label or recognize visually. These could be clearly defined objects like faces or intellectual property such as trademarks [6], or abstract concepts such as textures. Textures are particularly challenging because they often do not have corresponding labels in written language. Concept detection is essential because it gives the computer the ability to understand our notion of an object or concept. Instead of requiring all users to understand low level feature queries, we are asking the computer to understand the high level queries posed by humans. For instance, if we want to find an image with a beach under a blue sky, most systems require the user to translate the concept of beach to a particular color and texture. Ideally, the user should have the option to pose the query visually as a beach under a blue sky.

6.3 Complex Backgrounds

In many applications, it is assumed that the visual concept can be segmented from the image trivially by exploiting application specific knowledge such as uniform backgrounds, known backgrounds, or motion information. However, for the general problem of content-based retrieval, the segmentation of the visual concept from complex backgrounds is not merely necessary, but it is often the most difficult part of the process. An example of segmenting human faces from an image is shown in Fig. 6.1. Note that the face detector [17] finds a false positive from the sweater pattern on the person in the center of the image. Furthermore, Fig. 6.1 illustrates issues such as detecting partial examples of the visual concept. Specifically, along the bottom edge of the image are three partially occluded faces which are not found by the face detection algorithm.

Fig. 6.1 Locating faces in an image.

In the next section, we introduce the current body of work in feature selection regarding visual learning. This foundation is used for clarifying the current state of the art in visual concept detection.

6.4 Feature Selection

Detecting and segmenting visual concepts from imagery is one of the long-standing puzzles in artificial intelligence, psychology, computer vision, and biology. Human vision is gratifying because we are able to do it effortlessly, reliably, and accurately. However, it is also puzzling because we do not know the process of how it happens. From a computational perspective, one might expect that a process which is nearly instantaneous should also be simple and transparent. However, a simplistic solution has so far eluded us.

One paradigm for detecting visual patterns occurs when we have a classifier (i.e., nearest neighbor or neural network), and a large set of candidate features. In principle, we wish to use the subset of the features which minimizes the misdetection rate of the classifier. The first method which springs to mind is to try all of the possibilities. However, brute-force search may be impractical for even moderately large feature sets because the number of possible subsets grows exponentially.

Another method which may seem appealing is to use all of the features. In fact it is well known that in a statistical sense, the classification error should decrease when additional measurements are used. However, this is true when the samples sets have infinite extent. In most practical applications, only sparse training sets are available which means there could be insufficient data to arrive at statistically representative values for the unknown parameters associated with the features. Specifically, the classifier would be well tuned for the training set, but would lack the ability to generalize to new instances. Consequently, the probability of correct classification would decrease as the number of features increases, which has been called the curse of dimensionality. Therefore, we usually wish to minimize the size of the selected feature subset, which should minimize the number of unknown parameters, and consequently require a smaller training set.

This section gives a brief review of classical feature selection followed by more recent techniques from the research literature. Classifying patterns requires an assumption that classes occupy distinct regions in the pattern space. When the classes are more distant, the probability of successful recognition of class membership increases. Thus, the general approach is to select the feature subspace which maximally separates the classes. Formally, the problem of feature selection lies in selecting the best subset U from the set V which contains D candidate feature classes:
Select $d \le D$ from

$$V = \{v_j | j = 1, 2, \ldots, D\}$$

arriving at

$$U = \{u_i | i = 1, 2, \ldots, d\}$$

where each u_i is an element of V, and U maximizes a criterion function $J(U)$, which ideally is the probability of correct classification.

$$J(U) = \max_{Z} J(Z)$$

where $Z \subseteq V$ and $|Z| = d$.

Brute force feature selection is typically not feasible. For instance, selecting ten feature classes out of a hundred would necessitate evaluation of more than 10^{13} feature sets. Thus, in practical situations computationally feasible methods must be employed.

In practice, feature selection is typically done off-line, so it has been argued that the execution time of a particular algorithm is less important than the classification accuracy. Although this is true for feature sets of moderate size, there are important applications such as data mining where the data sets have hundreds of features, and running a feature selection algorithm even once may be impractical.

6.4.1 Class Separability

The notion of interclass distance is the most basic concept of class separability. It can be used to assess discriminatory potential of pattern representations in a given space. Figure 6.2 depicts an example of interclass distance, showing that when the interclass distance is small, it will be more difficult to separate the circles and the crosses. Furthermore, we note that the squares are easily separable from the other two classes due to the large interclass distance between the squares and circles, and between the squares and crosses. Intuitively, as the distance between the classes increases, the probability of successful recognition of class membership increases. From this perspective, it is reasonable to select as the feature space that d-dimensional subspace in which the classes are maximally separated.

From a theoretical viewpoint, an ideal measure of class separability in the space spanned by random variable ε for classes w_i is the error probability defined as

$$E(\Xi) = \int \left[1 - \max_{i}\{P(w_i|\varepsilon)\} \right] p(\varepsilon)d\varepsilon. \tag{6.1}$$

If for a class w given an instance ε, the probability is 1, then the error probability is 0, which is ideal in the sense of trying to minimize classification error since we will always know that the instance ε belongs to class w. Unfortunately, it is very difficult to evaluate, and we often must resort to other measures of class separability.

6.4.2 Probabilistic Separation

A well-known formulation of class separability uses the complete information about the probabilistic structure of classes. For instance, such information is provided by the class conditional probability density functions (PDFs), $p(\varepsilon|w_i)$, $i = 1, 2, \ldots, c$ and a priori class probabilities P_i.

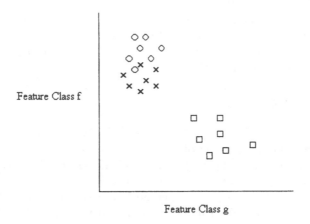

Fig. 6.2 Interclass distance as related to class membership.

If we have two classes, w_1 and w_2, then the classes will be fully separable if $p(\varepsilon|w_1)$ is zero for all ε such that $p(\varepsilon|w_2)$ does not equal zero. In addition, when $p(\varepsilon|w_1)$ equals $p(\varepsilon|w_2)$ for all ε, it is impossible to distinguish elements of class w_1 from those belonging to w_2. In this case, the classes are completely overlapping.

The overlap of density functions can be assessed by measuring the distance between $p(\varepsilon|w_1)$ and $p(\varepsilon|w_2)$. In general any function $J(\cdot)$,

$$J(\cdot) = \int g[p(\varepsilon|w_1), p(\varepsilon|w_2), P_1, P_2]d\varepsilon \tag{6.2}$$

satisfying:

$J(\cdot)$ is non-negative
$J(\cdot)$ attains a maximum when the classes in the ε space are disjoint, i.e.,
$\qquad J(\cdot) = \max$ if $\forall \varepsilon$, $p(\varepsilon|w_1) = 0$ when $p(\varepsilon|w_2) \neq 0$
$J(\cdot)$ equals zero when the probability density functions are equal, i.e.,
$\qquad J(\cdot) = 0$ if $\forall \varepsilon$, $p(\varepsilon|w_1) = p(\varepsilon|w_2)$

can be used as a probabilistic distance measure of class separability.

6.4.3 Mahalanobis Distance

In many applications, the class conditional densities are Gaussian distributions. An appropriate measure of class separation is the Mahalanobis distance:

$$d(u_1, u_2) = (u_1 - u_2)^T \Sigma^{-1}(u_1 - u_2) \tag{6.3}$$

where Σ is the common covariance matrix and u_1 and u_2 are the mean vectors of the instances of the two classes. When the class conditional densities of the features are Gaussian, the probability of error is inversely proportional to the Mahalanobis distance.

6.4.4 Optimal Search

We introduced the notion of class separability above, which can be used to assess the discriminatory power of individual feature sets which must be evaluated to determine the best combination of d features out of D measurements. Next, we discuss the process of selecting the feature subset.

One intuitive feature selection method is to select the features which individually have the best discriminatory power. It turns out that this method may lead to a suboptimal (in the sense of minimizing the classification error) feature set even if all of the features are statistically independent. The only situation where it will retain optimality is when

$$J(\boldsymbol{U}) = \sum_{i=1}^{d} J(u_i). \tag{6.4}$$

Such a situation could happen if the Mahalanobis distance were used with normally distributed and independent pattern vectors.

The overall strategy of the following search algorithms is to construct the best feature set by adding to and/or removing from the current feature set a small number of features until the required feature set is obtained. Specifically, the starting point can be either an empty set, which is then gradually built up, or we can start from the complete set of feature classes, and successively eliminate unnecessary feature classes. The former approach is known as the *bottom-up* method, while the latter is referred to as the *top-down* search.

Let S_k be a set containing k elements, $s_1, s_2, \ldots, s_k$, from the set of available feature classes, V. In addition let the criterion function $J(\cdot)$ be a measure of class separability chosen by the user. This notation will be used in the following sections.

The *branch* and *bound* algorithm is the only optimal search method in which all of the possible subsets, S_k, of d out of D feature classes are implicitly inspected without the exhaustive search. Basically, it is a top down search method but with a backtracking facility which allows all of the possible combinations of feature classes to be examined. The computational efficiency of the method lies in an effective organization of the search process. By virtue of this process, detailed enumeration of many candidate feature sets can be avoided without undermining optimality of the feature selection procedure. The branch and bound algorithm relies on the monotonicity property shared by the majority of the feature selection criterion functions. This property states that for nested sets of feature classes $S_i, i = 1, 2, \ldots, k$, i.e.,

$$S_1 \supset S_2 \supset \cdots \supset S_k, \tag{6.5}$$

the criterion functions, $J(S_i)$, satisfy

$$J(S_1) \geq J(S_2) \geq \cdots \geq J(S_k). \tag{6.6}$$

By a straightforward application of this property, many combinations of feature classes can be rejected from the set of candidate feature sets. Specifically, suppose we start the search with all feature classes, and we are deciding which feature classes to delete. Furthermore, if we are at state S_1 in our search, where our best $J(\cdot)$ is J_B, then we do not have to search in the branches of the search tree where we would subtract feature classes, $S_1 - S_2$, because we already know that $J(S_1) \geq J(S_2)$. This pruning of the solution tree is the principle behind the reduction of computational expense by using branch and bound.

6.4.5 Sequential Forward Selection (SFS)

SFS is essentially a bottom-up, hill climbing search procedure, where one feature class is added at a time to the current feature set. At each iteration, the feature class to be included in the feature set is selected from among the remaining feature classes such that the new feature set yields the greatest possible value of the criterion function.

The SFS algorithm selects new feature classes with reference to the current set of feature classes. In the D-dimensional feature space, we are finding a maximum of the criterion function. Since SFS is essentially a hill climbing method, it has the drawbacks associated with hill climbing methods. The two major drawbacks of hill climbing methods are the *local peak problem* and the *ridge problem*. The local peak problem occurs when there are multiple maxima in the problem space. The ridge problem occurs when the direction to a local maximum is not within the available directions. In mountain climbing, we could imagine a man who could only walk north, east, west, and south. If the direction of the peak was along a ridge to the north-east, any of the four directions available to him would cause him to decrease his altitude. Consequently, he would be at a false local maximum. With respect to SFS, the local peak problem is manifested by stopping at a local maximum. In the literature, this is also referred to as the nesting effect. The ridge problem occurs because the direction for an increase in the criterion function would require the addition of multiple feature classes instead of one at a time.

6.4.6 Generalized Sequential Forward Selection (GSFS)

GSFS is a generalization of the SFS algorithm where r feature classes are added to the current feature set at each stage of the algorithm. This method has greater potential of overcoming the ridge problem because the possibility of adding r feature classes will usually allow for more directions to move through feature space. The disadvantage of this method is that the computational expense for evaluating all of the possible directions increases.

6.4.7 Sequential Backward Selection (SBS)

SBS is the top down counterpart to the SFS algorithm. We begin with the complete set of feature classes as our initial feature set. At each stage, we remove one feature class at a time until $D - d$ feature classes have been deleted.

6.4.8 Generalized Sequential Backward Selection (GSBS)

By analogy, the GSBS algorithm is essentially the same as the SBS procedure, except that more than one feature is removed at a time.

6.4.9 Plus L Take Away R Selection (LRS)

In this method, we first enlarge the current feature set by L feature using the SFS method. From the resulting set, R features are then discarded using the SBS method. This algorithm still suffers from other drawbacks of the SFS and SBS algorithms. For instance, groups of feature classes are added and removed from the current set irrespective of their mutual relationship. This method has the advantage that it can find other local peaks as well as it allows more directions to overcome the ridge problem.

6.4.10 Floating Methods

In the previous method, L and R were fixed. Alternatively, we can let the values "float" which means dynamically change them so as to approximate the optimal solution as much as possible. Consequently, the dimensionality of the subset is not changing monotonously but is "floating" up an down. The difficulty is that there is no theoretical way of predicting the values of L and R to achieve the best feature set.

Let $X_k = \{x_i : 1 \leq i \leq k, x_i \in Y\}$ be the set of k features from the set $Y = \{y_i : 1 \leq i \leq D\}$ of D candidate features. Define the least significant feature as the feature which when removed gives the maximum class separation. The sequential forward floating selection algorithm (SFFS) [29] can be defined as follows.

Assume that k features have already been selected to form X_k with the corresponding criterion $J(X_k)$. Also, the values of $J(X_i)$ for all preceding subsets of size $i = 1, 2, \ldots, k - 1$, are stored. Furthermore, initialize k to 0, X_0 to the empty set, and use SFS until cardinality 2 is reached, then go to Step 1.

Step 1. Using SFS, select x_{k+1} from the set $Y - X_k$ to form X_{k+1}.

Step 2. Let z_j be the least significant feature in X_{k+1}.

    If z_j is x_{k+1}, then set $k = k + 1$ and go to Step 1, but if z_j is an element of X_k, then exclude z_j from X_{k+1} which consequently forms a new set Z_k. So, $Z_k = X_{k+1} - z_j$

    Note that now $J(Z_k) > J(X_k)$. If k equals 2, then set $X_k = Z_k$ and go to Step 1.

Step 3. Let z_s be the least significant feature in Z_k.

    If $J(Z_k - z_s) \leq J(X_{k-1})$, then set $X_k = Z_k$ and go to Step 1.

    If $J(Z_k - z_s) > J(X_{k-1})$, then exclude z_s from Z_k which consequently forms a new set Z_{k-1}.

    Set $k = k - 1$.

    If k equals 2, then set $X_k = Z_k$ and go to Step 1.

    Go to Step 3.

6.4.11 Stochastic Methods

Siedlecki and Sklansky [35] were the first to apply genetic algorithm methodology to feature selection. Briefly, a genetic algorithm (GA) is an optimization technique which has successfully attacked even NP-hard problems [11, 13]. In genetic algorithm terminology, a point in the search space is represented by a finite sequence of 0's and 1's, called a chromosome. Regarding feature selection, each chromosome represents a subset of features where the nth bit denotes the presence or absence of the nth feature. The chromosomes are allowed to crossover (mate) and to mutate. A crossover between two chromosomes produces a pair of chromosomes which are mixtures of the parents. A mutation produces a near copy which has some elements altered.

During a generation, a set of new chromosomes is created either through crossover or mutation. These new chromosomes are evaluated and only a predefined number of the best chromosomes survives to the next cycle of reproduction.

Three parameters which are present in most GA techniques are the population size, crossover rate, and mutation rate. The population size is the number of chromosomes which survives to the next generation. These are selected by a fitness function. The crossover rate is the probability of accepting an eligible pair of chromosomes for crossover. The mutation rate is the probability of switching bits in the chromosomes.

The crossover function is implemented by choosing a point (crossover point) at which two chromosomes exchange their parts to create two new chromosomes. For example, the crossover of two binary strings 0011 and 1010 in the middle would yield 0010 and 1011.

To incorporate the error of the classifier into the GA so that it would be constrained optimization, they also added a penalty function:

$$p(e) = [\exp((e - t)/m) - 1]/[\exp(1) - 1] \tag{6.7}$$

where e is the error rate, t is the feasibility threshold and m is a scale factor. The fitness function is defined as

$$f(a_i) = (1 + \varepsilon) \max_{a_j} J(a_j) - J(a_i) \qquad (6.8)$$

where ε is a small positive constant and

$$J(a) = l(a) + p(e(a)) \qquad (6.9)$$

where $l(a)$ is the number of features in the subset.

6.4.12 Neural Network Methods

In this paradigm, which was introduced by Mao et al. [22], a multilayer feedforward neural network is trained with a back-propagation learning algorithm, and then the nodes are pruned to reduce the complexity of the network. The pruning of input nodes effectively removes the corresponding features from the feature set, and develops both the optimal feature set and the optimum classifier.

The saliency of a node is defined as the sum of the increase in error over the set of training instances as a result of removing the node. Toward providing a computationally efficient solution, Mao et al. [22] approximated the saliency with a second order expansion. Then in a back-propagation fashion, they computed the value by finding the appropriate derivatives.

6.4.13 Direct Methods

Novovicova et al. [24] and Pudil et al. [28] suggest that it is possible to find a near optimal feature subset without involving any search procedure. In the approach by Pudil et al. [28], they proposed a feature selection method based on approximating the unknown class conditional probability density functions in the sense of minimizing the Kullback–Liebler [16] distances between the true and the postulated densities. The primary goal of the approach was toward finding the feature which are best from the point of view of approximating the unknown class distributions. However, it is not necessary that the best features for approximating a class are the best for separating it from another class.

In the work by Novovicova et al. [24], they attempted to employ a criterion more directly linked to the concept of separability between classes. Their algorithm has the advantage that it is suited for multimodal data. They transform the problem of selecting a subset of d features from D possible feature to the problem of choosing the vector Φ_d which satisfies

$$J(\Phi'_d) = \max_{\Phi_d} J(\Phi_d) \qquad (6.10)$$

where Φ_d is a vector which specifies which d of D features will be used, and $J(\cdot)$ is the Kullback J–divergence between the two classes, w_1 and w_2 [1]:

$$J(\cdot) = \sum_{\nu \in \{w_1, w_2\}} P(\nu) E_\nu \left\{ \log \frac{p(x|\nu)}{p(x|\{w_1, w_2\} - \nu)} \right\} \tag{6.11}$$

where E_ν is the mathematical expectation with respect to the class-conditional PDF $p(x|\nu)$.

6.4.14 Summary of the Feature Selection Algorithms

In the field of mathematical feature selection, the pioneering work is usually associated with Sebestyen [34], Lewis [20], and Marill and Green [21]. Sebestyen [34] was the early proponent of the use of interclass distance. Marill and Green [21] advocated a probabilistic distance measure. Lewis [20] and Marill and Green [21] were also among the first to address the problem of feature set search, and were the first to suggest the sequential backward selection of feature classes. SFS was first used by Whitney [38]. LRS was developed by Stearns [36]. SFS, SBS, and LRS were generalized by Kittler [15]. In optimal feature selection, the branch and bound algorithm was proposed by Narendra and Fukunaga [23]. The potential of any suboptimal feature selection algorithm to select the worst subset of features was discussed in Cover and Van Campenhout [5]. An overview of the search methods is given in Table 6.1.

Note that all of the search methods (i.e., SFS, GSFS, but not necessarily including the neural network methods) except for Branch and Bound are suboptimal. Only Branch and Bound implicitly searches all of the combinations and guarantees a globally optimum feature set.

6.5 Case Studies on Performance of Feature Selection Methods

Jain and Zongker [14] conducted a classification test where they evaluated 15 promising feature selection algorithms on a 20-dimensional, two-class data set. The "goodness" of the feature subset was evaluated using the Mahalanobis distance between the two class means. Two conclusions were:

- SFS and SBS algorithms have comparable performance, but the SFS is faster because it starts with small subsets and enlarges them, while the SBS starts with large subsets and reduces them. From a computational perspective, it is more efficient to evaluate small subsets than large subsets.
- The floating selection methods were for the most part faster than the branch and bound algorithm and gave comparable results.

Siedlecki and Sklansky [35] compared their genetic algorithm approach with SFS and SBS and showed that the GA was able to outperform the

Table 6.1 Search methods for finding optimal and suboptimal feature class sets

Branch and bound	A top-down search procedure with backtracking which allows all of the combinations to be implicitly searched without an exhaustive search
Sequential forward selection (SFS)	A bottom-up search procedure where the feature which will yield a maximum of the criterion function is added one at a time to the null set
Generalized sequential forward selection (GSFS)	Similar to SFS, but instead of adding one feature at a time, r features are added
Sequential backward selection (SBS)	Starting from the complete set of features, we discard one feature until $D - d$ features have been deleted
Generalized sequential backward selection (GSBS)	Similar to SBS, but instead of discarding one feature, multiple features are discarded
Floating methods	Similar to GLRS, but instead of having fixed l and r, the size of the feature set can increase and decrease
Stochastic methods	Applying genetic-algorithm-based optimization to feature selection
Neural network methods	Prune the least salient nodes from the neural network
Direct methods	Finding a feature subset without involving a search procedure

feature selection methods. However, Jain and Zongker [14] point out that there are several parameters in the GA algorithm which do not have a clear mechanism for setting their values.

Ferri [7] compared SFS, SFFS, and GA methods on high-dimensional data sets. They showed that the performance of GA and SFFS is comparable but GA degrades with increasing dimensionality.

6.6 Content-Based Retrieval

In content-based retrieval [8, 12, 25, 26], the number of features is quite large (> 100), so it is often impractical to perform optimal feature selection. Several researchers have shown the efficacy of interactive and heuristic feature selection for detecting visual concepts, i.e., Picard [27], Forsyth et al. [9], Chang et al. [4], Buijs and Lew [2], Fung and Loe [10], and Ratani [30]. Furthermore, there has been significant activity in detecting human faces in complex backgrounds. These methods can theoretically be applied to other visual concepts.

Picard [27] focuses on a class of interesting visual concepts which can be identified "at-a-glance." Specifically, she asks how people can identify classes of photos without taking time to look at the precise content of each photo. Her work begins by identifying a "society of models" which consists of a set of texture descriptions. These texture descriptions (such as wavelets, Gabor filters, etc.) can model any signal, but each has different signals and weaknesses. Their goal was to develop a system which could:

– Select the best model when one is best, and figure out how to combine models when that is best
– Learn to recognize, remember, and refine best model choices and combinations by analysis of both the data features and user interaction

Promising examples were given for detecting buildings, cars and streets as shown in Fig. 6.3.

In the report by Forsyth et al. [9], he discusses the importance of using several grouping strategies in combination with pixel- and template-based features such as color and texture. Promising results were found for the following:

– Fusing texture and geometry to represent trees
– Fusing color, texture, and geometry to find people and animals

An example of their system for finding unclothed people first locates images containing large areas of skin-colored regions. Within the skin-colored regions, it finds elongated regions and groups them into plausible skeletal possibilities of limbs and groups of limbs. More recent work by Carson et al. [3] shows the potential of using an ensemble of "blobs" representing image regions which are roughly homogeneous regarding color and texture. Good results were obtained for retrieving images containing tigers, cheetahs, zebras, and planes.

Pentland et al. [25] reported success in using variations on the Karhunen-Loeve transform to find optimally-compact representations for either appearance or shape. They also showed the utility of the Wold transform for decomposing signals into perceptually salient texture descriptions. The general idea of reducing images to a small set of perceptually significant coefficients was denoted as *semantics-preserving image compression* .

Vailaya et al. [37] showed promising results in classifying images as city vs. landscape. They found that the edge direction features have sufficient discriminatory power for accurate classification of their test set.

Buijs and Lew [2] and Lew et al. [19] use a direct feature selection method in conjunction with the Kullback discriminant [16] for selecting the best feature subset. They report encouraging results on detecting visual concepts such as sky, trees, sand, stone, and people. Furthermore, they show that rules for determining when a feature has low discriminating power can reduce the misdetection rate. An example of a visual query and results is shown in Fig. 6.4.

Several methods for detecting faces in complex backgrounds have been proposed, evaluated and benchmarked [17, 18, 32, 33]. The current methods are template based and have significant computational requirements because they pass a template window over copies of the image at varying resolutions.

Fig. 6.3 An example of using the society of models for the detection of visual concepts: building, car, and street.

For frontal images, the methods give surprisingly accurate results, but the techniques are fragile to large changes in orientation. For example, none of the methods detects side views of faces. Sung and Poggio [33] deal with the question of how many classes the frontal face detection problem should be split into. They show that clustering the face space into both positive and negative classes can yield low misdetection rates. Lew and Huijsmans [17] describe a direct method of feature selection which maximizes the Kullback

Fig. 6.4 A query (left) for a person with trees/grass above and below him, and the results (right).

discriminant. Their work shows that it is important to take into account correlated features. Other interesting findings include the significance of the eyes as the most important facial feature, and the nose as the least important feature. Moreover, it turns out that psychological findings of the relative importance of facial features in face perception tasks supports their results.

6.7 Discussion

In many ways, feature selection is a mature field. For example, there are extensive test sets for benchmarking the effectiveness of various methods. Techniques exist for finding optimal and near-optimal solutions. However, currently there are no broad scale content-based retrieval systems which can detect thousands of visual concepts automatically from images with complex backgrounds. Several major challenges exist and are elaborated upon in this section.

First, finding the optimal feature subset for very large (> 100) sets of candidate features is computationally expensive. If we instead focus on near-optimal solutions, then there are several possibilities. In particular, the direct methods appear to yield good results without requiring the computationally expensive search process.

Even if the feature selection process is computationally efficient, for very large sets of candidate features, it may impractical to gather sufficient training examples. The second challenge is therefore to determine how to overcome the curse of dimensionality, that is, select large feature subsets using sparse training sets. In the case where the Mahalanobis distance is used to measure class separation, the error from estimating the covariance matrix can lead to inferior results even if the selected feature subset is optimal for the training data [31].

Suppose an optimal feature subset is found for the class of visual concepts which include frontal and side views of human faces. The work from Sung and Poggio [33] and Buijs and Lew [2] indicates that better performance might be found from splitting the class of human faces into clusters: i.e., the frontal-view class and the side-view class, and then finding the appropriate feature subset for each class. The third challenge is to automatically determine when to split the training examples into multiple classes and find feature subsets for each of them.

In visual matching, there are frequently situations where part of the visual concept is occluded by another object as shown in Fig. 6.1. This gives rise to the fourth challenge which is to design the feature subset so that the performance degrades gracefully when a greater percentage of the visual concept is occluded.

6.8 Summary

In this chapter, we have described the state of the art in feature selection and the related methods in the content-based retrieval literature. Some interesting points:

- Floating selection methods have extensive experimental support regarding their efficacy for real-world problems.
- Direct methods have the significant advantage of avoiding the computationally expensive search procedure. Furthermore, they have the potential to be effective when the Mahanalobis distance measure is ineffectual.
- Including the user in the feature selection process is a promising paradigm for learning visual features.
- Grouping methods can be powerful in conjunction with pixel- and template-based features.
- Clustering and accounting for feature correlations can reduce misdetection rates.

In practice, there have been a wide variety of promising content-based search systems which could detect limited classes of visual concepts with low misdetection rates. It remains to be seen which methods can be generalized to a large number of interesting visual concepts.

References

1. Boekee, D and Van der Lubbe, J, "Some Aspects of Error Bounds in Feature Selection," Patt Recogn, 11, pp. 353–360, 1979.
2. Buijs, J and Lew, M, "Learning Visual Concepts," ACM Multimedia'99, 2, pp. 5–8, 1999.
3. Carson, C, Thomas, M, Belongie, S, Hellerstein, J, and Malik, J, "Blobworld: A System for Region-Based Image Indexing and Retrieval," Proc. VISUAL'99, Amsterdam, Netherlands, pp. 509–516, June, 1999.
4. Chang, SF, Chen, W, and Sundaram, H, "Semantic Visual Templates – Linking Visual Features to Semantics," IEEE Int. Conf. on Image Processing, Chicago, IL, October, 1998.
5. Cover, T and Van Campenhout, J, "On the Possible Orderings in the Measurement Selection Problem," IEEE Trans Syst Man Cybern, 7, pp. 657–661, 1977.
6. Eakins, J, "Similarity Retrieval of Trademark Images," IEEE Multimed, 5(2), pp. 53–63, 1998.
7. Ferri, F, Pudil, P, Hatef, M, and Kittler, J, "Comparative Study of Techniques for Large Scale Feature Selection," Pattern Recognition in Practice IV, E. Gelsema and L Kanal, eds, pp. 403–413, Elsevier Science, 1994.
8. Flickner, M, Sawhney, H, Niblack, W, Ashley, J, Huang, Q, Dom, B, Gorkani, M, Hafner, J, Lee, D, Petkovic, D, Steele, D, and Yanker, P, "Query by Image and Video Content: The QBIC System," Computer, IEEE Computer Society, pp. 23–32, September, 1995.
9. Forsyth, D, Malik, J, Fleck, M, Leung, T, Bregler, C, Carson, C, and Greenspan, H, "Finding Pictures of Objects in Large Collections of Images," International Workshop on Object Recognition, Cambridge, April, 1996.
10. Fung, C and Loe, K, "Learning Primitive and Scene Semantics of Images for Classification and Retrieval," ACM Multimedia'99, Orlando, Part 2, pp. 9–12, 1999.
11. Goldberg, DE, Genetic Algorithms in Search, Optimization and Machine Learning, Addison-Wesley, Reading, MA, 1989.
12. Gudivada, VN and Raghavan, VV, "Finding the Right Image, Content-Based Image Retrieval Systems," Computer, IEEE Computer Society, pp. 18–62, September, 1995.
13. Holland, J, Adaptation in Natural and Artificial Systems, University of Michigan Press, Ann Arbor, Michigan, 1975.
14. Jain, A and Zongker, D, "Feature Selection: Evaluation, Application, and Small Sample Performance," IEEE Trans Patt Anal Mach Intel, 19(2), pp. 153–158, 1997.
15. Kittler, J, "Une generalisation de quelques algorithmes sous-optimaux de recherche d'ensembles d'attributs," Reconnaissance des Formes et Traitement des Images, Paris, pp. 678–686, 1978.
16. Kullback, S, Information Theory and Statistics, Wiley, New York, 1959.
17. Lew, M and Huijsmans, N, "Information Theory and Face Detection," Int. Conf. on Pattern Recognition, Vienna, Austria, pp. 601–605, August 25-30, 1996.
18. Lew, M and Huang, TS, "Optimal Supports for Image Matching," IEEE Digital Signal Processing Workshop, Loen, Norway, pp. 251–254, September, 1996.
19. Lew, M, Lempinen, K, and Huijsmans, DP, "Webcrawling Using Sketches," VISUAL'97, San Diego, pp. 77–84, December, 1997.
20. Lewis, P, "The Characteristic Selection Problem in Recognition Systems," IRE Trans on Inform Theory, 8, pp. 171–178, 1962.

21. Marill, T and Green, DM, "On the Effectiveness of Receptors in Recognition Systems," IEEE Trans Inform Theory, 9, pp. 11–17, 1963.

22. Mao, J, Mohiuddin, K, and Jain, A, "Parsimonious Network Design and Feature Selection Through Node Pruning," Int. Conf. on Pattern Recognition, Jerusalem, pp. 622–624, 1994.

23. Narendra, PM and Fukunaga, K, "A Branch and Bound Algorithm for Feature Subset Selection," IEEE Trans Comput, 26, pp. 917–922, 1977.

24. Novovicova, J, Pudil, P, and Kittler, J, "Divergence Based Feature Selection for Multimodal Class Densities," IEEE Trans Patt Anal Mach Intel, 18(2), pp. 218–223, February, 1996.

25. Pentland, A, Picard, R, and Sclaroff, S, "Photobook: Content-Based Manipulation of Image Databases," Int J Computer Vision, 18, pp. 233–254, 1996.

26. Petkovic, D, "Challenges and Opportunities for Pattern Recognition and Computer Vision Research in Year 2000 and Beyond," Int. Conf. on Image Analysis and Processing, September, Florence, 2, pp. 1–5, 1997.

27. Picard, R, "A Society of Models for Video and Image Libraries," IBM Syst J, 1996.

28. Pudil, P, Novovicova, J, and Kittler, J, "Automatic Machine Learning of Decision Rule for Classification Problems in Image Analysis," BMVC '93, Fourth British Machine Vision Conf, Vol. 1, pp. 15–24, 1993.

29. Pudil, P, Novovicova, J, and Kittler, J, "Floating Search Methods in Feature Selection" Patt Recogn Lett, pp. 1119-1125, 1994.

30. Ratan, A, Maron, O, Grimson, W, and Perez, T, "A Framework for Learning Query Concepts in Image Classification," IEEE Conf. on Computer Vision and Pattern Recognition, Fort Collins, Colorado, pp. 423-431, 1999.

31. Raudys, S and Jain, A, "Small Sample Size Effects in Statistical Pattern Recognition: Recommendations for Practitioners," IEEE Trans Patt Anal Mach Intell, 13, pp. 252–264, March, 1991.

32. Rowley, H and Kanade, T, "Neural Network Based Face Detection," IEEE Trans Patt Anal Mach Intell, 20, pp. 23–38, 1998.

33. Sung, KK and Poggio, T, "Example-Based Learning for View-Based Human Face Detection," IEEE Trans. Patt Anal Mach Intell, 20, pp. 39–51, 1998.

34. Sebestyen, G, Decision Making Processes in Pattern Recognition, Macmillan, New York, 1962.

35. Siedlecki, W and Sklansky, J, "A Note on Genetic Algorithms for Large-Scale Feature Selection," Patt Recogn Lett, 10, pp. 335–347, 1989.

36. Stearns, S, "On Selecting Features for Pattern Classifiers," Int. Conf. on Pattern Recognition, pp. 71–75, 1976.

37. Vailaya, A, Jain, A, and Zhang, H, "On Image Classification: City vs. Landscape," IEEE Workshop on Content-Based Access of Image and Video Libraries, Santa Barbara, June, 1998.

38. Whitney, A, "A Direct Method of Nonparametric Measurement Selection," IEEE Trans Comput, 20, pp. 1100–1103, 1971.

7. Video Indexing and Understanding

Michael S. Lew, Nicu Sebe, and Paul C. Gardner

7.1 Introduction

More and more video is generated every day. While today much of this data is produced and stored in analog form, the tendency is to use the digital form. The digital form allows processing of the video data in order to generate appropriate data abstractions that enable content-based retrieval of video. In the future, video databases will be able to be searched with combined text and visual queries. Additionally, video clips will be retrieved from longer sequences in large databases on the basis of the semantic video content. Ideally, the video will also be automatically annotated as a result of the machine interpretation of the semantic content of the video.

Rowe et al. [32] characterized the types of video queries a user may ask and identified the following three types of metadata indexes that should be associated with the video data in order to satisfy a query:

Bibliographic data. This category includes information about the video (e.g., title, abstract, subject, and genre) and the individuals involved in the video (e.g., producer, director, and cast).

Structural data. Video and movies can be described by a hierarchy of movie, segment, scene, and shot where each entry in the hierarchy is composed of one or more entries at a lower level (e.g., a segment is composed of a sequence of scenes and a scene is composed of a sequence of shots [9]).

Content data. Users want to retrieve videos based on their content (audio and visual content). In addition, because of the nature of video the visual content is a combination of static content (frames) and dynamic content. Thus, the content indexes may be: (1) sets of keyframes that represent key images (e.g., frames that show each actor in the video or a sequence of images that depict the major segments and scenes in the video); (2) keyword indexes built from the soundtrack, and (3) object indexes that indicate entry and exit frames for each appearance of a significant object or individual.

Some of the current research is addressing the problem of manual video annotation [9, 19] and the types of annotation symbols needed. These annotations are added by an individual, often with the assistance of user interfaces. The problem is that when databases of video are involved, video processing to automatically extract content information may be crucial.

The interactive query process consists of three basic steps: formulating the query, processing the query, and viewing the results returned for a query. This requires an expressive method of conveying what is desired, the ability to match what is expressed with what is present, and ways to evaluate the outcome of the search. Conventional text-based query methods which rely on keyword lookup and string pattern matching are not adequate for all types of data, particularly not for auditory and visual data. Therefore, it is not reasonable to assume that all types of multimedia data can be sufficiently described with words alone, nor as metadata when it is first entered in the database, nor as queries when it is retrieved.

7.1.1 Video Query Formulation

In general, a video query is more complicated than a traditional query of text databases. In addition to text (manual annotations and closed captions) a video clip has visual and audio information. A primary question is how a query can be formulated across these multiple information modalities. Text can be used to formulate queries for visual data (image, video, graphs), but such queries are not very efficient and cannot encompass the hierarchical, semantic, spatial, and motion information. Describing visual data by text can be difficult because keywords are chosen by each user based on his subjective impression of the image and video. Thus, it is difficult to know which keywords were originally used when the target video shot was indexed. In addition, keywords must be entered manually, they are time consuming, error prone, and thus they may have prohibitive costs for large databases. For these reasons, visual methods should be considered in query formulation. A query system which allows retrieval and evaluation of multimedia data should be highly interactive to facilitate easy construction and refinement of queries.

The user can formulate diverse types of queries, which may require processing of multiple modalities. For example, a user may have seen a particular piece of video and wants to retrieve it for viewing or reusing. Another user may be only looking for a specific video without having seen it. Finally, a user may have only some vague idea of what he is looking for. Additionally, the system should accommodate queries of a user who just wants to browse the videos without having a specific goal. Ideally, the query formulation process should accommodate all these types of queries [6]. In order to achieve these goals, the different information modalities should be fully used and exploited in the query formulation and search process.

7.1.2 Video Categorization

In this stage the user or the search engine decides which category of video is to be searched (see Fig. 7.1). Categorization is the capability to use metadata, for example to direct the search to a specific topic. A form of classifying

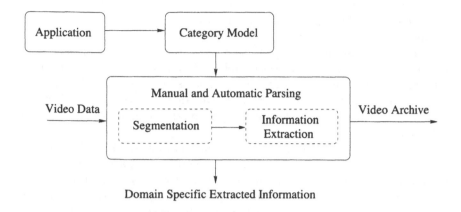

Fig. 7.1 Video data processing.

video on the basis of applications and purpose is discussed in Refs [14, 19]. The nature of the information captured and the purpose for which the video is used greatly affects the structure of the video. Four main purpose-based classes are extracted:

Entertainment. The information presented in this class is highly stylized depending on the particular sub-category: fiction (motion pictures, TV programs), nonfiction (sports, documentaries), and interactive video (games).

Information. The purpose of videos here is to convey information to the viewer (news).

Communication. The purpose of videos in this category is communication of information (video conferences).

Data analysis. Scientific video recording, like medical and psychology experiments may be used for data analysis purposes.

Such a classification of the video could create problems for certain videos, for example the ones presenting news about a sporting event. These videos will have to be placed into multiple categories (e.g., entertainment, information) or the user will have to select more than one category to search on.

In general, each video has to be classified using a combination of available manual textual annotation and visual and audio content. Considering that most of the current videos have some text associated with them, it is quite reasonable that the initial categorization is formulated in textual form.

7.1.3 Searching

Suppose that we are searching for a particular video shot. The result of the search is a list of candidate units that satisfy the constraints of the query. The ultimate goal of this stage is to minimize the number of candidates without missing the video(s) of interest.

The categorization stage limits to some extent the scope of data to be searched but there is still a need for efficient and effective means of further filtering the data of interest. Data-object models are used for organizing data in databases in order to allow fast retrieval. In video, the data objects developed have to cope with the various information modalities such as text, audio, and visual content.

In order to make the search possible, information contained in the video needs to be extracted (Fig. 7.1). This can be done by parsing the data automatically, manually, or a combination of the two (hybrid). Automatic extraction depends heavily on techniques used in computer vision. Content of unstructured data such as imagery or sound is easily identified by human observation. However, few attributes lead to reliable machine identification. Therefore, we have to rely more on hybrid extraction techniques.

Text-based search could be a first step in the searching stage. It serves as a good search filter and is an extremely important search tool that should not be neglected in video-retrieval problems. The text descriptors associated with the video should describe the content sufficiently to help the users to locate segments of interest or, at least, not exclude potential segments from the candidate list. These descriptors, such as the date of creation, the title, and the source, are commonly available at the time of creation. On the other hand, attributes such as the number of shots, audiovisual events such as dialogs and music, images, etc., can be derived from automatic analysis of the video.

Audiovisual attributes beyond the text are important and augment the text-query process. The difficult problem is to define audiovisual properties which can be extracted from the video and will reduce the size of the candidate list.

Query by image content techniques have been proposed in recent years. Here the queries are formulated using sample images or videos, rough sketches, or component features of an image (outline of objects, color, texture, shape, layout). In the QBIC [12] system, thumbnail images are stored in the database along with text information. Object identification is performed based on the images and the objects. Features describing their color, texture, shape, and layout are computed and stored. Queries can consist of individual or multiple features which are compared to the stored database features. The user can also use objects in the queries by drawing outlines around the object in an image. Another approach is to use iconic search [23, 32]. These systems take advantage of a user's familiarity with the world. An icon represents an entity/object in the world, and users can easily recognize the object from the icon image. The query is formulated by selecting the icons that are arranged in relational or hierarchical classes. A user refines a query by traversing through these classes. Iconic queries reduce the flexibility in a query formulation as queries can only utilize the icons provided, i.e., iconic databases tend to be rigid in their structure. There is also active ongoing research in audio indexing, to extract special audio features.

Another aspect of video search is that a user may not be interested in entire clips, but rather in portions of the clips. Thus, the search should be able to return as a result data objects (video portions) consisting of relevant footage. Furthermore, this implies that the data objects should be built in a hierarchical manner. A meaningful hierarchy is described in the film books: a shot is the fundamental unit, a scene is a collection of shots unified by the same dramatic incident, a segment is a collection of scenes, and a movie or clip is a collection of segments. Consequently, the attributes of a clip should be designed to include the most general features (attributes) which can be inherited by the segments, scenes, and shots it contains.

To enable video search, the video clips have to be properly segmented into semantic units. Automatic analysis of video content can achieve meaningful segmentation, provide valuable characterization of video, and offer features that are depictable with words.

7.1.4 Browsing

The result of the search stage is a collection of video candidates, of which the total duration can be long. Therefore, all candidates from the list should be browsable. This means that high-level representations of the content of the candidate videos should be displayed. A user, by looking at these representations, can quickly understand the video content and browse through many videos in a short time. On the other hand, the user should have random access to any point of any video and should be able to get an overview of each candidate video by viewing only a small area of a computer-display screen. As a consequence, high-level representations of video, in the form of visual summaries are necessary.

The area of visual summaries is very important in the context of the video query process. The problem is to derive or compute a mapping from the video data, considered as a three-dimensional object (the three dimensions are the two geometrical dimensions of the individual frames and the time dimension), to the two-dimensional plane for screen representation. The video structure represented by the video summary should be highly correlated with the semantic content, and allow simple semantic interpretation for sufficient understanding. This involves automatically finding the semantically most important dramatic elements within a video clip.

The development of visual summaries depends heavily on the video category. It is obvious that story structures are prominent in news or films, while programs of sporting events do not have such story structure. Therefore, different video categories will require different forms of visual summaries. For example, the VISION [13] system allows users to dynamically filter the contents of the category-based browsing structures using multiple criteria (source, date, and keywords). This provides a hybrid of browsing and searching.

7.1.5 Viewing

After one or more candidate videos are selected as more likely, the user has to be able to see parts of the candidate. The usual functions of today's video players should be available in this stage: play, pause, fast-forward, reverse, etc. More sophisticated capabilities like semantic fast-forward (the ability to move forward in the video on the basis of semantic video content) should also be available.

The huge amount of data present in the video requires compression in order to achieve efficient storage and transmission. One of the most common standard video compression formats is MPEG which uses predictive and differential coding techniques. This implies that random access to any frame of the video cannot be implemented in a straightforward way. Therefore, efficient algorithms should be implemented for manipulating MPEG videos without the need of a complete decompression.

7.2 Video Analysis and Processing

Most of the videos (movies, TV programs, documentaries, etc.) are structured such that they convey the underlying narrations and messages. This means that the shots are combined in a specific way and according to a special orders in order to form the montage in telling the story. Because of this, certain temporal features can be recognized and associated information can be extracted by automatically analyzing the visual contents and temporal associations of the video.

7.2.1 Video Shots

The fundamental unit of a film and video is a shot. The content of a shot depicts continuous action captured by a single camera during some interval. The importance of the shot as the fundamental video unit has been realized by many researchers, and the computational detection of video-shot boundaries has received much attention. Usually, two types of scene changes are considered: scene cuts (or camera breaks) and gradual transitions. An *abrupt transition* or *cut* (Fig. 7.2(a)) is an instantaneous change from one shot to another where the transition boundary is between two consecutive frames. *Gradual* scene changes are frequently used for editing techniques to connect two scenes together and can be classified into three common types: fade-in/out, dissolve, and wipe. A *fade* is a continuous, monotonic change in luminance (Fig. 7.2(b)). A *dissolve* can be viewed as a fade-out and fade-in with some overlap (Fig. 7.2(c)). A dissolve is more general than a fade because the chrominance can also change during the transition and each pixel can change independently. A *wipe* is a transition from one scene to another wherein the new scene is revealed by a moving boundary in the form of a line or pattern. Figure 7.3 shows some examples of wipes.

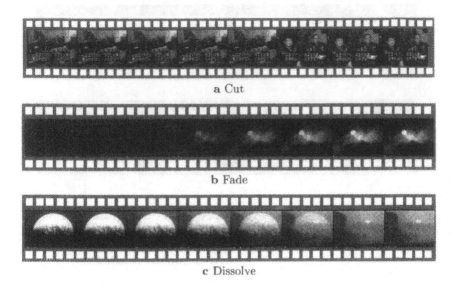

a Cut

b Fade

c Dissolve

Fig. 7.2 Examples of shot transitions.

Fundamentally, most approaches use the concept of partitioning the continuous video data into sequences for indexing. Videos are segmented primarily on the basis of camera breaks or shots, therefore, each sequence is a segment of data having a frame or multiple consecutive frames. Automatic isolation of camera transitions requires support of tools that provide accurate and fast detection. Abrupt camera transitions can be detected quite easily as the difference between two consecutive frame is so large that they cannot belong to the same shot. A problem arises when the transition is gradual, meaning that the shot does not change abruptly but just over a period of few frames. In this case, the difference between two shots is not large enough to declare it a camera break.

Recently, many video segmentation algorithms have been proposed. Ahanger and Little [3], Idris and Panchanathan [17], and Mandal et al. [25] have presented detailed surveys on some of the existing video segmentation methods. A large number of techniques have been reported in the literature for temporal segmentation. A number of metrics have been suggested for video scene segmentation for both the raw and compressed data. The metrics used to detect the difference between two frames can be divided broadly into

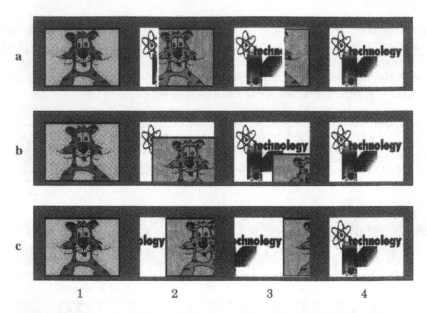

Fig. 7.3 Wipe examples. **a** piecewise replacement, neither images moves; **b** old image moves out, revealing the new one; **c** both images move, similar to panning.

four classes: (1) pixel or block comparison; (2) histogram comparison (of gray levels or color codes); (3) using the DCT coefficients in MPEG encoded video sequences; and (4) the sub-band feature comparison method.

Consider a video stream as a sequence of video frames $\{f_1, f_2, \ldots, f_N\}$, where N is the total number of frames and each f_i, $1 \leq i \leq N$, is a 24-bit RGB color image with $X \times Y$ pixels. Different image features such as intensity, color histogram, etc., extracted from each video frame are usually used to compute a quantitative measure of the content difference between two frames. We denote the inter-frame difference (or frame difference) between consecutive frames f_i and f_{i+1} by $D_i(f_i, f_{i+1})$ for $1 \leq i \leq N - 1$. In the following subsections, we describe some of the existing methods for computing the content difference for consecutive video frames.

Pixel Difference Method. Nagasaka and Tanaka [28] use the sum of absolute pixel-wise intensity differences between two frames as a frame difference. Let $f_i(x, y)$ denote the intensity of pixel (x, y) in a frame f_i. The difference between frame f_i and frame f_{i+1} is defined as:

$$D_i(f_i, f_{i+1}) = \frac{1}{X \cdot Y} \sum_{x=0}^{X-1} \sum_{y=0}^{Y-1} |f_i(x, y) - f_{i+1}(x, y)|. \tag{7.1}$$

Otsuji et al. [29] and Zhang et al. [49] count the number of the changed pixels, and a camera shot break is declared if the percentage of the total number of changed pixels exceeds a threshold. Mathematically, the differences in pixels and threshold are:

$$DP_i(x,y) = \begin{cases} 1 & \text{if } |f_i(x,y) - f_{i+1}(x,y)| > t, \\ 0 & \text{otherwise,} \end{cases} \tag{7.2}$$

$$\frac{100}{X \cdot Y} \sum_{x=0}^{X-1} \sum_{y=0}^{Y-1} DP_i(x,y) > T. \tag{7.3}$$

If the difference between the corresponding pixels in the two consecutive frame is above a certain minimum intensity value t, then $DP_i(x,y)$ is set to one. In Eq. 7.3, the difference percentage between the pixels in the two frames is calculated by summing the number of different pixels and dividing by the total number of pixels in a frame. If this percentage is above a certain threshold T, a camera break is declared.

Camera movement, e.g., pan or zoom, can change a large number of pixels and hence in such conditions a false shot break will be detected. Fast moving objects also have the same effect. Zhang et al. [49] propose the use of a smoothing filter (e.g., 3×3 window) to reduce the sensitivity to camera movement. To make cut detection more robust to illumination changes, Aigrain and Joly [4] normalize the video frame through histogram equalization.

Likelihood Ratio. Detecting changes at pixel level is not a very robust approach. A *likelihood ratio* approach is suggested based on the assumption of uniform second order statistics over a region [49]. The frames are first divided into blocks and then the blocks are compared on the basis of the statistical characteristics of their intensity blocks. Let μ_i and μ_{i+1} be the mean intensity values for a given region in two consecutive frames, and σ_i and σ_{i+1} be the corresponding variances. The frame difference is defined as the percentage of the regions whose likelihood ratios (Eq. 7.4) exceed a predefined threshold t:

$$\lambda = \frac{[\frac{\sigma_i + \sigma_{i+1}}{2} + (\frac{\mu_i + \mu_{i+1}}{2})^2]^2}{\sigma_i \cdot \sigma_{i+1}}, \tag{7.4}$$

$$DP_i(x,y) = \begin{cases} 1 \text{ if } \lambda > t, \\ 0 \qquad \text{otherwise.} \end{cases} \tag{7.5}$$

This approach is better than the previous one as it increases the tolerance against noise associated with camera and object movement. However, it is possible that even though the two corresponding blocks are different, they can have the same density function. In such cases no change is detected.

Shahraray [36] proposed dividing a frame into K non-overlapping blocks and finding the "best" matching blocks between frames for comparison, similar to the block matching technique of MPEG. Typically, $K = 12$ is used and block matching is performed on the image intensity values. A non-linear digital order-statistic filter *L-filter* is used in computing the block differences. Let $L = \{l_1, l_2, \ldots, l_K\}$ be an ordered list of the K block match values. Then the match coefficient between the two images is:

$$D(f_i, f_{i+1}) = \sum_{i=1}^{K} c_i l_i \qquad (7.6)$$

where c_i is a predetermined coefficient whose value depends on the order of the match values l_i in the list L.

Histogram Comparison. The sensitivity to camera and object motion can be further reduced by comparing the gray-level histograms of the two frames [28, 49]. This is due to the fact that two frames, which have minimal changes between backgrounds and some amount of object motion have almost the same histograms.

Let H_i be the gray-level histogram of a frame f_i, then the histogram difference between two consecutive frames is defined as:

$$D_i(f_i, f_{i+1}) = \sum_{k=0}^{G-1} |H_i(k) - H_{i+1}(k)| \qquad (7.7)$$

where G is the number of gray levels. If the sum is greater than a given threshold t a transition is declared.

Histogram comparison using color code is suggested in Refs [28, 49]. The authors use a 6-bit color code of a 24-bit RGB pixel obtained by combining the two most significant bits extracted from each of the 8-bit red, green, and blue color components. Let HC_i be the histogram of the color codes in a frame f_i then, the frame difference is defined as:

$$D_i(f_i, f_{i+1}) = \sum_{k=0}^{63} |HC_i(k) - HC_{i+1}(k)|. \qquad (7.8)$$

Nagasaka and Tanaka [28] propose using the χ^2 test as a measure of similarity between color histograms. It computes the square of the difference between the two histograms in order to strongly reflect the difference. Let H_{C_i} be the color histogram of a frame f_i having G bins then, the frame difference is defined as:

$$D_i(f_i, f_{i+1}) = \sum_{k=0}^{G-1} \frac{|H_{C_i}(k) - H_{C_{i+1}}(k)|^2}{H_{i+1}(k)}. \qquad (7.9)$$

The χ^2-test enhances the difference between the camera breaks but also small changes due to camera or object motion. Therefore, this method may not be more effective than the gray and color histogram comparison techniques.

They also propose a robust approach to accommodate momentary noise such as a the light from a flashlight, which only brightens part of the frame. The video frame is divided into 4×4 rectangular regions, each denoted by a region number r. Histogram comparison is made on each pair of corresponding regions between two frames. The frame difference is defined as the sum of the eight smallest sub-frame histogram differences calculated using Eq. 7.9 for each $0 \leq r \leq 15$.

Twin Comparison. The previous techniques are based on a single threshold and lack the power of detecting gradual transitions. This is because the threshold is higher than the difference found between the frames in which the transition takes place due to special effects. Lowering the threshold does not solve the problem because the difference value due to the special effects can be smaller than the ones which take place within the shot. For example, object motion and/or camera motion might contribute more changes than the gradual transition. Making the value of the threshold even smaller will lead to false detections due to camera and object motions. When the beginning and end frames for the gradual transitions are detected, the frames in the transition can be declared as a separate segment.

The twin comparison method [49] takes into account the cumulative differences between the frames for gradual transitions. This method (Fig. 7.4) requires two cutoff thresholds, one higher threshold (T_h) for detecting abrupt transitions and a lower one (T_l) for gradual transitions. In the first stage, the higher threshold is employed to detect abrupt shot boundaries. In the next stage, the lower threshold is used on the rest of the frames. Any frame that has a difference more than this threshold is declared as a potential start (F_s) of the transition. Once F_s is identified, it is compared with subsequent frames measuring the accumulated difference instead of frame-to-frame difference. Usually, this difference value increases and when this value increases to the level of the higher threshold, a camera break is declared at that frame (F_e). If the value falls between the consecutive frames, then the potential frame is dropped and the search starts all over. Although the twin comparison approach is effective in detecting gradual transitions, the type of transition cannot be identified.

The gradual transitions might include special effects due to camera panning and zooming. Optical flow is used to detect camera motions and motion vectors are computed to detect the changes due to pan and zoom.

Model-Based Segmentation. It is not only important to identify the gradual transition position, but also the type of transition. In a video sequence, a gradual transition from one scene to another is the result of the editing

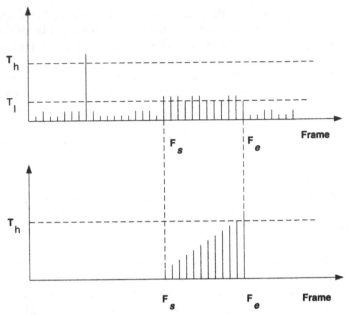

Fig. 7.4 Twin comparison.

process. Therefore, in model-based techniques the problem of video segmentation is viewed as locating the edit boundaries within the video sequence. Here, different edit types, such as cuts, translations, fades, wipes, and dissolves are modeled by mathematical functions.

Let the symbol S denote a single continuous shot which is a set of consecutive 2D images. The individual images of the set are denoted $I(x, y, t)$, where x and y are the pixel position and t is the discrete time index. A shot containing N images is represented as follows:

$$S = \{I(x, y, 1), I(x, y, 2), \ldots, I(x, y, N)\}. \tag{7.10}$$

Abrupt cuts are formed by concatenating two different sequences as $S = S_1 \circ S_2$. They are the easiest to detect since the two sequences S_1 and S_2 are usually dissimilar. Due to the abrupt nature of this transition, it is expected that it will produce significant changes in the scene lighting and geometric structure.

Two types of fades are considered. The first is a fade-out, where the luminance of the scene is decreased over multiple frames, and the second is a fade-in, where the luminance of the scene is increased from some base level, usually black, to full scene luminance. A simple way to model a fade-out is to take a single frame in the scene, $I(x, y, t_1)$, and gradually decrease the luminance. One way to do this is by scaling each frame in an N frame edit sequence as:

$$S(x,y,l) = I(x,y,t_1) \times \left(1 - \frac{l}{N}\right), \qquad l = 1,\ldots,N. \qquad (7.11)$$

The shape of the intensity histogram remain fixed (ideally) for each frame in the sequence, but the width of the histogram is scaled by the multiplicative factor $(1 - \frac{l}{N})$. The intensity mean, median, and standard deviation values in each frame are also scaled by this factor (relative to their value in frame t_1). Another way to implement a fade-out is to shift the luminance level as:

$$S(x,y,l) = I(x,y,t_1) - m_1 \times \left(\frac{l}{N}\right), \qquad l = 1,\ldots,N \qquad (7.12)$$

where m_1 is the maximum intensity value in the frame $I(x,y,t_1)$. In this model, the mean and median intensity values are shifted downward in each consecutive frame, but the standard deviation remains constant. In practice, a non-linear limiting operation is applied to the sequence because intensity values are non-negative, and the resulting negative intensity values are set equal to 0. The limiting operation decreases the width of the histogram and likewise decreases the standard deviation. A general expression for the change in standard deviation cannot be determined because it depends on the shape of the histogram. If the limiting operation is applied to any of the inputs, the standard deviation will decrease, otherwise it will remain constant.

Two analogous models for fade-ins are:

$$S(x,y,l) = I(x,y,t_1) \times \left(\frac{l}{N}\right), \qquad l = 1,\ldots,N \qquad (7.13)$$

and

$$S(x,y,l) = I(x,y,t_1) + m_N \times \left(\frac{l}{N}\right), \qquad l = 1,\ldots,N \qquad (7.14)$$

where m_N is the maximum intensity value in the frame $I(x,y,t_N)$.

The scaling model of Eqs 7.11 and 7.13 was employed for detecting chromatic edits in video sequences [15] (a chromatic edit is achieved by manipulating the color or intensity space of the shots being edited). Experimental results indicate that some, but not all, fades follow this model. The shifting model of Eqs 7.12 and 7.14 is proposed as one alternative. Both fade models are too simple because they model a sequence which is a single static image whose brightness is varied. In reality, this operation is applied to non-static sequences where inter-frame changes due to object motion occur.

During a fade it is assumed that the geometric structure of the scene remains fairly constant between frames, but that the lighting distribution changes. For example, during a fade-out that obeys Eq. 7.13 the mean, median, and standard deviation all decrease at the same constant rate, but the content of the scene remains constant. If the fade obeys Eq. 7.14 the mean and the median decrease at the same rate, but the standard deviation may not. The converse is true for fade-ins.

Dissolves are a combination of two scenes, and can simply be a combination of a fade-out of one shot (I_1) and a simultaneous fade-in of another (I_2). The model for this is [15]:

$$S(x,y,l) = I_1(x,y,t_1) \times \left(1 - \frac{l}{N}\right) + I_2(x,y,t_2) \times \left(\frac{l}{N}\right). \tag{7.15}$$

Dissolves are difficult to detect since they can occur gradually, and often do not obey a simple mathematical model such as Eq. 7.15. This is because the fade rates do not have to be equal as modeled. For example, there may be object/camera motion during the transition and complex special effects can be applied. There are also other common sources of problems in detecting fades and dissolves: relatively small areas of change, e.g., title sequences where the frame is mostly black with only a few lines of text that actually fade in or out; and persistent unchanging regions, e.g., overlay graphics and station identifiers/logos.

During a dissolve the scene structure and lighting characteristics may change simultaneously. In addition, the expected changes in the sequence are difficult to predict unless the final image in the sequence is known (i.e., the edit is already detected). However, if the dissolve obeys Eq. 7.15 the frames comprising the edit will be a linear combination of the initial and final scenes. In practice, one does not need a priori knowledge of the start and end frames of a complete dissolve in order to model the expected changes. A solution is to measure the pixel differences between three consecutive frames A, B, and C. The task is to determine if B is a dissolving (or fading) frame. The A/B differences are computed and compared to the B/C differences. Only if there is enough consistency is an attempt made to verify the effect by synthesizing the average of A and C and comparing it to B. A and C might also be dissolving (fading) frames but that does not affect the AC/B comparison, and identifying where the overall beginning or end of the effect is something a higher level of logic can determine afterward. Note that one cannot just always compute the AC/B comparison to discover dissolves (fades) because a static sequence also satisfies that condition. There must be a pixel-wise consistent trend to the change in luminance or chrominance over at least three frames. More frames may be necessary if false positives due to the transient pixel jitter and brief flashes are to be minimized.

A similar approach is used by Meng et al. [26] for detecting dissolves. Consider that $I_1(t)$ and $I_2(t)$ are two ergotic sequences with intensity variance σ_1^2 and σ_2^2. The model used for dissolve is given by the Eq. 7.15. In the ideal case when $I_1(t)$ and $I_2(t)$ are ergotic with σ_1^2 and σ_2^2, the variance curve in the dissolve region shows a parabolic shape. In real sequences with motions, the variances of $I_1(t)$ and $I_2(t)$ are not constant but the dissolve region still demonstrates the parabolic shape.

Corridoni and Del Bimbo [8] used the $L^*u^*v^*$ color space representation, which separates the brightness feature from the color. Considering that in a fade sequence there is a linear variation in the luminance values (l) and constancy in the chrominance values (u and v), they introduced the following measure for fade detection:

$$D_{fade} = \frac{l_i}{u_i v_i (\varepsilon + |C_i^L - C_i|)} \tag{7.16}$$

where l_i is the number of pixels whose change in l is linear; u_i and v_i are the number of pixels whose change in u and v is lower than a threshold imposed by Gaussian noise; C_i is the spatial center of video frame f_i and

$$C_i^L = \frac{1}{l_i}(x_i^L, y_i^L) \tag{7.17}$$

where x_i^L and y_i^L are the number of pixels whose change in luminance along x and y is linear. The right term in Eq. 7.16 is a spatial constraint to impose uniform variation of brightness on the video frame. The measure is averaged over a wide temporal window (e.g., 16) and whenever the temporal average over the window overcomes a threshold, a fade is detected.

The dissolve is described by the composition of a fade-in and a fade-out for the brightness value. Although the u and v values do not remain constant, the amount of changing pixels is high and is statistically distributed uniformly over the image. The dissolve detection is based on the following measures:

$$D_{diss}^1 = \frac{l_i}{\varepsilon + |C_i^L - C_i|}, \tag{7.18}$$

$$D_{diss}^2 = \frac{u_i + v_i}{\varepsilon + |C_i^{UV} - C_i|} \tag{7.19}$$

where C_i^{UV} is defined similar to C_i^L (see Eq. 7.17).

A dissolve is characterized by two consecutive maxima in the D_{diss}^1 measure and by a maximum in the D_{diss}^2 measure, in correspondence with the valley between the two maxima.

Aigrain and Joly [4] developed a model for the probability density function of intershot pixel differences in successive frames for various transition effects. In their model, pixel differences in a cut transition follow the linear law with density function defined as:

$$Q(s) = \frac{2(a - |s|)}{a(a - 1)} \tag{7.20}$$

where a is the number of gray levels.

Pixel differences in a wipe transition also follow a linear law with density function defined as:

$$Q(s) = \frac{2(a - |s|)}{da(a - 1)} \tag{7.21}$$

where d is the duration of the wipe transition in a number of sampled images.

In the case of a cross-dissolve transition, the pixel differences follow the law:

$$Q(s) = \begin{cases} \frac{2d(a-d|s|)}{a(a-1)} & \text{for } d|s| \leq a, \\ 0 & \text{otherwise} \end{cases} \qquad (7.22)$$

where d is the duration of the cross-dissolve.

Pixel differences in a fade to black, fade to white, fade from black or fade from white transition follow a linear law:

$$Q(s) = \begin{cases} \frac{d}{a} & \text{for } d|s| \leq a, \\ 0 & \text{otherwise} \end{cases} \qquad (7.23)$$

for $s > 0$ (fade from black or fade to white) or $s < 0$ (fade from white or fade to black).

Shot transitions are detected by fitting these models to pixel difference data obtained from the sequence.

Detection of Camera Motion. To detect the camera motion, optical flow techniques are utilized. Optical flow gives the distribution of velocity with respect to an observer over the points in an image. Optical flow is determined by computing the motion vector of each pixel in the image. The fields generated by zoom, pan and tilt are shown in Fig. 7.5. Detecting these fields helps in separating the changes introduced due to the camera movements from the special-effects such as wipe, dissolve, fade-in, fade-out.

As seen in Fig. 7.5 most of the motion vectors between consecutive frames due to pan and tilt point in a single direction thus exhibiting a strong modal value corresponding to the camera movement. Disparity in direction of some of the motion vectors will result from object motion and other kinds of noise. Thus, a single modal vector is exhibited with respect to the camera motion. As given by Eq. 7.24, a simple comparison technique can be used to detect pan and tilt. Pan or tilt are detected by calculating the differences between the modal vector and the individual motion vectors [49]. Let θ_l be the direction of the motion vectors and θ_m the direction of the modal vectors, then:

$$\sum_{l=1}^{N} |\theta_l - \theta_m| \leq \Theta_p. \qquad (7.24)$$

If the sum of the differences of all vectors is less than or equal to a threshold variation Θ_p then a camera movement is detected. This variation should be zero if no other noise is present.

Motion vectors for zoom have a center of focus, i.e., a focus of expansion (FOE) for zoom-in, and a focus of contraction (FOC) for zoom-out. Due to the absence of noise, it is easy to detect the zoom because the sum of the motion vectors around the FOC/FOE will be zero. However, it is difficult to find

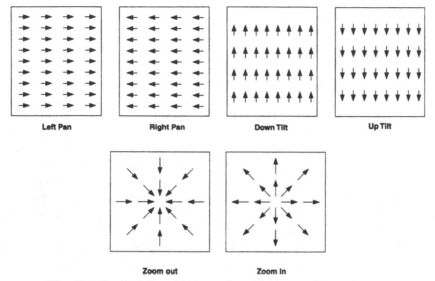

Fig. 7.5 Optical flow field produced by pan, tilt, and zoom.

the center of focus of zoom since it could be present across two consecutive frames. Zhang et al. [49] assume that the FOE/FOC lies within a frame, therefore, it is not necessary to locate it for vector comparison. A simple comparison technique can be used, the vertical vectors from the top (V_k^{top}) and the bottom row (V_k^{bottom}) can be compared for magnitude and horizontal vectors from the left-most (U_k^{top}) and right-most (U_k^{bottom}) vectors at the same row can be compared. In the case of zoom, the vectors will have opposite signs, and the magnitude of the difference of these components should exceed the magnitude of the highest individual component. This is due to the fact that in every column the magnitude of the difference between these vertical components will exceed the magnitude of both components:

$$|V_k^{top} - V_k^{bottom}| \geq (|V_k^{top}|, |V_k^{bottom}|), \tag{7.25}$$

$$|U_k^{top} - U_k^{bottom}| \geq (|U_k^{top}|, |U_k^{bottom}|). \tag{7.26}$$

When both Eqs 7.25 and 7.26 are satisfied, a zoom is declared. Thus, camera movements can be separated from the gradual transitions.

Optical flow works well for detecting motion, but it can be difficult to distinguish between object and camera motion without understanding the video content. Moreover, pure pans and tilts are exceptions in real-world videos, mixed-axis motion is more common. In general, it is more likely that pure X or Y motion is attributable to the camera, whereas mixed-axis motion is more likely attributable to objects. However, there are too many exceptions to this generalization to make it very useful.

Motion vector information can also be obtained from MPEG compressed video sequences. However, the block matching performed as a part of MPEG encoding selects vectors based on compression efficiency and thus often selects inappropriate vectors for image processing purposes.

DCT-Based Method. In this category of methods, the discrete cosine transform (DCT) coefficients are used in comparing two video frames. The DCT is commonly used for reducing spatial redundancy in an image in different video compression schemes such as MPEG and JPEG. Compression of the video is carried out by dividing the image into a set of 8×8 pixel blocks. Using the DCT the pixels in the blocks are transformed into 64 coefficients which are quantized and Huffman entropy encoded. The DCT coefficients are analyzed to find frames where camera breaks take place. Since the coefficients in the frequency domain are mathematically related to the spatial domain, they can be used in detecting the changes in the video sequence.

A DCT correlation method is described by Arman et al. [5]. Given 8×8 blocks of a single DCT-based encoded video frame f_i, a subset of blocks is chosen a priori. The blocks are chosen from n connected regions in each frame. For each block, 64 randomly distributed AC coefficients are chosen. By taking coefficients from each frame a vector $V_i = \{c_1, c_2, c_3, \dots\}$ is formed. The frame difference between two successive frames f_i and f_{i+1} is defined as the inner product of the two corresponding vectors V_i and V_{i+1}:

$$D_i(f_i, f_{i+1}) = 1 - \frac{V_i V_{i+1}}{|V_i||V_{i+1}|}. \tag{7.27}$$

A transition is detected when $D_i > T$, where T is the threshold.

Zhang et al. [51] proposed a method which uses the pairwise difference of DCT coefficients. Let $c_{i,b}(m,n)$ be the DCT coefficient at (m,n) of a block b in video frame f_i. A binary function for deciding whether there is a change between corresponding DCT blocks of video frames f_i and f_{i+1} is defined as follows:

$$DB_{i,b}(f_i, f_{i+1}) = \begin{cases} 1 & \text{if } \frac{1}{64} \sum_{m=0}^{7} \sum_{n=0}^{7} \frac{|c_{i,b}(m,n) - c_{i+1,b}(m,n)|}{\max[c_{i,b}(m,n), c_{i+1,b}(m,n)]} > t, \\ \\ 0 & \text{otherwise} \end{cases} \tag{7.28}$$

where t is a user-defined threshold. Further, the frame difference between f_i and f_{i+1} is defined as:

$$D_i(f_i, f_{i+1}) = \frac{1}{B} \sum_{b=0}^{B-1} DB_{i,b}(f_i, f_{i+1}) \tag{7.29}$$

where B is the total number of 8×8 blocks. So, if in Eq. 7.28 $DB_{i,b}$ is larger than the threshold, the block b is considered to be changed. If the number of changed blocks Eq. 7.29 exceeds a threshold, a shot boundary between f_i and f_{i+1} frames is declared. Compared with the previous technique [5], the processing time of this technique is smaller; however, it is more sensitive to gradual changes.

Yeo and Liu [43] use the pixel differences of the luminance component of DC frames. The DC image is a spatially reduced version (1/64) of the original video frame. Each pixel of a DC image is equal to the average intensity of each 8×8 block of the frame. Under these conditions the frame difference between f_i and f_{i+1} is defined as:

$$D_i(f_i, f_{i+1}) = \sum_{b=0}^{B-1} |dc_i(b) - dc_{i+1}(b)| \qquad (7.30)$$

where B is the total number of 8×8 blocks and $dc_i(b)$ is the DC coefficient of block b in frame f_i. Although this technique is fast, cuts may be misdetected between two frames which have similar pixel values but different density functions.

Patel and Sethi [30] have experimented with various statistics and have found that the χ^2 statistic gives the best performance. They use intensity histograms obtained from the entire frame. The histograms are found using DC coefficients of MPEG video for only I frames. Taskiran and Delp [40] use a *generalized trace* (GT). This high dimensional feature vector is composed of a set of features extracted from each DC frame. The GT is then used in a binary regression tree to determine the probability that each frame is a shot boundary. These probabilities can then be used to detect frames corresponding to shot boundaries.

Vector Quantization. Idris and Panchanathan [16] use vector quantization to compress a video sequence using a codebook of size 256 and 64-dimensional vectors. The histogram of the labels obtained from the codebook for each frame is used as a frame similarity measure and a χ^2 statistic is used to detect cuts. Consider the histogram of the labels of a frame f_i, $\{H_i(k); k = 1, 2, \ldots, N\}$ where $H_i(k)$ is the number of labels k in the compressed frame and N is the number of codewords in the codebook. The difference between two frames f_i and f_j is:

$$D_i(f_i, f_j) = \sum_{k=1}^{N} \frac{(H_i(k) - H_j(k))^2}{|H_i(k) - H_j(k)|}. \qquad (7.31)$$

A large value of $D_i(f_i, f_j)$ indicates that f_i and f_j belong to different scenes. An abrupt change is declared if the difference between two successive frames exceeds a threshold and a gradual transition is detected if the difference between the current frame and the first frame of the current shot is greater than a threshold.

Color-Ratio-Based Method. A color-ratio-based method for uncompressed and compressed videos is introduced by Adjeroh et al. [2]. Using the fact that neighborhood-based color ratios are invariant under various changes in viewing conditions, including both spectral and intensity variations of the illuminant, their aim is to obtain illumination invariant and motion invariant features for window-based video partitioning.

Let $h(x, y)$ be the red, green, or blue intensity of pixel (x, y), then the four-neighbor color ratio for the pixel (x, y) of a video frame f_i is defined as:

$$\Phi_i(x, y) = \frac{h(x, y - 1) + h(x, y + 1) + h(x - 1, y) + h(x + 1, y)}{4h(x, y)}. \qquad (7.32)$$

The window-based difference between frames f_i and f_{i+1} is computed:

$$\mathcal{J}_{i,w}(f_i, f_{i+1}) = \begin{cases} 1 - \xi_{i,w}(f_i, f_{i+1}) & \text{if } \xi(\cdot) \leq 1, \\ 1 - \frac{1}{\xi_{i,w}(f_i, f_{i+1})} & \text{otherwise} \end{cases} \qquad (7.33)$$

where w denotes a window, and

$$\xi_{i,w}(f_i, f_{i+1}) = \frac{\Psi_i(w)}{\Psi_{i+1}(w)}, \text{ and } \Psi_i(w) = \prod_{(x,y) \in w} \Phi_i(x, y). \qquad (7.34)$$

The frame difference between f_i and f_{i+1} is further defined as:

$$DW_{i,w}(f_i, f_{i+1}) = \begin{cases} 1 & \text{if } \mathcal{J}_{i,w}(f_i, f_{i+1}) > t, \\ 0 & \text{otherwise,} \end{cases} \qquad (7.35)$$

$$D_i(f_i, f_{i+1}) = \frac{1}{W} \sum_{w=0}^{W-1} DW_{i,w}(f_i, f_{i+1}) \qquad (7.36)$$

where t is a user-defined threshold for deciding whether there is a change between corresponding windows of two video frames and W is the total number of windows.

In the case of compressed video, similar to the case of uncompressed video, an n-neighbor color ratio for a DCT coefficient $H_{i,b}(u, v)$ of a DCT image is defined as:

$$\Phi_i(u, v) = \frac{\sum\limits_{r \in \{n-\text{neighbor of } b\}} H_{i,r}(u, v)}{nH_{i,b}(u, v)} \qquad (7.37)$$

where n is the number of neighboring blocks involved, $H_{i,b}(u, v)$ is the transform coefficient at the location (u, v) in a block b, and $H_{i,r}(u, v)$ is the coefficient at the corresponding (u, v) location in neighboring block r. The block-based difference between f_i and f_{i+1} is computed as:

$$\mathcal{J}_{i,b}(f_i, f_{i+1}) = \begin{cases} 1 - \xi_{i,b}(f_i, f_{i+1}) & \text{if } \xi(\cdot) \leq 1, \\ 1 - \frac{1}{\xi_{i,b}(f_i, f_{i+1})} & \text{otherwise} \end{cases} \tag{7.38}$$

where b denotes a block, and

$$\xi_{i,b}(f_i, f_{i+1}) = \frac{\Psi_i(b)}{\Psi_{i+1}(b)}, \qquad\qquad \Psi_i(b) = \prod_{(u,v)\in b} \Phi_i(u, v). \tag{7.39}$$

Since the DC and AC coefficients usually carry different types of information about the scene, a weighting factor $\phi, 0 \leq \phi \leq 1$, can be defined to vary the block-based difference adaptively, depending on the scene.

1. DC represents an average of the sample value in the spatial domain.
2. AC components provide information on the level of "scene activity" or detail:

$$\mathcal{J}_{i,b}(f_i, f_{i+1}) = \phi \mathcal{J}_{i,b}^{DC}(f_i, f_{i+1}) + (1 - \phi)\mathcal{J}_{i,b}^{AC}(f_i, f_{i+1}). \tag{7.40}$$

The frame difference between f_i and f_{i+1} is finally defined as:

$$DB_{i,b}(f_i, f_{i+1}) = \begin{cases} 1 & \text{if } \mathcal{J}_{i,b}(f_i, f_{i+1}) > t, \\ 0 & \text{otherwise}, \end{cases} \tag{7.41}$$

$$D_i(f_i, f_{i+1}) = \frac{1}{B} \sum_{b=0}^{B-1} DB_{i,b}(f_i, f_{i+1}) \tag{7.42}$$

where t is a user-defined threshold for deciding whether there is a change between corresponding windows of two video frames and B is the total number of blocks.

Motion-Based Segmentation. Motion analysis is an important step in video processing. A video stream is composed of video elements constrained by the spatiotemporal piecewise continuity of visual cues. The normally coherent visual motion becomes suddenly discontinuous in the event of shot boundaries. Hence, motion discontinuities may be used to mark the change of a scene, the occurrence of occlusion, or the inception of a new activity.

In the MPEG standard, to reduce temporal redundancy, video frames are coded with reference to the contents of the other previous or following frames by motion compensation prediction. Figure 7.6 illustrates an MPEG sequence of frames with a GOP (group of pictures) size of 15.

There are three types of frames, namely I-frame, P-frame, and B-frame. Every I-frame is encoded without referencing to other frames. In this case no motion compensation is performed. A P-frame is predictively coded with motion compensation from past I- or P-frames. Both these frames are used as a basis for bidirectional motion compensated B-frames. The prediction is

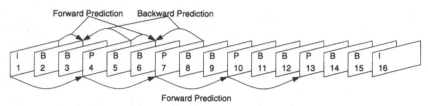

Fig. 7.6 A typical MPEG frame sequence in display order.

based on matching a 16×16 pixel macroblock of a frame as closely as possible to another macroblock of a preceding or succeeding frame. For those predicted frames (i.e., P or B) a number of motion vectors indicating the types and the distances will be generated. The number of motion vectors associated with a frame depends on the residual error after motion compensation. If two frames are quite different, motion compensation will be abandoned and most macroblocks will be intra-coded, resulting in a smaller number of motion vectors. Thus, a scene cut can be detected based on the counts of motion vectors.

Zhang et al. [51] have proposed a technique for cut detection using motion vectors in MPEG. In P-frames, M is the number of motion vectors and in B-frames, M is the smaller of the counts of the forward and backward non-zero motion. Then $M < T$ will be an effective indicator of a camera boundary before or after the B- and P-frame, where T is a threshold value close to zero. However, this method yields false detection where there is no motion. This is improved by applying the normalized inner product metric (Eq. 7.27) to the two I-frames on the sides of the B-frame where a break has been detected.

Kuo et al. [20] define two types of reference ratios (RRs) for P- and B-frames. Forward reference ratio (FRR) is defined as the ratio between the number of the forward-predicted macroblocks (R_f) and the total number of macroblocks (N). On the other hand, backward reference ratio (BRR) is defined as R_b/N, where R_b is the number of backward predicted macroblocks. RRs of frames within the same shot will be high (compared with a pre-defined threshold) and when a shot boundary occurs, since the contents of frames are assumed to be different from the preceding frames, the RRs will be low. The conditions for a scene change to occur are described as follows:

Scene cut at I-frame. The $BRRs$ of previous B-frames are low.

Scene cut at P-frame. The $BRRs$ of previous B-frames and the FRR of this P-frame are low.

Scene cut at B-frame. The $BRRs$ of previous B-frames (if any), the FRR of this B-frame, and the $FRRs$ of following B- and P-frames (if any) are all low.

They propose a mask matching approach to detect the above different ratio patterns. For example, let the frame pattern be $IBBPBBPBBI\ldots$, then $\{I : (B_b, B_b, @I)\}$ is a mask which specifies the condition that the $BRRs$

of B-frames before the current I-frame should be low, if a scene change occurs at the current I-frame. A frame marked with a symbol @ denotes the frame under consideration. Similarly, $\{P : (B_b, B_b, @P_f)\}$ is a mask designed for detecting scene cuts at the current P-frame. P_f denotes a criterion that the FRR of the P-frame should be low for a scene change to occur. For a B-frame, either $\{B : (@B_f, B_f, P_f)\}$ or $\{B : (B_b, @B_f, P_f)\}$ is used depending on whether or not the current B-frame is immediately preceded by an I- or P-frame. A function is defined to convert the results of mask matching to shot change probabilities from which the global threshold method is used to detect scene cuts:

$$P_i = 1 - \frac{RR_{f_1}^2 + RR_{f_2}^2 + \cdots + RR_{f_n}^2}{RR_{f_1} + RR_{f_2} + \cdots + RR_{f_n}} \tag{7.43}$$

where $f_1, f_2, \ldots, f_n$ are mask frames of a mask i and RR_f is the RR of mask frame f.

The global threshold T_b is defined as $(F + F')/2$, where F is the average of the 97.5% smallest scene change probabilities and F' is the average of the 2.5% largest scene change probabilities. This is based on the assumption that on average there is a scene change for every 40 frames.

To reduce the effects of illumination and fast motions, and to avoid the situations of two or more shot changes in a short period of time, an adjustment to shot change probabilities can be defined as follows:

$$F(P_i) = \begin{cases} P_i - \min(P_k) & \text{if } P_i = \max(P_k), \quad i - j \le k \le i + j, \\ 0 & \text{otherwise} \end{cases} \tag{7.44}$$

where j is a parameter of the window size.

Liu and Zick [24] have presented a technique based on the error signal and the number of motion vectors. A scene cut between a P-frame P_m and a past reference P-frame P_n increases the error energy. Hence, the error energy provides a measure of similarity between P_m and the motion compensated frame P_n:

$$S(P_m, P_n) = \frac{\sum_{i=1}^{R_f} E_i}{R_f^2} \tag{7.45}$$

where R_f is the number of forward predicted macroblocks and E_i is the error energy of macroblock i:

$$E_i = \sum_m \sum_{j=1}^{64} c_{i,m,j}^2 \tag{7.46}$$

where $c_{i,m,j}$ is the jth DCT coefficient in block m of macroblock i.

For the detection of scene changes based on B-frames, the difference between the number of forward predicted macroblocks R_f and backward predicted R_b is used. A scene change between a B-frame and its past reference B-frame will decrease R_f and increase R_b. A scene change is declared if the difference between R_f and R_b changes from positive to negative.

Meng et al. [26] define three ratios R_p, R_b, and R_f based on the MPEG coding characteristics. Let X be the number of forward predicted macroblocks, Y the number of backward predicted macroblocks, and Z the number of intra-coded blocks. Then $R_p = Z/X$, $R_b = Y/X$, and $R_f = 1/R_b$. Within the same scene, motion vector ratios tend to be similar. However, different scenes may have very different motion vector ratios. An adaptive local window threshold setting technique for detecting peaks in R_p, R_b, and R_f is proposed. Ratio values in a window of two GOPs are used in calculating a ratio histogram with 256 bins. If the peak-to-average ratio is greater than a pre-defined threshold T_d, then the peak frame is declared as a suspected scene change. Note that the peak values are not included in calculating the average and the typical value for T_d is 3. If a scene change happens on a B-frame, the ratio of the immediately subsequent B-frame will also be high. Both ratios will be considered as peaks and only the first B-frame will be marked as a suspected scene change. The summary of the method is as follows:

Detect R_p peaks in P-frames. Mark them as suspected shot boundaries.

Detect R_b peaks in B-frames. Mark them as suspected shot boundaries.

Detect R_f peaks in B-frames. Detect all $|\Delta\sigma^2|$ peaks in I- and P-frames. Mark I-frames as a suspected shot boundary if they have $|\Delta\sigma^2|$ peaks and the B-frames immediately before them have R_f peaks.

All the marked frames are then examined. If the difference between the current marked frame and the last shot boundary exceeds a threshold, then the current marked frame is a true shot boundary.

We should note some shortcomings of the techniques involving MPEG. The most severe one is the dependence of the algorithms on the characteristics of a particular encoder. Different MPEG encoders are more or less aggressive in their exploitation of motion compensation and can vary widely in the size of their GOP and selection of I-, P-, and B-frames. As a consequence, compressed domain shot boundary detection that works well with video produced by one MPEG encoder could require different parameters settings or might not even find the information it requires in video from another encoder.

Segmentation Based on Object Motion. Video data can also be segmented based on the analysis of encapsulated object motion [22]. The dynamic-scene analysis is based on recognizing the objects and then calculating their motion characteristics. Objects in video data are recognized either by their velocity or by their shape, size, and texture characteristics.

The motion which we perceive is the effect of both an object and camera motion. Camera motion information can be exploited in the scene segmentation process based on object motion. For example, if the camera is in motion, then the image has nonzero relative velocity with respect to the camera. The relative velocity can be defined in terms of the velocity of the point in the image and the distance from the camera. For segmenting moving-camera scenes, the component of the motion due to camera motion should be removed from the total motion of an object. Accumulative-difference pictures [18] can be used to find areas that are changing (usually due to object motion) and hence imply scene segmentation. Velocity components of the points in an image can be used for segmenting a scene into different moving objects. This is due to the fact that points of an object optimistically have the same velocity components under certain camera motions and image surfaces. Equation (7.47) can be used for calculating the image intensity velocity, where $I(x, y, t)$ is the image intensity, $\partial I/\partial x$ and $\partial I/\partial y$ denote the image intensity gradient, and dx/dt and dy/dt denote the image velocity:

$$dI/dt = \partial I/\partial x \times dx/dt + \partial I/\partial y \times dy/dt + \partial I/\partial t. \tag{7.47}$$

If the image intensity velocity changes abruptly, a shot boundary can be declared. This method is much more computationally intensive because the object must be traced through a sequence of frames.

Segmentation Based on Subband-Coded Video Data. Lee and Dickinson [21] have presented a scene detection algorithm where the temporal segmentation is applied on the lowest subband in subband-compressed video. Four different metrics for segmentation of video data have been developed as follows:

1. Difference of histograms: measures the absolute sum of the histograms of two frames. This metric is insensitive to object motion; however, it is sensitive to camera operations such as panning and zooming.
2. Histogram of difference frame: measures the histogram of the pixel to pixel difference frame. The degree of change is large if there are more pixels distributed away from the origin. It is more sensitive to object motion that the previous measure.
3. Block histogram difference: block histogram difference is obtained by computing histograms of each block and summing the absolute difference of these block histograms between the two frames. This metric is sensitive to local object motion.
4. Block variance difference: instead of using the histogram the variance of the block is used. Since it is a block-based metric, it is sensitive to local object motion.

Segmentation Based on Features. Work on feature-based segmentation is being done by Zabih et al. [48]. The segmentation process involves analyzing intensity edges between two consecutive frames. During cut and dissolve operations, new intensity edges appear far from the old ones due to change in content. In addition, old edges disappear far from the location of new edges. The authors define an edge pixel that appears far from an existing edge pixel as an *entering* edge pixel, and an edge pixel that disappears far from an existing edge pixel as an *exiting* edge pixel. By counting the entering and exiting edge pixels, cuts, fades, and dissolves are detected and classified. By analyzing the spatial distribution of entering and exiting edge pixels, wipes are detected and classified.

The algorithm takes as input two consecutive images I and I' and using an edge detection step, two binary images E and E' are obtained. Let ρ_{in} denote the fraction of edge pixels in E' which are more than a fixed distance r from the closest edge pixel in E. ρ_{in} measures the proportion of the entering edge pixels. It should assume a high value during a fade-in, or a cut, or at the end of a dissolve. Similarly, let ρ_{out} be the fraction of edge pixels in E which are farther than r away from the closest edge pixel in E'. ρ_{out} measures the proportion of exiting edge pixels. It should assume a high value during a fade-out, or a cut, or at the beginning of a dissolve.

The basic measure of similarity introduced is: $\rho = \max(\rho_{in}, \rho_{out})$. This represents the fraction of changed edges. Scene breaks can be detected by looking for peaks in ρ. Cuts are easy to distinguish, because a cut is the only scene break that occurs entirely between two consecutive frames. As a consequence, a cut will lead to a single isolated frame with a high value of ρ, while the other scene breaks will lead to an interval where ρ's value is elevated.

Fades and dissolves can be distinguished from each other by looking at the relative values of ρ_{in} and ρ_{out} in a local region. During a fade-in, ρ_{in} will be much higher than ρ_{out}, since there will be many entering edge pixels and few exiting edge pixels. Similarly, at a fade-out, ρ_{out} will be higher than ρ_{in}. A dissolve, on the other hand, consists of an overlapping fade-in and fade-out. During the dissolve, there is a initial peak in ρ_{in} followed by a peak in ρ_{out}.

During a wipe, each frame will have a portion of the old scene and a portion of the new scene. Between adjacent frames, a single strip of the image will change from the old scene to the new scene. For an horizontal wipe there is a vertical strip that passes either left–right or right–left, depending on the direction of the wipe. Since the between-scene transition occurs in this strip, the number of edge pixels that either enter or exit should be higher inside the strip and lower in the other areas of the image. The edge pixel that is either entering or exiting is called *changing* pixel. When computing ρ the location of the changing edge pixels can be recorded and their spatial distribution analyzed. There are different types of wipes that are detected. For vertical and horizontal wipes, the percentage of changing pixels in the top half and the left half of the image is calculated. For a left-to-right horizontal wipe, the majority changing pixels will occur in the left of the image during the first half of the wipe, then in the right half of the images during the rest of the wipe. The other cases can be handled similarly.

Motion compensation is also provided in the algorithms to give better results in the presence of camera and object motion. Shen et al. [37] applied this technique to MPEG sequences using a multi-level Hausdorff distance histogram.

Video Shot Representation. Various researchers observed that the representation of a video shot as a single still image is a significant step towards retrieving pieces of video from large collections of video clips. If there is no independent motion (only camera motion is presented) in a shot, the transformation between subsequent frames can be determined, and the frames can be "pasted" together into a larger image, possibly of higher resolution. This technique is called mosaicking [33]. On the other hand, if there is independent motion, the moving objects must be separated from the background, and the image of the background can still be reconstructed along with images and trajectories of moving objects. Techniques proposed for obtaining such still images are salient stills [38, 41], mosaicking [33, 39] and video space icons [42]. One of the reasons for this extracting process is that once a complete shot is transformed into a single still image, then all image database browsing and retrieval techniques can be applied. Along with providing the information about this static image (color, texture, shape) this process also provides information about the camera motion and independent object motions, which permits motion queries [33].

A simpler way to obtain a single image to represent a shot is to select a keyframe or, if there is much motion and action in the shot, to select multiple keyframes [44, 51]. In this way the problem of video query is reduced to the problem of image query. However, in this case usually one will obtain a large collection of shots from a normal video clip. Thus, searching large video databases may be transformed to searching large image databases.

In summary, video search requires more information to be used than just static images derived from video. In the case of video, the search has to go beyond the static features of images and incorporate the dynamics within a shot and between the shots.

7.2.2 Video Stories

During a film, a story is told using the presentation of the images in time; the sequence of such a presentation forms the montage or the edit of the film. Technically, [1] the montage is "the most essential characteristic of the motion picture. Montage refers to the editing of the film, the cutting and piecing together of exposed film in a manner that best conveys the intent of the work." Properly edited video has a continuity of meaning that overwhelms the inherent discontinuity of presentation. The idea is that intrinsically a film has two levels of discontinuity. One is the frame-to-frame discontinuity which is mostly imperceptible; the other one is the shot-to-shot transition which can be easily detected and percepted. In essence, a shot in a video is very much like a sentence in a piece of text; it has some semantic meaning, but taken out of context, it may be ambiguous. Thus, there is a need to concatenate groups of shots in order to form a three-dimensional event which is continuous in time. This concatenation forms a story unit which often represents a time and place in the narration. The continuity in time is not as important as the continuity in meaning. Therefore, one important challenging problem of video analysis is to find the discontinuities in meaning or, equivalently, establishing from shot to shot whether there is a continuity in the subject of video.

Some work towards such achievements was carried out in the HyperCafe Project [34]. The authors described a formal approach for the design of video-centric hypermedia applications. Their framework defines a way to structure video-based scenes into narrative sequences that can be dynamically presented as spatial and temporal opportunities. Some other efforts were directed towards creation of storyboards [7, 31, 47]. Here, high-level video structure is extracted and presented to the user. In such conditions, the user can conveniently browse through the video on a story-unit basis. Typically, there is an order of magnitude reduction from the number of shots to the number of scenes [47]. This represents a significant reduction of information to be conveyed or presented to the user, thus making it easy for the user to identify segments of interest.

Furthermore, the ability to automatically label a story unit as a "dialogue" or "action" means that one can query on the basis of such semantic characteristics. Video summaries can be labeled with such semantic descriptions, allowing video content viewing and fast random access.

7.2.3 Video Parsing

Video parsing has been successfully applied to recover individual news items from news programs [50]. A priori models are constructed through the use of state transitions, where each state corresponds to a phase of a news broadcast such as news anchor speaking, a reporter speaking and a weather forecast. In the recognition stage, visual and temporal elements are used together with domain knowledge of spatial layouts of objects.

After the different items of the news are recovered the user can view any part of a news program. This offers a high-level random access into a different segment of the news according to what the user is interested in. If he is interested only in the weather forecast he can directly jump into the segment corresponding to weather forecast without having to look into all of the video. In work by Mohan [27] a shot-boundary detection algorithm in combination with the detection of audio silences is used to segment the news into individual news items. Additionally, the closed caption text is synchronized with the visual news item and used for text-based retrieval of the news items.

A different approach is used in the system by Shahraray and Gibbon [35]. The authors use the closed caption information for news archive retrieval. The pictorial transcript system they developed transcribes news programs into HTML-based summaries. Each HTML page contains several news items, each being represented by keyframes with detailed text derived from closed captions.

7.2.4 Video Summaries

Visual queries give the end user the opportunity to have a quick idea of what a particular candidate is about. The summaries allow the user to refine his precedent query or to have an idea whether the query has been posed correctly. Using a summary the user can have rapid non-linear access in viewing the video and can get an idea of the visual content.

In Yeung et al. [45] a compact representation of video content called *scene-transition graph* can be built on clusters of similar video shots. This representation is a form of visual summary that maps a sequence of video shots onto a two-dimensional display on the screen. The graph representation has nodes, which capture the temporal video content information, and edges, which depict the temporal flow of the story.

Another way to provide the user with a video summary is to emulate a video player that displays some extracted keyframes. In Ding et al. [11], they propose a *slide show interface* that flips through computer-selected video keyframes in a temporally ordered sequence, displaying the "slides" at rates set by the user. A similar approach is described in the work by DeMethod et al. [10]. Their interface allows the user to change the summarization level on the fly. They also provide a way to automatically select the summarization level that provides a concise and representative set of keyframes.

Another form of visual summary is the so called *pictorial summary* [46]. The pictorial summary consists of a sequence of representative images arranged in temporal order. Each representative image, called a video poster, consists of one or more subimages, each of which represents a unique "dramatic element" of the underlying story, such that the layout of the subimages emphasizes the dramatic "importance" of individual story units. Note that in this case a kind of semantic VCR is emulated. The pictorial summary is thus a general scheme for visual presentation, and each video poster visually summarizes the dramatic events taking place in a meaningful story unit of the story.

7.2.5 Automatic Attributes Generation for Video

The features that are extracted by the described algorithms can provide high-level annotation of video. The annotations form the attributes that are attached to the different levels of the hierarchy of video representation described in Section 7.1.3. For a video sequence the story units partition the video into a small number of meaningful units, each representing a dramatic event. The number of dialogs and their duration can be an example of attributes for story units. At the lowest level of the hierarchy will be the attributes of each shot. The temporal ones can be derived from motion analysis, object track- ing, color changes, etc., while the static ones include the properties of the shot itself (duration) and those of its images (keyframes or mosaics) such as color histograms, texture patterns, etc.

Beyond visual features, the information from the audio track, biblio- graphic data and closed caption can also provide textual annotation. For example audio analysis can be used to classify the audio information of a video (silence, speech, music).

7.3 Discussion

A framework for video retrieval and query formulation that is based on a se- quence of categorization, searching, browsing, and viewing has been proposed. This framework considers that a video query system should incorporate cer- tain capabilities such as the search on dynamic visual properties of video and the ability for rapid nonlinear viewing of video. To achieve these capabili- ties, algorithms have to be developed for automatic extraction of dynamic visual video properties and for processing of video to extract visual features for compact presentation of the video content. Such representations facilitate nonlinear access into video and give quick views of the visual content of video.

Several techniques for detecting shot boundaries were discussed. Some of these techniques are suitable for detecting abrupt transitions where others were aimed at detecting gradual transitions. Simple methods based on pixel differences can be successfully used for detecting cuts only for particular videos where the shot boundaries are clearly delimited. In practice, these methods are not likely to work properly. Camera movement, e.g. pan or zoom, fast moving objects, illumination changes, etc., can change a large number of pixels and consequently can result in detection of a false shot break. Moreover, these techniques are based on a single threshold and therefore, lack the power of detecting gradual transitions. The twin comparison technique was used for detecting gradual transitions but this method cannot be used to detect the type of the transition. In order to achieve this, different transition types such as cuts, fades, wipes, and dissolves were modeled by mathematical functions. However, these models do not take into account perturbing factors such as object and camera motion. Optical flow techniques are utilized for detecting the camera motion. Optical flow works well but may not distinguish between object and camera motion without understanding the content of the scene. Velocity components of the points in an image can be used for segmenting a scene into different moving objects. Using the fact that neighborhood-based color ratios are invariant under various changes in view condition, including both spectral and intensity variations of the illuminant, a window-based video partitioning method based on color ratio was also proposed. Several other techniques were used for detecting shot boundaries in the compressed domain. A problem here is the dependence of the algorithms on the characteristics of a particular encoder. As a consequence, compressed domain shot boundary detection that works well with video produced by one MPEG encoder could require different parameter settings or might not even find the information it requires in video from a different encoder.

An important aspect is the evaluation of the search results of a video query. Accurate content processing and annotation, whether it is automatic or manual, may not result in the retrieval of the desired segments. Different types of queries can be expected from different types of end users: one who may have seen a particular piece of video and wants to find it or may not have seen it but knows that the video exists; and one who may not have seen the piece but has some vague idea what he or she is looking for. In either case, one could define what it would mean for a retrieved video to be appropriate to the query. For the first case, a candidate can be appropriate only if it is the particular video clip or if the candidate has the clip embedded in it in some form. For the second case, the concept of appropriateness is less clear, and the user must decide if the candidate is what he or she is interested in.

The evaluation of the performance of a video retrieval system is made difficult by the interactiveness of the query process. The performance has to be measured not only in terms of recall and precision, but also in terms of the average amount of time required to select the segments of interest. The discipline of text retrieval has already given some valuable thought to these issues, which will be useful for video-query performance evaluation. Nonetheless, the measures for performance of video query are still open research issues.

The audiovisual features, together with textual descriptions, are integral components of video queries. Analysis of visual content is one step beyond the traditional database query approach. Analysis of audio features is another step forward, and the integration of the available multiple-media features is the ultimate step toward the successful deployment of video queries in multimedia database systems.

References

1. "Britannica Online" http://www.eb.com:180.
2. Adjeroh, D, Lee, M, and Orji, C, "Techniques for Fast Partitioning of Compressed and Uncompressed Video," Multimed Tools Applic, 4(3), pp. 225–243, 1997.
3. Ahanger, A and Little, T, "A Survey of Technologies for Parsing and Indexing Digital Video," J Visual Commun Image Represent, 7(1), pp. 28–43, 1996.
4. Aigrain, P and Joly, P, "The Automatic Real-Time Analysis of Film Editing and Transition Effects and Its Applications," Computers Graphics, 18(1), pp. 93–103, 1994.
5. Arman, F, Hsu, A, and Chiu, M, "Image Processing on Encoded Video Sequences," Multimed Syst, pp. 211–219, 1994.
6. Bolle, R, Yeo, B, and Yeung, M, "Video Query: Research Directions," Multimed Syst, 42(2), pp. 233–252, 1998.

7. Chen, J, Taskiran, C, Delp, E, and Bouman, C, "ViBE: A New Paradigm for Video Database Browsing and Search," IEEE Workshop on Content-Based Access of Image and Video Database, 1998.
8. Corridoni, JM and Del Bimbo, A, "Structured Digital Video Indexing," Int. Conf. Pattern Recognition, 1996.
9. Davenport, G, Smith, TA, and Pincever, N, "Cinematic Primitives for Multimedia," IEEE Computer Graphics Applic, 6(5), pp. 67–74, 1991.
10. DeMethon, D, Kobla, V, and Doermann, D, "Video Summarization by Curve Simplification," ACM Multimed, 1998.
11. Ding, W, Marchionini, G, and Tse, T, "Previewing Video Data: Browsing Key Frames at High Rates," Int. Symp. on Digital Libraries, 1997.
12. Flicker, M, Sawhney, H, Niblack, W, Ashley, J, Huang, Q, Dom, B, Gorkani, M, Hafner, J, Lee, D, Petkovic, D, Steele, D, and Yanker, P, "Query by Image and Video Content: The QBIC System," IEEE Computer, 28(9), pp. 23–32, 1995.
13. Gauch, S, Gauch, J, and Pua, K, "VISION: A Digital Video Library," ACM Digital Libraries, 1996.
14. Hampapur, A, Digital Video Indexing in Video Databases, PhD Thesis, University of Michigan, 1994.
15. Hampapur, A, Jain, R, and Weymouth, T, "Digital Video Segmentation," ACM Multimedia, pp. 357–364, 1994.
16. Idris, F and Panchanathan, S, "Indexing of Compressed Video Sequences," Proc. SPIE 2670: Storage and Retrieval for Image and Video Databases IV, pp. 247–253, 1996.
17. Idris, F and Panchanathan, S, "Review of Image and Video Indexing Techniques," J Visual Commun Image Represent, 8(2), pp. 146–166, 1997.
18. Jain, RC, "Segmentation of Frame Sequence Obtained by a Moving Observer," IEEE Trans Patt Anal Mach Intell, 6(5), pp. 624–629, 1984.
19. Jain, R and Hampapur, A, "Metadata in Video Databases," ACM SIGMOD, 23(4), 1994.
20. Kuo, T, Lin, Y, Chen, A, Chen, S, and Ni, C, "Efficient Shot Change Detection on Compressed Video Data," Proc. Int. Workshop on Multimedia Database Management Systems, 1996.
21. Lee, J and Dickinson, BW, "Multiresolution Video for Subband Coded Video Databases," Proc. SPIE: Storage and Retrieval for Image and Video Databases, 1994.
22. Lee, SY and Kao, HM, "Video Indexing: An Approach Based on Moving Object and Track," Proc. SPIE 1908: Storage and Retrieval for Image and Video Databases, pp. 25–36, 1993.
23. Little, T, Ahanger, G, Folz, R, Gibbon, J, Reeves, F, Schelleng, D, and Venkatesh, D, "A Digital On-Demand Video Service Supporting Content-based Queries," ACM Multimedia, pp. 427–433, 1993.
24. Liu, HC and Zick, GL "Scene Decomposition of MPEG Compressed Video," Proc. SPIE 2419: Digital Video Compression, pp. 26–37, 1995.
25. Mandal, M, Idris, F, and Panchanathan, S, "A Critical Evaluation of Image and Video Indexing Techniques in the Compressed Domain," Image Vision Comput, 17, pp. 513–529, 1999.
26. Meng, J, Juan, Y, and Chang, SF, "Scene Change Detection in a MPEG Compressed Video Sequence," Proc. SPIE 2419: Digital Video Compression, pp. 267–272, 1995.
27. Mohan, R, "Text Based Indexing of TV News Stories," Proc. SPIE 2916: Multimedia Storage and Archiving Systems, 1996.

28. Nagasaka, A and Tanaka, Y, "Automatic Video Indexing and Full-Video Search for Object Appearances," Visual Database Systems II, pp. 113–127, 1992.
29. Otsuji, K, Tonomura, Y, and Ohba, Y, "Video Browsing using Brightness Data," Proc SPIE 1606: Visual Communications and Image Processing, pp. 980–989, 1991.
30. Patel, N and Sethi, I, "Video Shot Detection and Characterization for Video Databases," Patt Recogn, 30(4), pp. 583–592, 1997.
31. Poncelon, D, Srinivasan, S, Amir, A, and Petkovic, D, "Key to Effective Video Retrieval: Effective Cataloging and Browsing," ACM Multimed, 1998.
32. Rowe, LA, Boreczky, JS, and Eads, CA, "Indices for User Access to Large Video Databases," Proc. SPIE 2185: Storage and Retrieval for Image and Video Databases II, 1994.
33. Sawhney, H and Ayer, S, "Compact Representations of Videos Through Dominant and Multiple Motion Estimation," IEEE Trans Patt Anal Mach Intell, 18(8), pp. 814–830, 1996.
34. Sawhney, N, Balcom, D, and Smith, I, "Authoring and Navigating Video in Space and Time," IEEE Multimed J, 1997.
35. Shahraray, B and Gibbon, D, "Automatic Generation of Pictorial Transcripts of Video Programs," Proc. SPIE 2417: Multimedia Computing and Networking, 1995.
36. Shahraray, B, "Scene Change Detection and Content-Based Sampling of Video Sequence," Proc. SPIE 2419: Digital Video Compression, pp. 2–13, 1995.
37. Shen, B, Li, D, and Sethi, IK, "Cut Detection via Compressed Domain Edge Extraction," IEEE Workshop on Nonlinear Signal and Image Processing, 1997.
38. Szeliski, R, "Image Mosaicking for Tele-Reality Applications," ACM Multimedia, 1993.
39. Taniguchi, Y, Akutsu, A, and Tonomura, Y, "PanoramaExcerpts: Extracting and Packing Panoramas for Video Browsing," ACM Multimedia, 1997.
40. Taskiran, C and Delp, E, "Video Scene Change Detection Using the Generalized Trace," IEEE Int. Conf. on Acoustic, Speech and Signal Processing, pp. 2961–2964, 1998.
41. Teodosio, L and Bender, W, "Salient Video Stills: Content and Context Preserved," TR. DEC Cambridge Research Laboratory, 1994.
42. Tonomura, Y, Akutsu, A, Otsuji, K, and Sadakata, T, "VideospaceIcon: Tools for Anatomizing Video Content," ACM INTERCHI, 1993.
43. Yeo, BL and Liu, B, "Rapid Scene Analysis on Compressed Video," IEEE Trans Circuits Syst Video Technol, 5(6), pp. 533–544, 1995.
44. Yeung, MM and Liu, B, "Efficient Matching and Clustering of Video Shots," International Conf. Image Processing, pp. 338–341, 1995.
45. Yeung, MM, Yeo, B, Wolf, W, and Liu, B, "Video Browsing Using Clustering and Scene Transitions on Compressed Sequences," Proc. SPIE 2417: Multimedia Computing and Networking, pp. 399–413, 1995.
46. Yeung, MM and Yeo, B, "Video Visualization for Compact Presentation of Pictorial Content," IEEE Trans Circuits Syst Video Technol, 7(5), pp. 771–785, 1997.
47. Yeung, MM and Yeo, B, "Video Content Characterization and Compaction for Digital Libraries," Proc. SPIE 3022: Storage and Retrieval for Image and Video Databases V, pp. 45–58, 1997.
48. Zabih, R, Miller, J, and Mai, K, "A Feature-Based Algorithm for Detecting and Classifying Scene Breaks," ACM Multimed, 1995.
49. Zhang, HJ, Kankanhalli, A, and Smoliar, SW, "Automatic Partitioning of Full-Motion Video," Multimed Syst, 3(1), pp. 10–28, 1993,

50. Zhang, HJ, Tan, Y, Smoliar, SW, and Gong, Y, "Automatic Parsing and Indexing of News Video," Multimed Syst, 6(2), pp. 256–266, 1995.
51. Zhang, HJ, Low, CY, and Smoliar, SW, "Video Parsing and Browsing using Compressed Data," Multimed Tools Applic, 1, pp. 89–111, 1995.

PART II
Advanced Topics

8. Query Languages for Multimedia Search

Shi-Kuo Chang and Erland Jungert

8.1 Introduction

For multimedia databases, there are not only different media types, but also different ways to query the databases. The query mechanisms may include free text search, SQL-like querying, icon-based techniques, querying based upon the entity-relationship (ER) diagram, content-based querying, sound-based querying, as well as virtual-reality (VR) techniques. We will first discuss query languages for multimedia databases with emphasis on visual query languages, and then survey recent advances in query languages for heterogeneous multimedia databases and data sources. A powerful spatial/temporal query language called ΣQL is described, and we will illustrate this approach by data fusion examples.

With the rapid expansion of the wired and wireless networks, a large number of soft real-time, hard real-time and non-real-time sources of information need to be processed, checked for consistency, structured and distributed to the various agencies and people involved in an application [15]. In addition to spatial/temporal multimedia databases, it is also anticipated that numerous websites on the World Wide Web will become rich sources of spatial/temporal multimedia information. The retrieval and fusion of spatial/temporal multimedia information from diversified sources calls for the design of spatial/temporal query languages capable of dealing with both multiple data sources and databases in a heterogeneous information system environment.

Powerful query languages for multiple data sources and databases are needed in applications such as emergency management (fire, flood, earthquake, etc.), tele-medicine, digital library, community network (crime prevention, child care, senior citizens care, etc.), military reconnaissance and scientific exploration (field computing). These applications share the common characteristics that information from multiple sources and databases must be integrated. A typical scenario for information fusion in emergency management may involve a live report from a human observer, data collected by a heat sensor, video signal from a camera mounted on a helicopter, etc. Current systems often have preprogrammed, fixed scenarios. In order to enable the end user to effectively retrieve spatial/temporal multimedia information and to discover relevant associations among media objects, a flexible spatial/temporal multimedia query language for multiple data sources and databases should be provided.

Data sources such as cameras, sensors, or signal generators usually provide continuous streams of data. Such data need to be transformed into abstracted information, i.e., into various forms of spatial/temporal/logical data structures, so that the processing, consistency analysis, and fusion of data become possible. The abstracted information does not necessarily represent different levels of knowledge and can be various representations of common knowledge and therefore needs to be integrated and transformed into fused knowledge that is the common knowledge derivable from, and consistent with, the various abstractions.

As an example, information items such as time and number of people can be extracted from multiple data sources such as image, scenario document, and sound [27]. Cooperation among various media is carried out by exchanging these information items. In an experimental study involving a TV drama scene, this approach could successfully realize good synchronization between image, scenario, and sound, and moreover could also perform personal character identification [27].

As a second example, a recent study first identified known people's names from the text, and then tried to detect corresponding faces from the video stream [24]. As a third example, a video camera is a data source that generates video data. Such video data can be transformed into various forms of abstracted representations including: text, keyword, assertions, time sequences of frames, qualitative spatial description of shapes, frame strings, and projection strings [13]. To describe a frame containing two objects a and b, the text is a *is to the northwest of* b, the keywords are $\{a, b\}$, and the assertion is (a northwest b). The x-directional projection string is ($u : a < b$). The time sequence of three frames C_{t_1}, C_{t_2}, C_{t_3}, is ($t : C_{t_1} < C_{t_2} < C_{t_3}$). (These representations will be explained later in Section 8.4).

We will first discuss query languages for multimedia databases in Section 8.2. Then we will survey recent advances in query languages for heterogeneous multimedia databases and data sources in Section 8.3. To support the retrieval and fusion of multimedia information from multiple sources and databases, a powerful spatial/temporal query language called ΣQL will be described in Section 8.4. ΣQL is based upon the σ-operator sequence and in practice expressible in an SQL-like syntax. The natural extension of SQL to ΣQL allows a user to specify powerful spatial/temporal queries for both multimedia data sources and multimedia databases, eliminating the need to write different queries for each. A ΣQL query can be processed in the most effective manner by first selecting the suitable transformations of multimedia data to derive the multimedia static schema, and then processing the query with respect to this multimedia static schema.

8.2 Query Languages for Multimedia Databases

Multimedia databases, when compared to traditional databases, have the following special requirements [18, 19]:

- The size of the data items may be very large. The management of multimedia information therefore requires the accessing and manipulation of very large data items.
- Storage and delivery of video data requires guaranteed and synchronized delivery of data.
- Various query mechanisms need to be combined, and the user interface should be highly visual and should also enable visual relevance feedback and user-guided navigation.
- The on-line, real-time processing of large volumes of data may be required for some types of multimedia databases.

For multimedia databases, there are not only different media types, but also different ways to query the databases. The query mechanisms may include free text search, SQL-like querying, icon-based techniques, querying based upon the entity-relationship (ER) diagram, content-based querying, sound-based querying, as well as virtual-reality (VR) techniques.

Let us take the retrieval of books from a library as an example. Some of the above-described query mechanisms are based upon traditional approaches, such as free text search (**"retrieve all books authored by Einstein"**) and SQL query language (**"SELECT title FROM books WHERE author = 'Einstein'"**). Some are developed in response to the special needs of multimedia databases such as content-based querying for image/video databases (**"find all books containing the picture of Einstein"**) [17], and sound-based queries that are spoken rather than written or drawn [28]. Some are dictated by new software/hardware technologies, such as icon-based queries that use icons to denote query targets (**"a book icon"**) and objects (**"a sketch of Einstein"**), and virtual reality queries where the query targets and objects can be directly manipulated in a virtual reality environment (**"the book shelves in a Virtual Library"**).

Except for the traditional approaches and those relying on sound, the other techniques share the common characteristic of being highly visual. Therefore, in the rest of this section we will concern ourselves mainly with multiparadigmatic visual interfaces to accessing multimedia documents. The query languages to deal with heterogeneous databases and data sources will be discussed in the next section.

A visual interface to multimedia databases, in general, must support some type of visual querying language. Visual query languages (VQLs) are query languages based on the use of visual representations to depict the domain of interest and express the related requests. Systems implementing a visual query language are called Visual Query Systems (VQSs) [2, 4]. They include both a language to express the queries in a pictorial form and a variety of functionalities to facilitate human–computer interaction. As such, they are oriented to a wide spectrum of users, ranging from people with limited technical skills to highly sophisticated specialists.

In recent years, many VQSs have been proposed, adopting a range of different visual representations and interaction strategies. Most existing VQSs restrict human–computer interaction to only one kind of interaction paradigm. However, the presence of several paradigms, each one with different characteristics and advantages, will help both naive and experienced users in interacting with the system. For instance, icons may well evoke the objects present in the database, while relationships among them may be better expressed through the edges of a graph, and collections of instances may be easily clustered into a form. Moreover, the user is not required to adapt his perception of the reality of interest to the different views presented by the various data models and interfaces.

The way in which the query is expressed depends on the visual representations as well. In fact, in the existing VQSs, icons are typically combined following a spatial syntax [8], while queries on diagrammatic representations are mainly expressed by following links and forms are often filled with prototypical values. Moreover, the same interface can offer to the user different interaction mechanisms for expressing a query, depending on both the experience of the user and the kind of the query itself [5].

To effectively and efficiently access information from multimedia databases, we can identify the following design criteria for the user interface: (1) various query mechanisms need to be combined seamlessly, (2) the user interface should be visual as much as possible, (3) the user interface should enable visual relevance feedback, (4) the user interface should support user-guided navigation, and (5) the user interface should facilitate the user-controlled discovery of new associations among media objects.

8.3 Query Languages for Heterogeneous Multimedia Databases

Query language for heterogeneous multimedia databases is a new research area and therefore the body of related work only just begins to grow. There has been substantial research on query languages for images and spatial objects, and a survey can be found in [12]. Of these query languages, many are based upon extension of SQL, such as PSQL [26] and Spatial SQL [16].

After the SQL-like query languages, next come video query languages where the focus is shifted to temporal constraints [1] and content-based retrieval [9]. Recent efforts begin to address query languages involving images, video, audio and text. Vazirgiannis describes a multimedia database system for multimedia objects that may originate from sources such as text, image,

Table 8.1 Multimedia query languages and their characteristics

Query language name	Query language type	Reference	Application area	Special features and operators
GeoAnchor	Hypermap query	[3] (see Sect. 8.5)	Hypermapped GIS	Semantic filtering operators
MOQL	Object query language	[24]	Multimedia database	Temporal and spatial operators
PSQL	Extended SQL	[26]	Pictorial database	Spatial operators
QBIC	Query by color patches and texture patterns	[18]	Content-based image retrieval	Color and texture matching operators
QL/G	SQL-like	[6]	Geometric database	Geometric operators
Spatial SQL	Extended SQL	[16]	GIS	Spatial operators
Video SQL	Extended SQL	[25]	Video database	Emphasis on presentation
Visual SEEk	Query by color sketches	[9]	Content-based video retrieval	Similarity operators

video, [29]. The query language QL/G developed by Chan and Zhu supports the querying of geometric databases and is applicable to both geometric and text data [6], but does not handle temporal constraints. In [25], Oomoto and Tanaka describes an SQL-like query language for video databases, where the emphasis is more on presentation, rather than on retrieval.

A multimedia object query language MOQL that extends the object query language OQL is reported in Li et al. [24]. An interoperable multi-database platform in a client/server environment using a common object model is described in Xu et al. [31], which can provide inter-operations between popular database systems. A related approach is to provide a database integrator (DBI) for customers who have data stored in multiple data sources, typically heterogeneous and/or non-relational, and want to view those data sources as a single logical database from the data and/or metadata perspective [20].

Table 8.1 summarizes some representative multimedia/spatial query languages, to illustrate the different types of query languages, their application areas, and most importantly, their special features and operators. The intent of Table 8.1 is not to give an exhaustive survey, but to illustrate the diversity of databases/data sources, and the diversity of operators.

8.4 The σ-Query Language

While the above-described approaches each address some important issues, there is a lack of unified treatment of queries that can deal with both spatial and temporal constraints from both live data sources and stored databases. Since the underlying databases are complex, the user also needs to write complicated queries to integrate multimedia information. The proposed approach differs from the above in the introduction of a general powerful operator called

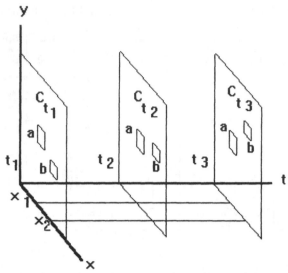

Fig. 8.1 Example of extracting three time slices (frames) from a video source.

the σ-operator, so that the corresponding query language can be based upon σ-operator sequences. The rest of the chapter is organized as follows. The basic concepts of the σ-query are explained in Section 8.4.1. Section 8.4.2 introduces elements of symbolic projection theory and the general σ-operator, and Section 8.4.3 describes the SQL query language. An illustration of data fusion using the σ-query is presented in Section 8.4.4. Extension to hyper-mapped virtual world query is discussed in Section 8.5.

8.4.1 Basic Concepts

As mentioned in Section 8.1, the σ-query language is a spatial/temporal query language for information retrieval from multiple sources and databases. Its strength is its simplicity: the query language is based upon a single operator – the σ-operator. Yet the concept is natural and can easily be mapped into an SQL-like query language. The σ-query language is useful in theoretical investigation, while the SQL-like query language is easy to implement and is a step towards a user-friendly visual query language. An example is illustrated in Fig. 8.1. The source R, also called a universe, consists of time slices of 2D frames. To extract three predetermined time slices from the source R, the query in mathematical notation is: $\sigma_t(t_1, t_2, t_3)R$.

The meaning of the σ-operator in the above query is select, i.e., we want to select the time axis and three slices along this axis. The subscript t in σ_t indicates the selection of the time axis. In the SQL-like language a ΣQL query is expressed as:

SELECT t
CLUSTER t_1, t_2, t_3
FROM R

A new keyword "CLUSTER" is introduced, so that the parameters for the σ-operator can be listed, such as t_1, t_2, t_3. The word "CLUSTER" indicates that objects belonging to the same cluster must share some common characteristics (such as having the same time parameter value). A cluster may have a sub-structure specified in another (recursive) query. Clustering is a natural concept when dealing with spatial/temporal objects. The mechanism for clustering will be discussed further in Section 8.3. The result of a ΣQL query is a string that describes the relationships among the clusters. This string is called a *cluster string*, which will also be discussed further in Section 8.4.2.

A cluster is a collection of objects sharing some common characteristics. The SELECT-CLUSTER pair of keywords in ΣQL is a natural extension of the SELECT keyword in SQL. In fact, in SQL implicitly each attribute is considered as a different axis. The selection of the attributes' axes defines the default clusters as those sharing common attribute values. As an example, the following ΣQL query is equivalent to an SQL query to select attributes' axes "sname" and "status" from the suppliers in Paris.

SELECT sname, status
CLUSTER *
FROM supplier
WHERE city = "Paris"

In the above ΣQL query, the * indicates any possible values for the dimensions sname and status. Since no clustering mechanism is indicated after the CLUSTER keyword the default clustering is assumed. Thus by adding the "CLUSTER *" clause, every SQL query can be expressed as a ΣQL query.

Each cluster can be open (with objects inside visible) or closed (with objects inside not visible). The notation is t_2^o for an open cluster and t_2^c or simply no superscript for a closed cluster. In the ΣQL language the keyword "OPEN" is used:

SELECT t
CLUSTER t_1, OPEN t_2, t_3
FROM R

With the notation described above, it is quite easy to express a complex, recursive query. For example, to find the spatial relationship of two objects a and b from the three time slices of a source R, as illustrated in Fig. 8.1, the ΣQL query in mathematical notation is:

$$\sigma_x(x_1, x_2)(\sigma_t(t_1^o, t_2^o, t_3^o)R).$$

In the ΣQL language the query can be expressed as:

SELECT x
CLUSTER x_1, x_2
FROM
    SELECT t
    CLUSTER OPEN t_1 , OPEN t_2 , OPEN t_3
    FROM R

The query result is a cluster string describing the spatial/temporal relationship between the objects a and b. How to express this spatial/temporal relationship depends upon the (spatial) data structure used. In the next section we explain symbolic projection as a means to express spatial/temporal relationships.

8.4.2 A General σ-Operator for σ-Queries

As mentioned above, the ΣQL query language is based upon a single operator – the σ-operator – which utilizes symbolic projection to express the spatial/temporal relationships in query processing. In the following, symbolic projection, the cutting mechanism and the general σ-operator are explained, which together constitute the theoretical underpinnings of ΣQL.

Symbolic projection [10, 23] is a formalism where space is represented as a set of strings. Each string is a formal description of space or time, including all existing objects and their relative positions viewed along the corresponding coordinate axis of the string. This representation is qualitative because it mainly describes sequences of projected objects and their relative positions. We can use symbolic projection as a means for expressing the spatial/temporal relationships extracted by a spatial/temporal query.

Continuing the example illustrated by Fig. 8.1, for time slice C_{t_1} its x-projection using the fundamental symbolic projection is:

$$\sigma_x(x_1, x_2)C_{t_1} = (u : C_{x_1,t_1} < C_{x_2,t_1})$$

and its y-projection is:

$$\sigma_y(y_1, y_2)C_{t_1} = (u : C_{y_1,t_1} < C_{y_2,t_1}).$$

In the above example, a time slice is represented by a cluster C_{t_1} containing objects with the same time attribute value t_1. A cluster string is a string composed from cluster identifiers and relational operators. The single cluster C_{t_1} is considered a degenerated cluster string. After the σ_y operator is applied, the resulting cluster C_{y_1,t_1} contains objects with the same time and space attribute values. In the above example, the cluster string $(v : C_{y_1,t_1} < C_{y_2,t_1})$ has the optional parentheses and projection variable $Òv : Ó$ to emphasize the direction of projection.

The query $\sigma_t(t_1, t_2, t_3)R$ yields the following cluster string α:

$$\alpha = (t : C_{t_1} < C_{t_2} < C_{t_3}).$$

When another operator is applied, it is applied to the clusters in a cluster string. Thus the query $\sigma_x(x_1, x_2)\sigma_t(t_1^o, t_2^o, t_3^o)R$ yields the following cluster string β:

$$\beta = (t : (u : C_{x_1,t_1} < C_{x_2,t_1}) < (u : C_{x_1,t_2} < C_{x_2,t_2}) < (u : C_{x_1,t_3} < C_{x_2,t_3})).$$

The above cluster string β needs to be transformed so that the relationships among the objects become directly visible. This calls for the use of a *materialization function* MAT to map clusters to objects. Since $C_{x_1,t_1} = C_{x_1,t_2} = C_{x_1,t_3} = \{a\}$ and $C_{x_2,t_1} = C_{x_2,t_2} = C_{x_2,t_3} = \{b\}$, the materialization MAT(β) of the above cluster-string yields:

$$\text{MAT}(\beta) = (t : (u : a < b) < (u : a < b) < (u : a < b)).$$

Returning now to the ΣQL query that is equivalent to an SQL query to select attributes (i.e., axes) "sname" and "status" from the suppliers in Paris.

SELECT sname, status
CLUSTER *
FROM supplier
WHERE city = "Paris"

The result of the above query is a cluster string α that describes the relationships among the clusters. Since each cluster corresponds to a unique (sname, status) pair, the query result α is:

$$\alpha = C_{sname1,status1} > C_{sname2,status2} > \cdots > C_{sname-n,status-n}$$

where $>$ denotes an ordering relation. When this cluster string α is materialized into objects using a materialization function MAT_R, the result $\text{MAT}_R(\alpha)$ is an ordered list of (sname, status) pairs from suppliers in Paris.

The query result in general depends upon the clustering that in turn depends upon the cutting mechanism. The cutting is an important part of symbolic projection because a cutting determines both how to project and also the relationships among the objects or partial objects in either side of the cutting line. In most of the examples presented in this chapter, the cuttings are ordered lists that are made in accordance with the fundamental symbolic projection. The cutting type, κ-type, determines which particular cutting mechanism should be applied in processing a particular σ-query.

The general σ-operator is defined by the following expression where, in order to make different cutting mechanisms available, the cutting mechanism κ-type is explicitly included:

$$\sigma_{axes,\kappa-type}^{\sigma-type}(\text{clusters})_\varphi < \text{cluster string} >= stype :< \text{cluster string} > .$$

The general σ-operator is of the type $\sigma\text{-}type$ and selects an *axis* or multiple *axes*, followed by a cutting mechanism of the type κ-type on (clusters)$_\varphi$ where φ is a predicate that objects in the clusters must satisfy. The σ-operator operates on a cluster string that either describes a data source (e.g., data from a specified sensor) or is the result of another σ-operator. The result of the σ-operator is another cluster string of type *stype*. Since the result of the σ-operator is always a cluster string, a materialization operator MAT is needed to transform the cluster string into real-world objects and their relationships for presentation to the user.

8.4.3 The ΣQL Query Language

ΣQL is an extension of SQL to the case of multimedia sources. In fact, it is able to query seamlessly traditional relational tables and multimedia sources and their combination. The ΣQL query language operates on the extended *multimedia static structure* MSS [7]. The syntax of ΣQL can be presented in BNF notation:

```
<query>::= <select_type> <dimension_list>
           CLUSTER <cluster_type> <cluster_values>
           FROM <source>
           WHERE <condition>
           PRESENT <presentation_description>
<select_type> ::= SELECT | MERGE_AND | MERGE_OR
<dimension_list> ::= <dimension>, <dimension_list> | <dimension>
<dimension>::= x | y | z | t | image_object | audio_object | video_object
              | type | attribute | object | ...
<cluster_type ::= ε | interval_projection | ...
<cluster_values> ::= * | <cluster_list>
<cluster_list> ::= <cluster_val>, <cluster_list> | <cluster_val>
<cluster_val >::= <val> | OPEN <val> | (<val> ALIAS <identifier>)
              | OPEN (<val> ALIAS <identifier>)
<val> ::= <variable_identifier> | <string_constant> | <numeric_constant>
<source>::= <query> | <source_name>
<condition>::= <string>
<presentation_description> ::= <string>
<source_name> ::= <source_identifier>
```

A template of an ΣQL query is given below:

```
SELECT dimension_list
CLUSTER [cluster_type] [OPEN] cluster_val1,..., [OPEN] cluster_valn
FROM source
WHERE conditions
PRESENT presentation_description
```

which can be translated as follows: "Given a source (FROM *source*) and a list of dimensions (SELECT *dimensions*), select clusters (CLUSTER) corresponding to a list of projection values or variables ([OPEN] *cluster_val*$_1$, ...) on the dimension axes using the default or a particular clustering mechanism ([cluster_type]). The clusters must satisfy a set of conditions (WHERE conditions) on the existing projection variables and/or on cluster contents if these are open ([OPEN]). The final result is presented according to a set of presentation specifications (PRESENT *presentation_description*)."

Each σ-query can be expressed as an ΣQL query. For example, the σ-query $\sigma_{s,k}(s_1, s_2, s_3, \ldots, s_n)_\phi R$ can be translated as follows:

SELECT s
CLUSTER κ s_1 , OPEN $s_2, s_3, \ldots, s_n$
FROM R
WHERE ϕ

A σ-query can be processed according to the following procedure:

Procedure σ-query_Processor($\sigma_{s,k}(s_1, s_2, s_3, \ldots, s_n)R$)
Input: (1) A cluster string representing R, and (2) a σ-query.
Output: The retrieval results.
Step 1: Apply cutting mechanism κ to R to find all of its sub-clusters according to the clustering $(s_1, s_2, s_3, \ldots, s_n)$.
Step 2: Apply $\sigma_{s,k}$ to all the clusters $C_{s_1}, C_{s_2}, \ldots, C_{s_n}$ and return a cluster string (a relational expression) on them.
Step 3: For each sub-cluster C_{s_i}, if s_i is closed, it is treated as a single object and $\sigma(C_{s_i}) = C_{s_i}$. If s_i is open, it is treated as a set of objects and σ can be applied to the constituent objects that may be sub-sub-clusters.

The above algorithm is recursive, i.e., each R may itself be the form $\sigma_{w,k}(w_1, w_2, w_3, \ldots, w_n)R'$ and can be evaluated recursively.

8.4.4 Querying for Multisensor Data Fusion

In this section, ΣQL will be illustrated with a query that uses data from two different sensors, i.e., a laser radar and a video. The data from these two sensors are heterogeneous with respect to each other. An example of a laser radar image is given in Fig. 8.2. This image shows a car park with a fairly large number of cars, which look like rectangles when viewed from the top. The only moving car can be seen in the lower right part of the image with a north–south orientation while all other cars in the image have an east–west orientation. The moving car and five of its parked neighbors can also be seen in Fig. 8.3. This image is a somewhat enlarged version of a part of the image in Fig. 8.2 and viewed in an elevation angle that shows the three dimensions of the image. The holes at the vehicles in this figure are due to the fact that no information from the sides of the vehicles has been registered.

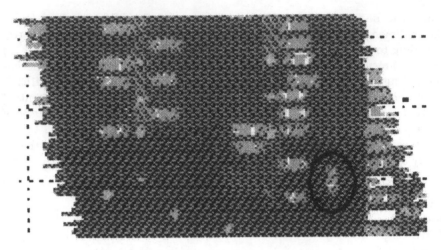

Fig. 8.2 An example of a laser radar image taken across a car park with a moving car in the lower right part of the image in north–south orientation.

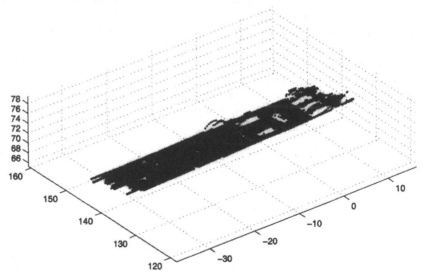

Fig. 8.3 A narrow slice of the laser radar image in Fig. 8.2 shows the moving car and some of its parked neighbors.

Laser radar images are characterized by being three-dimensional and in having geometric properties, that is, each image point is represented with x-, y-, and z-coordinate values. The particular laser radar used here is manufactured by SAAB Dynamics in Sweden, is helicopter-borne and generates image elements from a laser beam that is split into short pulses by a rotating mirror.

The laser pulses are transmitted to the ground, in a scanning movement, and when reflected back to the platform a receiver collects the returning pulses which are stored and analyzed. The results of the analysis are points represented with their three coordinates. There is also a time indication for each point. The resolution of a laser radar image is about 0.3 m.

The video, see Fig. 8.4(a) and (b), is of ordinary type and carried by the same platform. The two sensors are observing the same area at the same time. This means that most cars in the park can be seen in the images from both sensors. The moving car in Fig. 8.4(a) is outside the park and immediately to the left of the entrance and is white. In Fig. 8.4(b) it has entered the park and reached the first of the parked cars. It is quite simple to generate the various projection strings from both types of sensor images. Fig. 8.5 shows two symbolic images corresponding to the two video images in Fig. 8.4. Almost identical projection strings can be generated from the laser radar image.

a   b

Fig. 8.4 Two video frames showing a white car at the entrance in the middle of the image **a** and the white car between some of the parked cars **b**.

Basically, the query that is of concern here can be formulated as follows. Assume that we are interested in determining moving objects along a flight. This can theoretically be done by analyzing the video alone, but that requires hundreds and probably even more sequential video frames to be analyzed. This will take both a very long time and really large computational resources, which may not always be available. Furthermore, this problem cannot, at this time, be solved in real time. By using images from a laser radar, on the other hand, it is possible to recognize any type of vehicle in real time with respect to time and position. This has been shown by Jungert et al. in [21, 22]. However, it cannot be determined from this sensor whether the vehicle is moving. The solution to this problem is to analyze the laser radar image to first find occurring vehicles and determine their position in time and from this information in a second step identify a very limited set of video frames that includes the vehicles found in the laser radar image. From the now limited set of video frames it is possible to determine which of the vehicles are in motion. Finally, in a fusion process, it can be determined which of the vehicles are moving. This will be illustrated by the query below where the query first is split into two subqueries that correspond to queries concerned with data from just one of the sensors. In the final step, it will also be demonstrated how the information from the sensors is fused in order to answer the query.

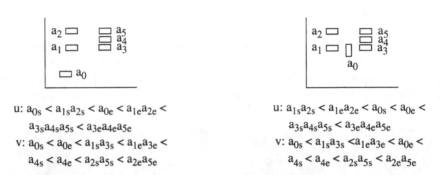

u: $a_{0s} < a_{1s}a_{2s} < a_{0e} < a_{1e}a_{2e} <$
     $a_{3s}a_{4s}a_{5s} < a_{3e}a_{4e}a_{5e}$
v: $a_{0s} < a_{0e} < a_{1s}a_{3s} < a_{1e}a_{3e} <$
     $a_{4s} < a_{4e} < a_{2s}a_{5s} < a_{2e}a_{5e}$

u: $a_{1s}a_{2s} < a_{1e}a_{2e} < a_{0s} < a_{0e} <$
     $a_{3s}a_{4s}a_{5s} < a_{3e}a_{4e}a_{5e}$
v: $a_{0s} < a_{1s}a_{3s} < a_{1e}a_{3e} < a_{0e} <$
     $a_{4s} < a_{4e} < a_{2s}a_{5s} < a_{2e}a_{5e}$

Fig. 8.5 Two symbolic images showing the situation of the two video frames in Fig. 8.4 with the moving car and its close neighbors and the corresponding interval projection strings [12].

An important problem that is not addressed in this work, but will be subject to future research, is the handling of uncertain sensor information. Clearly, this is a very important problem that cannot be excluded when designing a query language for sensor data fusion, where in particular all input data come from heterogeneous data sources. However, we have found it necessary to address the basic query techniques, the syntax of the query language, and the basic abstract spatial/temporal structures for reasoning first. In this perspective, the query is first represented as σ-sequences and then translated into ΣQL-syntax.

subquery1: Are there any moving objects in the video sequence in the time interval t_1 through t_2?

$$Q_1 = \sigma_{motion}(\text{moving})\sigma_{type}(\text{vehicle})\sigma_{xy,interval\_cutting}(*)$$
$$\sigma_t(T^o)_{T \bmod 10=0 \ and \ T>t_1 \ and \ T<t_2}$$
$$\sigma_{media\_sources}(\text{video}^o)\text{media\_sources}.$$

subquery2: Are there any vehicles in the laser radar image in the time interval t_1 through t_2?

$$Q_2 = \sigma_{type}(\text{vehicle})\sigma_{xyz,interval\_cutting}(*)\sigma_t(T^o)_{T>t_1 \ and \ T<t_2}$$
$$\sigma_{media\_sources}(\text{laset\_radar}^o)\text{media\_sources}.$$

The subquery Q_1 first selects the video source and then the video frames which all are opened. However, the selection of video frames also includes some conditions with respect to which frames to accept and in which time interval. In this case we have chosen to select each tenth video frame within the interval $[t_1, t_2]$. In the next selection the σ_{xy}-operator is applied to the video frames using the interval cutting mechanism [11, 12]. This operator generates the (u, v)-strings from which the object types are determined by the σ_{type}-operator. That is, in this particular case, the vehicles, in the selected frames. Eventually the vehicles in motion are determined by the application of the motion operator. The motion string (m) is generated from the time projection string (t) where the single video frames are opened with respect to the x- and y-dimensions, i.e.,

$$t : (u : a_{0s} < a_{1s}\, a_{2s} < a_{0e} < a_{1e}\, a_{2e} < a_{3s}\, a_{4s}\, a_{5s} < a_{3e}\, a_{4e}\, a_{5e},$$
$$v : a_{0s} < a_{0e} < a_{1s}\, a_{3s} < a_{1e}\, a_{3e} < a_{4s} < a_{4e} < a_{2s}\, a_{5s} < a_{2e}\, a_{5e}) <$$
$$(u : \cdots, v : \cdots) <$$
$$(u : a_{1s}\, a_{2s} < a_{1e}\, a_{2e} < a_{0s} < a_{0e} < a_{3s}\, a_{4s}\, a_{5s} < a_{3e}\, a_{4e}\, a_{5e},$$
$$v : a_{0s} < a_{1s}\, a_{3s} < a_{1e}\, a_{3e} < a_{0e} < a_{4s} < a_{4e} < a_{2s}\, a_{5s} < a_{2e}\, a_{5e}) <$$
$$(u : \cdots, v : \cdots) <$$
$$\cdots$$

From this string the motion string is generated by applying the σ-operator which generates a string similar to an implicit merge_or operation, i.e.,

$$m : t : (u : a_{0s} < a_{0e} < a'_{0s} < a'_{0e}, v : a_{0s} < a_{0e} < a'_{0s} < a'_{0e}).$$

The subquery Q_2 returns first the (u, v)-strings for the time interval $[t_1, t_2]$. This is sufficient, since that gives the relative position of the vehicles (the z-information is normally unnecessary for this purpose and will only be used for object type recognition). An intermediate result of the subquery will thus look like:

$$u : a_{1s}\, a_{2s} < a_{1e}\, a_{2e} < a_{0s} < a_{0e} < a_{3s}\, a_{4s}\, a_{5s} < a_{3e}\, a_{4e}\, a_{5e},$$
$$v : a_{1s}\, a_{3s} < a_{0s} < a_{1e}\, a_{3e} < a_{4s} < a_{4e} < a_{2s}\, a_{5s} < a_{0e} < a_{2e}\, a_{5e}.$$

The laser scanner determines each point in a sequential order so that an object that is in the lower left corner of the image is first registered by the sensor. Therefore it is possible to implicitly determine the t-string, if needed.

$$t : (u : a_1) < \cdots < (u : a_i) \cdots$$

This, however, requires a further application of σ_t but that has not been applied here. It can, nevertheless, be motivated because it may support the outcome of the data fusion process. In the final step of this subquery existing vehicles are determined by applying the σ_{type}-operator.

The two subqueries can now be fused with respect to equality. For this purpose another operator that can perform this is needed. However, this fusion operator is different from the σ-operators that have been used so far, since its input will be coming from multiple data sources that must be of equal type. For this reason, a fusion operator, called ϕ, is defined. This operator performs the data fusion operation, here called *merge-and*, with respect to the three dimensions x, y, and t; in other words, it fuses the vehicle information with respect to equality of type and position in time. The object type in question is in this case already determined. All object types are consequently equal in both subqueries. The final query with the fusion operator thus becomes:

$$Q_3 = \phi_{xyt}^{\text{merge-and}}(*)(Q_1, Q_2).$$

This means that a fusion operation is applied such that only those objects selected from the two subqueries and which can be *associated* with each other will remain in the output string, which here is called **mo** (motion objects). This gives us the following result:

mo:a_0.

The complete query now looks like:

$\phi_{xyt}^{\text{merge-and}}(*)$

$\quad(\sigma_{motion}(\text{moving})\sigma_{type}(\text{vehicle})\sigma_{xy,interval\_cutting}(*)$

$\quad\sigma_t(T^o)_{T\ mod\ 10=0\ and\ T>t_1\ and\ T<t_2}$

$\quad\sigma_{media\_sources}(\text{video}^o)\text{media\_sources},$

$\quad\sigma_{type}(\text{vehicle})\sigma_{xyz,interval\_cutting}(*)\sigma_t(T^o)_{T>t_1\ and\ T<t_2}$

$\quad\sigma_{media\_sources}(\text{laset\_radar}^o)\text{media\_sources}).$

The important problem here is, as always in data fusion, the association problem. In other words, the query must determine whether a certain object found in one of the two subqueries is the same as any of the vehicles found in the other subquery. This problem is generally very difficult and is discussed more deeply by Waltz and Llinas [30].

Translating the σ-query into ΣQL-syntax is now a fairly simple task and the result from this translation becomes:

```
MERGE-AND x,y,t
CLUSTER *,*,[t₁,t₂]
FROM (SELECT type
      CLUSTER vehicle
      FROM SELECT x,y,z
           CLUSTER interval, *
           FROM SELECT t
                CLUSTER OPEN (* ALIAS T)
```

FROM SELECT media_sources
      CLUSTER OPEN laser_radar
      FROM media_sources
    WHERE $T > t_1$ AND $T < t_2$,
SELECT motion
CLUSTER moving
FROM SELECT type
    CLUSTER vehicle
    FROM SELECT x,y
      CLUSTER interval *
      FROM SELECT t
        CLUSTER OPEN (* ALIAS T)
        FROM SELECT media_sources
        CLUSTER OPEN video
        FROM media_sources
      WHERE T mod 10 =0 AND $T>t_1$ AND $T<t_2$)

8.5 Discussion

As explained in previous sections, ΣQL can express both spatial and temporal constraints individually using the SELECT/CLUSTER construct and nested subqueries. A visual language version of this new ΣQL language may be suitable for the hypermapped virtual world (HVW) information model [15].

The HVW information model is a combination of hypermap with virtual reality, so that each hyperlink can lead to a virtual world. Hypermaps can be used advantageously as a metaphor for the representation of all the multimedia hyperbase elements. For example, in GeoAnchor [3] a map can be built dynamically as a view of the multimedia hyperbase. Each displayed geometry is an anchor to either a geographic node or to a related node. Hence, the map on the screen acts both as an index to the nodes and as a view to the multimedia hyperbase. As another example, in a virtual classroom a hypermap can also be used as a metaphor to link the most frequently accessed items such as reading rooms, book shelves, etc. to present different views to the end user. This combined metaphor of hypermapped virtual classroom (which is a combination of the VR information space and the logical information hyperspace [14]) may lead to efficient access of multimedia information in a distance learning environment. The σ-query may serve as the basis of a visual query language for the hypermapped virtual world.

References

1. Ahanger, G, Benson, D, and Little, TD, "Video Query Formulation," Storage and Retrieval for Images and Video Databases II, San Jose, pp. 280–291, February, 1995.

2. Batini, C, Catarci, T, Costabile, MF, and Levialdi, S, "Visual Query Systems," Technical Report N. 04.91, Dipartimento di Informatica e Sistemistica, Universita di Roma "La Sapienza," Italy, 1991 (revised in 1993).

3. Caporal, AJ and Viemont, AY, "Maps as a Metaphor in a Geographical Hypermedia System," J Visual Lang Comput, 8(1), pp. 3–25, February, 1997.

4. Catarci, T and Costabile, MF (Eds), "Special Issue on Visual Query Systems," J Visual Lang Comput, 6(1), 1995.

5. Catarci, T, Chang, SK, Costabile, MF, Levialdi, S, and Santucci, G, "A Graph-Based Framework for Multiparadigmatic Visual Access to Databases," IEEE Trans Knowledge Data Engin, 8(3), pp. 455–475, 1996.

6. Chan, E and Zhu, R, "QL/G – A Query Language for Geometric Databases," 1st Int. Conf. on GIS in Urban Regional and Environment Planning, Samos, Greece, pp. 271–286, April, 1996.

7. Chang, H, Hou, T, Hsu, A, and Chang, SK, "The Management and Applications of Tele-Action Objects," ACM J Multimed Syst, 3(5-6), pp. 204–216, 1995.

8. Chang, H, Hou, T, Hsu, A, and Chang, SK, "Tele-Action Objects for an Active Multimedia System," 2nd Int. IEEE Conf. on Multimedia Computing and Systems, pp. 106–113, Washington, D.C, May 15-18, 1995.

9. Chang, SF, Chen, W, Meng, HJ, Sundaram, H, and Zhong, D, "VideoQ: An Automated Content Based Video Search System Using Visual Cues," 5th ACM Multimedia, November, 1997.

10. Chang, SK, Shi, OY, and Yan, CW, "Iconic Indexing by 2D Strings," IEEE Trans Patt Anal Mach Intell, 9(3), pp. 413–428, 1987.

11. Chang, SK and Jungert E, "Pictorial Data Management Based Upon the Theory of Symbolic Projection," J Visual Lang Comput, 2(3), pp. 195–215, 1990.

12. Chang, SK and Jungert, E, Symbolic Projection for Image Information Retrieval and Spatial Reasoning, Academic Press, 1996.

13. Chang, SK and Jungert, E, "Human and System Directed Fusion of Multimedia and Multimodal Information Using the s-Tree Data Model," 2nd Int. Conf. on Visual information Systems, San Diego, CA, December 15-17, 1997.

14. Chang, SK and Costabile, MF, "Visual Interface to Multimedia Databases," in Handbook of Multimedia Information Systems, WI Grosky, R Jain, and R Mehrotra, eds, Prentice Hall, pp. 167–187, 1997.

15. Chang, SK, "Content-Based Access to Multimedia Information," Aizu International Student Forum-Contest on Multimedia (N Mirenkov and A. Vazhenin, eds.), The University of Aizu, Aizu, Japan, pp. 2–41, 1998. (The paper is available at www.cs.pitt.edu/~chang/365/cbam7.html)

16. Egenhofer, M, "Spatial SQL: A Query and Presentation Language," IEEE Trans Knowledge Data Eng, 5(2), pp. 161–174, 1991.

17. Faloutsous, C, Barber, R, Flickner, M, Niblack, W, Petkovic, D, and Equitz, W, "Efficient and Effective Querying by Image Content" IBM Research Division Almaden Research Center Technical Report RJ9543 (83074), August, 1993.

18. Faloutsous, C, Barber, R, Flickner, M, Hafner, J, Niblack, W, Petkovic, D, and Equitz, W, "Efficient and Effective Querying by Image Content" J Intell Inform Syst, 3, pp. 231–262, 1994.

19. Fox, EA, "Advances in Interactive Digital Multimedia Systems," IEEE Computer, 24(10), pp. 9–21, October, 1991.

20. Holden, R, "Digital's DB Integrator: a Commercial Multi-Database Management System," 3rd Int. Conf. on Parallel and Distributed Information Systems, Austin, TX, USA, pp. 267–268, September, 1994.

21. Jungert, E, Carlsson C, and Leuhusen, C, "A Qualitative Matching Technique for Handling Uncertainties in Laser Radar Images," SPIE Conf. on Automatic Target Recognition VIII, Orlando, Florida, pp. 62–71, April, 1998.
22. Jungert, E, "A Qualitative Approach to Recognition of Man-Made Objects in Laser-Radar Images," Conf. on Spatial Data Handling, Delft, The Netherlands, Vol II, pp. A15–A26, August, 1998.
23. Lee, SY and Hsu, FS, "Spatial Reasoning and Similarity Retrieval of Images Using 2D Cstring Knowledge Representation," Patt Recogn, 25, pp. 305–318, 1992.
24. Li, JZ, Ozsu, MT, Szafron, D, and Oria, V, "MOQL: A Multimedia Object Query Language," 3rd Int. Workshop on Multimedia Information Systems, Como, Italy, pp. 19–28, September, 1997.
25. Oomoto, E and Tanaka, K, "Video Database Systems – Recent Trends in Research and Development Activities," in Handbook of Multimedia Information Management (Grosky, WI, Jain, R, and Mehrotra, R, eds), Prentice Hall, pp. 405–448, 1997.
26. Roussopoulos, N, Faloutsos, C, and Sellis, T, "An Efficient Pictorial Database System for PSQL," IEEE Trans Soft Engin, 14(5), pp. 639–650, May, 1988.
27. Srihari, RK and Zhang, ZF, "Finding Pictures in Context," MINAR'98, Hong Kong, Lecture Notes in Computer Science, 1464, (Ip, H and Smeulders, A, eds), pp. 109–123, August, 1998.
28. Tabuchi, "Hyperbook," Int. Conf. on Multimedia Information Systems, Singapore, 1991.
29. Vazirgiannis, M, "Multimedia Data Object and Application Modelling Issues and an Object Oriented Model," Multimedia Database Systems: Design and Implementation (Nwosu, KC, Thuraisingham, B, and Berra PB, eds), Kluwer Academic Publishers, pp. 208–250, 1996.
30. Waltz, E and Llinas, J, Multisensor Data Fusion, Artect House, Boston, 1990.
31. Xu, XB, Shi, BL, and Gu, N, "FIMDP: an Interoperable Multi-Database Platform," 8th Int. Hong Kong Computer Society Database Workshop, Data Mining, Data Warehousing and Client/Server Databases, Hong Kong, pp. 166–176, July, 1997.

9. Relevance Feedback Techniques in Image Retrieval

Yong Rui and Thomas S. Huang

9.1 Introduction

Despite the extensive research effort, the retrieval techniques used in content-based image retrieval (CBIR) systems lag behind the corresponding techniques in today's best text search engines, such as Inquery [2], Alta Vista, and Lycos. One reason is that the information embedded in an image is far more complex than that in text. To better understand the history and methodology of CBIR and how we can improve CBIR's performance, we will first introduce an image object model before we go into the details of the discussions. An image object (O) can be modeled as a function of the image data (D), features (F), and representations (R). This is described below and also shown in Fig. 9.1:

$$O = O(D, F, R). \tag{9.1}$$

- D is the raw image data, e.g., a JPEG image.
- $F = \{f_i\}$, $i = 1, \ldots, I$ is a set of visual features associated with the image object, such as color, texture, and shape.
- $R_i = \{r_{ij}\}$, $j = 1, \ldots, J_i$ is a set of representations for a given feature f_i, e.g., both color histogram and color moments are representations for the color feature [33]. Note that, each representation r_{ij} itself is normally a vector consisting of multiple components, i.e.,

$$r_{ij} = [r_{ij1}, \ldots, r_{ijk}, \ldots, r_{ijK_{ij}}] \tag{9.2}$$

where K_{ij} is the length of the vector r_{ij}.

This image model has three information abstraction levels (object, feature, and representation), increasing in the information granularity. Furthermore, different weights (U at the object level, V_i at the feature level, and W_{ij} at the representation level) exist to reflect a particular entity's importance to its upper level. Most of the existing research in CBIR can fit in this comprehensive image object model, even though most research only explores part of it. We next briefly review the history and methodology of CBIR with respect to this model.

At the early stage of CBIR, research primarily focused on exploring what is the "best" feature (f_i) for a given image or the "best" representation (r_{ij}) for a given feature. For example, for the texture feature alone, almost a dozen

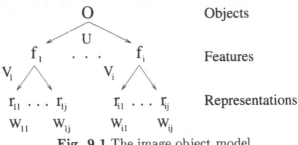

Fig. 9.1 The image object model.

representations have been proposed [11], including Tamura [8], MSAR [19], Word decomposition [13], Fractal [4], Gabor Filter [15,18], and Wavelets [3,16,30]. Once the "best" features and representations are found, their corresponding weights (V_i and W_{ij}) are further fixed by the system designer based on his or her judgment of the importance of each r_{ij} and r_{ijk}. In addition, at the query stage, the user is asked to specify the weights at the object level (U). Based on the user's query image and specified weights, the retrieval system then tries to find similar images to the user's query.

In this type of approach, there is no human–computer interaction except that the user has to give a set of weights (U). We therefore refer to this type of approach as an *isolated* approach. While this approach establishes the basis of CBIR, its performance is not satisfactory because of the following two reasons:

– Specifying the object level weights (U) imposes a burden on the user, as it requires the user to have comprehensive knowledge of the visual features used in the retrieval system and how they related to his or her information need. This is not easy for an expert to do, let alone a normal user.

– Specifying feature and representation levels' weights (V_i and W_{ij}) imposes a burden on the system designer as well. Not knowing beforehand which r_{ij} and r_{ijk} can best match a particular user's perception of image content, it is almost impossible for the system designer to give accurate values for V_i and W_{ij}. Furthermore, these weights can not be easily obtained by training on typical users, because human perception of image content is subjective. That is, different persons, or the same person under different circumstances, may perceive the same visual content differently. This is called *human perception subjectivity* [25]. The subjectivity exists at various levels. For example, one person may be more interested in an image's color feature, and another may be more interested in the texture feature. Even if both people are interested in texture, how they perceive the similarity of texture may be quite different. This is illustrated in Fig. 9.2.

Among the three texture images, some may say that (a) and (b) are more similar if they do not care for the intensity contrast, while others may say that (a) and (c) are more similar if they ignore the local property on the

a   b   c

Fig. 9.2 Subjectivity in perceiving the texture feature.

seeds. No single texture representation (r_{ij}) can capture everything. Different representations capture the visual feature from different perspectives.

The above situation makes the *isolated* approach difficult to use. The burden imposed on both the user and system designer boils down to one thing. That is, in this approach, all the weights are *static* (fixed). For static weights, it is not possible to correctly model a user's information need.

Motivated by the limitations of the *isolated* approach, recent research in CBIR has moved to an interactive mechanism that involves a human as part of the retrieval process [11, 22, 23, 28]. Examples include *interactive* region segmentation [6], interactive image database annotation [20, 22], usage of *supervised* learning before the retrieval [14, 17], and *interactive* integration of keywords and high level concepts to enhance image retrieval performance [31, 32].

Based on the above analysis, in this chapter we explore *interactive* approaches that address the difficulties faced by the *isolated* approach. In particular, we will focus on techniques for interactive image retrieval based on *relevance feedback*, in which both user and computer interact to refine the user's true information need. Relevance feedback is a powerful technique first introduced in traditional text-based information retrieval systems. It is the process of automatically adjusting an existing query using the information fed back by the user about the relevance of previously retrieved objects such that the adjusted query is a better approximation to the user's information need [1, 27, 29]. In a *interactive* system, neither the user nor the system designer needs to specify any weights. The user only needs to mark which images he or she thinks are relevant to his or her query. The weights associated with the query object are *dynamically* updated to model the user's information need and perception subjectivity. In general, there are three approaches to relevance feedback in image retrieval. One is based on artificial intelligence (AI) learning techniques [22], one on a Bayesian framework [5], and the last on information retrieval (IR) techniques [23, 26]. The last one is the most widely used and is the topic of this chapter.

The rest of the chapter is organized as follows. Section 9.2 describes a retrieval model based on the proposed image object model (Fig. 9.1) and discusses how to compute the distance between two image objects. How to estimate weights U, V_i, and W_{ij}, as well as the query vector, is essential to the *interactive* approach. We therefore explore various techniques in parameter (weights and queries) updating, ranging from a heuristic-based approach to an optimization approach. They are discussed in great detail in Sections 9.3–9.5. Experimental results of the proposed approaches over multiple data sets are given in Section 9.6. Concluding remarks are given in Section 9.7.

9.2 The Retrieval Process Based on Relevance Feedback

In order to compare the distance between two images, we need to define a retrieval model. The object model $O(D, F, R)$ together with a set of distance measures specifies a retrieval model.

Let the distance measures at the three levels be $\Phi()$, $\Theta()$, and $\Psi()$. How many arguments they have will depend on which forms (e.g., linear or quadratic) they are taking (see Eqs 9.3, 9.26, and 9.47). Let r_{mij}, $m = 1, \ldots, M$; $i = 1, \ldots, I$; $j = 1, \ldots, J_i$ be the the ijth representation vector for the mth image in the database, where M is the total number of images in the database, I the number of features, and J_i the number of representations for feature i. Let q_{ij}, $i = 1, \ldots, I$; $j = 1, \ldots, J_i$ be the query vector for the ijth representation. The retrieval process can be described as follows and is illustrated in Fig. 9.3.

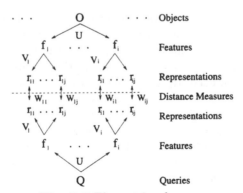

Fig. 9.3 The retrieval process.

1. Initialize the values of the weights U, V_i, and W_{ij}.
2. The distance between an image and a query in terms of the ijth representation is

$$d_m(\boldsymbol{r}_{ij}) = \Psi_{ij}(\boldsymbol{r}_{mij}, \boldsymbol{q}_{ij}, W_{ij}),$$
$$m = 1, \ldots, M,$$
$$i = 1, \ldots, I,$$
$$j = 1, \ldots, J_i$$

where $\boldsymbol{r}_{mij}$ is the ijth representation vector for the mth image in the database, and $d_m(\boldsymbol{r}_{ij})$ denotes the distance between the mth image and a query in terms of representation ij.

3. The distance between the image and the query in terms of feature i is then

$$d_m(f_i) = \Theta_i(d_m(\boldsymbol{r}_{ij}, V_i) = \Theta_i(\Psi_{ij}(\boldsymbol{r}_{mij}, \boldsymbol{q}_{ij}, W_{ij}), V_i).$$

4. The overall distance is then

$$d_m = \Phi(d_m(f_i, U) = \Phi(\Theta_i(\Psi_{ij}(\boldsymbol{r}_{mij}, \boldsymbol{q}_{ij}, W_{ij}), V_i), U).$$

5. The images in the database are ordered by their overall distances to the query (d_m). The N_{RT} most similar ones are returned to the user, where N_{RT} is the number of images the user wants to retrieve.
6. For each of the retrieved images, the user provides a degree-of-relevance score, according to the user's information need and perception subjectivity.
7. The system dynamically updates the parameters (described in Sections 9.3–9.5) according to the user's feedback such that the adjusted query $\boldsymbol{q}_{ij}$ and weights U, V_i, W_{ij} better match the user's information need.
8. Go to Step 2 with the adjusted parameters and start a new iteration of retrieval until the user is satisfied.

In Fig. 9.3, the information need embedded in the query Q flows up while the content of image object O flows down. They meet at the dashed line, where the distance measures are applied to calculate the distance values between Q and O.

Whether a retrieval model can update its parameters (weights or query vectors or both) distinguishes the *interactive* approach from the *isolated* approach. In the *isolated* approach, all the parameters (weights and query vectors) are fixed. Because of the fixed parameters, this approach cannot effectively model the user's information need and perception subjectivity. This approach also places a burden on both the user and the system designer. On the other hand, for the *interactive* approach, weights and query vectors are *dynamically* updated via relevance feedback. The burden of specifying the weights is removed from the user.

It is clear that the most essential part of the retrieval model is parameter updating. We next present two such approaches, a heuristic one, and an optimal one. The former is faster to compute but has less accuracy, and the latter is optimal for the given objective function but needs more time and computing power.

9.3 Parameter Updating: a Heuristic Approach

In this approach, we restrict the distance measures $\Phi()$ and $\Theta_i()$ to be linear functions in both their corresponding entities and their weights, and $\Psi_{ij}()$ to be linear in their weights but arbitrary in corresponding entities, that is,

$$d_m = \Phi(\Theta_i(\Psi_{ij}(\boldsymbol{r}_{mij}, \boldsymbol{q}_{ij}, W_{ij}), V_i), U) = \sum_{i=1}^{I} u_i \sum_{j=1}^{J_i} v_{ij} \Psi_{ij}(\boldsymbol{r}_{mij}, \boldsymbol{q}_{ij}, \boldsymbol{w}_{ij})$$

where U takes the form of a vector $\boldsymbol{u}^T = [u_1, \ldots, u_i, \ldots, u_I]$ and u_i is the weight for the ith feature; V_i takes the form of a vector $\boldsymbol{v}_i^T = [v_{i1}, \ldots, v_{ij}, \ldots, v_{iJ_i}]$ and v_{ij} is the weight for the ijth representation; W_{ij} takes the form of a vector $\boldsymbol{w}_{ij}^T = [w_{ij1}, \ldots, w_{ijk}, \ldots, w_{ijK_{ij}}]$ and w_{ijk} is the weight for the kth component for representation ij; and T denotes transpose of a matrix or a vector.

The reason that we can assume $\Phi()$ and $\Theta_i()$ to be linear functions is that the weights are proportional to the entities' relative importance [9]. For example, if a user cares twice as much about one feature (color) as he does about another feature (shape), the overall similarity would be a linear combination of the two individual similarities with the weights being 2/3 and 1/3, respectively [9]. Note that, at the representation level, the distance measure Ψ_{ij} normally cannot be simplified to linear functions. For example, intersection distance is used in histogram comparison and Euclidean is used for wavelet texture comparison.

Because $\Phi()$ and $\Theta_i()$ are both linear functions, we can combine the object level and feature level into a single level, that is,

$$d_m = \sum_{i=1}^{I} u_i d_m(\boldsymbol{r}_i), \tag{9.3}$$

$$d_m(\boldsymbol{r}_i) = \Psi_i(\boldsymbol{r}_{mi}, \boldsymbol{q}_i, \boldsymbol{w}_i) \tag{9.4}$$

where I is now redefined as the total number of representations; i is the index for a particular representation; and u_i redefined as the weight directly from the image object to the ith representation. We next discuss how to update these two levels' weights, as well as how to refine the query vector, in the following three subsections.

9.3.1 Update of the Query Vector $\boldsymbol{q}_i$

It is clear that the specification of the query vector $\boldsymbol{q}_i$ is critical, because the computed distance values (d_m's) are based on them. However, it is usually difficult for a user to map his or her information need into a query vector precisely. Relevance feedback is a way of refining the query vectors [1, 27, 29].

The mechanism of this method can be described elegantly in the vector space. If the sets of relevant images (D_R) and non-relevant images (D_N) are known, the optimal query can be proven to be [1, 27, 29]:

$$q_i^* = \frac{1}{N_R} \sum_{n \in D_R} x_{ni} - \frac{1}{N_T - N_R} \sum_{i \in D_N} x_{ni} \qquad (9.5)$$

where N_R is the number of documents in D_R, N_T the number of the total documents, and x_{ni} is the nth training sample in the sets D_R' and D_N'.

In practice, D_R and D_N are not known in advance. However, the relevance feedback obtained from the user furnishes approximations to D_R and D_N, which are referred to as, D_R' and D_N'.

The original query q_i can be modified by putting more weights on the relevant components and fewer weights on the non-relevant ones:

$$q_i' = \alpha q_i + \beta \left(\frac{1}{N_{R'}} \sum_{n \in D_R'} x_{ni} \right) - \gamma \left(\frac{1}{N_{N'}} \sum_{n \in D_N'} x_{ni} \right), \qquad i = 1, \ldots, I$$

where α, β, and γ are suitable constants [27, 29], and $N_{R'}$ and $N_{N'}$ are the numbers of objects in the relevant set D_R' and non-relevant set D_N'.

9.3.2 Update of w_i

The vector w_i consists of K_i components, associated with r_{mi} and q_i. Each component w_{ik}, $k = 1, \ldots, K_i$ reflects the different contribution of a component to the representation vector r_i. For example, in the wavelet texture representation, we know that the mean of a subband may be corrupted by the lighting condition, while the standard deviation of a subband is independent of the lighting condition. Therefore, more weight should be given to the standard deviation component, and less weight to the mean component. The support of different weights for r_i's enables the system to have more reliable feature representation and thus better retrieval performance.

A standard-deviation-based weight updating approach has been proposed in our previous work [23]. Out of the N_{RT} returned objects, for those objects that are marked *highly relevant* or *relevant* by the user, stack their representation vector r_i's to form a $N \times K_i$ matrix, where N is the number of objects marked *highly relevant* or *relevant*. In this way, each column of the matrix is a length-N sequence of r_{ik}'s ($k = 1, \ldots, K_i$). Intuitively, if all the relevant objects have similar values for the component r_{ik}, it means that the component r_{ik} is a good indicator of the user's information need. On the other hand, if the values for the component r_{ik} are very different among the relevant objects, then r_{ik} is not a good indicator. Based on this analysis, the inverse of the standard deviation of the r_{ik} sequence is a good estimation of the weight w_{ik} for component r_{ik}, that is, the smaller the variance, the larger the weight and vice versa:

$$w_{ik} = \frac{1}{\sigma_{ik}} \tag{9.6}$$

where σ_{ik} is the standard deviation of the length-N sequence of r_{ik}'s. Here we assume that the user will mark more than two images as relevant or highly relevant, such that σ_{ik} is computable. Even though simple, we later show that this heuristic-based weight updating approach is actually optimal under some conditions (Section 9.4.1).

9.3.3 Update of u

The vector u consists of I elements, associated with each representation. Each u_i, $i = 1, \ldots, I$, reflects the user's different emphasis of a representation in the overall distance measure. The support of different weights enables the user to specify his or her information need more precisely. We next develop a heuristic approach to update u_i's according to the user's relevance feedback.

Let RT be the set of the most similar N_{RT} image objects according to the overall distance value d_m (Eq. 9.3):

$$RT = [RT_1, \ldots, RT_l, \ldots, RT_{N_{RT}}]. \tag{9.7}$$

Let $Score$ be the set containing the degree-of-relevance scores fedback by the user for RT_l's (see Section 9.2):

$$Score_l = \begin{cases} 3 & \text{if highly relevant,} \\ 1 & \text{if relevant,} \\ 0 & \text{if no-opinion,} \\ -1 & \text{if non-relevant,} \\ -3 & \text{if highly non-relevant.} \end{cases} \tag{9.8}$$

The choice of 3, 1, 0, -1, and -3 as the scores (degree of relevance) is arbitrary. Experimentally we find that the above scores capture the semantic meaning of *highly relevant, relevant,* etc. In Eq. 9.8, we provide the user with five levels of relevance. Although more levels may result in more accurate feedback, it is less convenient for the user to interact with the system. Experimentally we find that five levels is a good trade-off between convenience and accuracy.

For representation i, let RT^i be the set containing the N_{RT} image objects most similar to the query Q, according to the distance values $d_m(r_i)$ (Eq. 9.3):

$$RT^i = [RT_1^i, \ldots, RT_l^i, \ldots, RT_{N_{RT}}^i]. \tag{9.9}$$

To calculate the weight for representation i, first initialize $u_i = 0$, and then use the following procedure:

$$u_i = u_i + Score_l, \quad \text{if } RT_l^i \text{ is in } RT, \tag{9.10}$$

$$= u_i + 0, \quad \text{if } RT_l^i \text{ is not in } RT, \tag{9.11}$$

$$l = 0, \dots, N_{RT}. \tag{9.12}$$

Here, we consider all the images outside set RT as marked with *no opinion* and having the score of 0. After this procedure, if $u_i < 0$, set it to 0 (weights are non-negative). Intuitively, the above procedure means the more the overlap of relevant objects between RT and RT^i, the larger the weight of u_i. That is, if a representation reflects the user's information need, it receives more emphasis.

9.4 Parameter Updating: Optimal Approaches

In the previous section, we have discussed a heuristic approach to parameter updating, which is both easy to implement and quick to compute. However, because it is heuristic based, there is no optimality guarantee of any kind. In this and next sections we will explore optimization-based approach for parameter updating. For simplicity, in the remaining discussion, we will combine the object level and feature level into a single level. That is, the distance between an image and a query is calculated as

$$d_m = \Phi(\Psi_i(r_{mi}, q_i, W_i), U) \tag{9.13}$$

where i is the index for a particular representation and U is redefined as the weight directly from the image object to the ith representation.

The remainder of the chapter is organized as follows. The parameter update for the representation level has attracted a lot of attention from researchers. The rest of this section is dedicated to this topic, where we review, analyze, and compare various approaches. How to update the parameters at the object level is much less addressed in the literature. To the authors' best knowledge, the only approach seen is the one we discussed in Section 9.3. Even worse, there is no approach which can update both the object level and the representation level simultaneously. In the Section 9.5, we will develop a few approaches to address this problem.

9.4.1 Parameter Update at the Representation Level

Among the three levels in the image object model (Fig. 9.1), the representation level is the most concrete, because the distance between an image and a query can be directly computed at this level. Because of this, researchers have developed many techniques for updating parameters (weights and query vectors) at this level.

Rui et al. [23] discussed two approaches for query refinement. One approach is to move the query vector:

$$q'_i = \alpha q_i + \beta \left(\frac{1}{N_{R'}} \sum_{n \in D'_R} x_{ni} \right) - \gamma \left(\frac{1}{N_{N'}} \sum_{n \in D'_N} x_{ni} \right), \quad i = 1, \ldots, I$$

where α, β, and γ are suitable constants [27, 29]; $N_{R'}$ and $N_{N'}$ are the numbers of objects in the relevant set D'_R and non-relevant set D'_N; and x_{ni} is the nth training sample in the sets D'_R and D'_N.

The other approach is to update the weights associated with the query vector q. As we have discussed in Section 9.3.2, a standard deviation approach was devised to compute w_{ik}:

$$w_{ik} = \frac{1}{\sigma_{ik}} \tag{9.14}$$

where σ_{ik} is the standard deviation of the length-N sequence of r_{ik}'s (see Section 9.3.2).

Since the appearance of the above approaches to relevance feedback in image retrieval, many improved versions have been proposed. One of the most elegant is MindReader, developed by Ishikawa et al. [12]. After some revisions, the MindReader algorithm can be summarized as follows.

The distance function used for $\Psi_i()$ is a generalized Euclidean function. That is, the distance between an image and a query in terms of representation i is defined as

$$d_m(r_i) = (r_{mi} - q_i)^T W_i (r_{mi} - q_i) \tag{9.15}$$

where T denotes transpose, and W_i is a $(K_i \times K_i)$ weight matrix associated with the ith representation. Note that both r_{mi} and q_i are vectors of length K_i.

Let N denote the number of training samples (the images fed back by the user). Let $\pi = [\pi_1, \ldots, \pi_n, \ldots, \pi_N]$, $\pi_n > 0$ be the vector of degree-of-relevance for the corresponding training samples. They proposed to solve the following optimization problem:

$$\min J = \sum_{n=1}^{N} \pi_n (x_{ni} - q_i)^T W_i (x_{ni} - q_i),$$

$$s.t. \quad det(W_i) = 1$$

where x_{ni} denotes the nth training sample and is a vector of length K_i.

By forming the Lagrange multiplier as

$$L = J + \lambda \left(det(W_i) - 1 \right), \tag{9.16}$$

we can obtain the optimal solutions for q_i and W_i: [12]

$$q_i^{T*} = \frac{\pi^T X_i}{\sum_{n=1}^{N} \pi_n}, \tag{9.17}$$

$$W_i^* = (det(C_i))^{\frac{1}{J_i}} C_i^{-1} \tag{9.18}$$

where X_i is the training sample matrix for representation i, obtained by stacking the N training vectors (x_{ni}) into a matrix. It is therefore an $(N \times K_i)$ matrix. The term C_i is the $(K_i \times K_i)$ weighted covariance matrix of X_i. That is,

$$C_{i_{rs}} = \frac{\sum_{n=1}^{N} \pi_n (x_{nr} - q_r)(x_{ns} - q_s)}{\sum_{n=1}^{N} \pi_n}, \qquad r, s = 1, \ldots, K_i.$$

Both solutions match our intuition nicely. The optimal query vector is the weighted average of the training samples. The optimal weight matrix is inversely proportional to the covariance matrix of the training samples.

MindReader formulated an optimization problem to solve for best q_i and W_i. We can arrive at similar and even simpler solutions by formulating a maximum likelihood estimation (MLE) problem. Assume the training samples obey the normal distribution of $N(\mu_i, \Sigma_i)$, where μ_i and Σ_i are the mean vector and covariance matrix respectively, and they are independent and identically distributed (IID). The training samples therefore will form a single cloud or cluster and the centroid of the cluster is the ideal query vector, which achieves minimum distance distortion and maximum likelihood. That is, $q_i = \mu_i$. We further define the distance between a training sample and the query vector to be the *Mahalanobis distance*:

$$d_n(r_i) = (x_{ni} - \mu_i)^T \Sigma_i^{-1} (x_{ni} - \mu_i) \tag{9.19}$$

where Σ_i^{-1} plays the role of W_i.

Assume the training samples are IID, the likelihood function and log likelihood function can be written as

$$L(\mu_i, \Sigma_i) = \prod_{n=1}^{N} p(x_{ni}|\mu_i, \Sigma_i),$$

$$LL(\mu_i, \Sigma_i) = \sum_{n=1}^{N} p(x_{ni}|\mu_i, \Sigma_i).$$

The degree-of-relevance vector π comes into play as the relative frequency of a particular training sample. That is, the higher its degree of relevance, the more frequently the training sample will appear in the training set. By taking this into account, the log likelihood function can be written as

$$LL(\mu_i, \Sigma_i) = \sum_{n=1}^{N} \pi_n \, p(x_{ni}|\mu_i, \Sigma_i). \tag{9.20}$$

It is easy to prove the MLE for μ_i, Σ_i is [7]:

$$q_i^{T^*} = \mu_i^{T^*} = \frac{\pi^T X_i}{\sum_{n=1}^{N} \pi_n}, \tag{9.21}$$

$$W_i^* = \Sigma_i^{-1^*} = C_i^{-1}. \tag{9.22}$$

Comparing Eqs 9.17–9.18 and 9.21–9.22, we can see that from different ways, we have arrived at similar solutions. The physical meanings of Eqs 9.21 and 9.22 are even more intuitive and satisfactory.

After deriving the optimal solutions for W_i, if we look back at the heuristic approach (Eq. 9.6) proposed in Ref. [23], it is actually optimal under some assumptions. That is, if we restrict W_i to be a diagonal matrix (i.e., we use Euclidean distance rather than generalized Euclidean distance), the solution in Ref. [23] is optimal. This is very obvious, since from both Eqs 9.18 and 9.22 the diagonal weights are inversely proportional to the standard deviation.

9.4.2 Practical Considerations

We have shown the optimality of the solutions in the previous subsection. In this subsection, we will further discuss some real-world issues.

In both solutions for W_i (Eqs 9.18 and 9.22), we need to compute the inverse of the covariance matrix C_i. It is clear that, if $N < K_i$, C_i is not invertible, and we thus cannot have W_i. In MindReader, the authors proposed a solution to this by using a pseudo-inverse defined below.

The singular value decomposition (SVD) of C_i is

$$C_i = A \, \Lambda \, B^T \tag{9.23}$$

where Λ is a diagonal matrix: $diag(\lambda_1, \ldots, \lambda_j, \ldots, \lambda_{K_i})$. The λ's are either positive or zero. Suppose there are L non-zero λ's, the pseudo-inverse of C_i is defined as

$$C_i^+ = A \, \Lambda^+ B^T,$$
$$\Lambda^+ = diag\left(\frac{1}{\lambda_1}, \ldots, \frac{1}{\lambda_L}, 0, \ldots, 0\right)$$

where $+$ denotes the pseudo-inverse of a matrix. The approximation solution of W_i^* is then [12]

$$W_i^* = \left(\prod_{l=1}^{L} \lambda_l\right)^{\frac{1}{L}} C_i^+. \tag{9.24}$$

Even though, in theory, we can get around the *singular* problem by using the above procedure, in reality this solution does not give satisfactory results. This is especially true when N is far less than K_i. Remember, we need to use $(N-1) \times K_i$ numbers from the training samples to estimate $\frac{K_i(K_i+1)}{2}$ parameters in C_i. In MindReader, the authors used a $K_i = 2$ example to show the performance of the algorithm. However, in real image retrieval systems, $K_i = 2$ is not very realistic. For example, in HSV color histogram, an $(8 \times 4 \times 4 = 32)$ vector is normally used. In other shape and texture representations, we also have high-dimensionality vectors [11].

Considering this, in practice it is not bad to use the standard deviation approach (Eq. 9.6), especially when $N < K_i$. This approach is much more robust, as it requires only two different training samples to estimate the parameters.

9.5 Parameter Update at Both Levels: Optimal Approaches

As we can see from the previous section, there are many approaches to parameter updating at the representation level. Unfortunately, there is almost no research in optimizing both levels simultaneously.

One natural thought from the previous section is to stack all the representation vectors r_i to form a huge overall vector r and directly use the results from the previous section. In theory, this will work, as long as we have enough training samples. Unfortunately, as we discussed in Section 9.4.2, even for each individual r_i we have the risk of not being able to invert C_i, let alone deal with the stacked huge vector r. This suggests that using a flat model (stacking all r_i's together) will not work in practice. We therefore need to explore approaches that can take advantage of the hierarchical image object model (Fig. 9.1). Recall that the overall distance between a training sample and a query is

$$d_n = \Phi(\Psi_i(x_{ni}, q_i, W_i), U). \tag{9.25}$$

The goal of parameter updating is to minimize the weighted distance between all the training samples and the query. Because Φ and Ψ_i can be non-linear functions in general, a direct optimization of J over U, W_i, q_i is intractable. Depending on which distance functions we use, there are different algorithms. We next examine two of them.

9.5.1 Quadratic in both $\Phi()$ and $\Psi_i()$

In this subsection, we restrict ourselves to use the generalized Euclidean distance for both $\Phi()$ and $\Psi_i()$. That is,

$$d_n = g_n^T U g_n, \tag{9.26}$$

$$g_n = [g_{n1}, \ldots, g_{ni}, \ldots, g_{nI}], \tag{9.27}$$

$$g_{ni} = (x_{ni} - q_i)^T W_i (x_{ni} - q_i). \tag{9.28}$$

The parameter updating process can therefore be formulated as a constrained optimization problem:

$$\min \ J = \boldsymbol{\pi}^T \times \boldsymbol{d}, \tag{9.29}$$

$$d_n = \boldsymbol{g}_n^T \ U \ \boldsymbol{g}_n, \tag{9.30}$$

$$\boldsymbol{g}_n = [g_{n1}, \ldots, g_{ni}, \ldots, g_{nI}], \tag{9.31}$$

$$g_{ni} = (\boldsymbol{x}_{ni} - \boldsymbol{q}_i)^T W_i (\boldsymbol{x}_{ni} - \boldsymbol{q}_i), \tag{9.32}$$

$$s.t. \quad det(U) = 1, \tag{9.33}$$

$$det(W_i) = 1, \tag{9.34}$$

$$n = 1, \ldots, N, \tag{9.35}$$

$$i = 1, \ldots, I. \tag{9.36}$$

We use Lagrange multipliers to solve this constrained optimization problem:

$$L = \boldsymbol{\pi}^T \times \boldsymbol{d} - \lambda(det(U) - 1) - \sum_{i=1}^{I} \lambda_i(det(W_i) - 1). \tag{9.37}$$

Optimal Solution for q_i.

$$\frac{\partial L}{\partial q_i} = \boldsymbol{\pi}^T \times \begin{bmatrix} \frac{\partial d_1}{\partial q_i} \\ \cdots \\ \frac{\partial d_n}{\partial q_i} \\ \cdots \\ \frac{\partial d_N}{\partial q_i} \end{bmatrix} = \boldsymbol{\pi}^T \times \begin{bmatrix} 2\boldsymbol{g}_1^T \ U \ \frac{\partial \boldsymbol{g}_1}{\partial q_i} \\ \cdots \\ 2\boldsymbol{g}_n^T \ U \ \frac{\partial \boldsymbol{g}_n}{\partial q_i} \\ \cdots \\ 2\boldsymbol{g}_N^T \ U \ \frac{\partial \boldsymbol{g}_N}{\partial q_i} \end{bmatrix} \boldsymbol{\pi}^T \times \begin{bmatrix} 2\boldsymbol{g}_1^T \ U \ \times \begin{bmatrix} \frac{\partial g_{11}}{\partial q_i} \\ \cdots \\ \frac{\partial g_{1i}}{\partial q_i} \\ \cdots \\ \frac{\partial g_{1I}}{\partial q_i} \end{bmatrix} \\ \cdots \\ 2\boldsymbol{g}_n^T \ U \ \times \begin{bmatrix} \frac{\partial g_{n1}}{\partial q_i} \\ \cdots \\ \frac{\partial g_{ni}}{\partial q_i} \\ \cdots \\ \frac{\partial g_{nI}}{\partial q_i} \end{bmatrix} \\ \cdots \\ 2\boldsymbol{g}_N^T \ U \ \times \begin{bmatrix} \frac{\partial g_{N1}}{\partial q_i} \\ \cdots \\ \frac{\partial g_{Ni}}{\partial q_i} \\ \cdots \\ \frac{\partial g_{NI}}{\partial q_i} \end{bmatrix} \end{bmatrix}$$

$$= \boldsymbol{\pi}^T \times \begin{bmatrix} -4 \, \boldsymbol{g}_1^T \, U \, \times \begin{bmatrix} 0 \\ \cdots \\ (\boldsymbol{x}_{1i} - \boldsymbol{q}_i)^T \, W_i \\ \cdots \\ 0 \end{bmatrix} \\ \cdots \\ -4 \, \boldsymbol{g}_n^T \, U \, \times \begin{bmatrix} 0 \\ \cdots \\ (\boldsymbol{x}_{ni} - \boldsymbol{q}_i)^T \, W_i \\ \cdots \\ 0 \end{bmatrix} \\ \cdots \\ -4 \, \boldsymbol{g}_N^T \, U \, \times \begin{bmatrix} 0 \\ \cdots \\ (\boldsymbol{x}_{Ni} - \boldsymbol{q}_i)^T \, W_i \\ \cdots \\ 0 \end{bmatrix} \end{bmatrix}$$

$$= \boldsymbol{\pi}^T \times \begin{bmatrix} -4 \, \boldsymbol{h}_1^T \times \begin{bmatrix} 0 \\ \cdots \\ (\boldsymbol{x}_{1i} - \boldsymbol{q}_i)^T \, W_i \\ \cdots \\ 0 \end{bmatrix} \\ \cdots \\ -4 \, \boldsymbol{h}_n^T \times \begin{bmatrix} 0 \\ \cdots \\ (\boldsymbol{x}_{ni} - \boldsymbol{q}_i)^T \, W_i \\ \cdots \\ 0 \end{bmatrix} \\ \cdots \\ -4 \, \boldsymbol{h}_N^T \times \begin{bmatrix} 0 \\ \cdots \\ (\boldsymbol{x}_{Ni} - \boldsymbol{q}_i)^T \, W_i \\ \cdots \\ 0 \end{bmatrix} \end{bmatrix} = \boldsymbol{\pi}^T \times \begin{bmatrix} -4 \, h_{1i}^T \times (\boldsymbol{x}_{1i} - \boldsymbol{q}_i)^T \, W_i \\ \cdots \\ -4 \, h_{ni}^T \times (\boldsymbol{x}_{ni} - \boldsymbol{q}_i)^T \, W_i \\ \cdots \\ -4 \, h_{Ni}^T \times (\boldsymbol{x}_{Ni} - \boldsymbol{q}_i)^T \, W_i \end{bmatrix}$$

where $\boldsymbol{h}_n^T = \boldsymbol{g}_n^T \, U$, and h_{ni} is the ith component of $\boldsymbol{h}_n$.

Setting the above equation to zero, we have

$$\pi^T \times \begin{bmatrix} -4\, h_{1i}^T \times (x_{1i} - q_i)^T \ W_i \\ \cdots \\ -4\, h_{ni}^T \times (x_{ni} - q_i)^T \ W_i \\ \cdots \\ -4\, h_{Ni}^T \times (x_{Ni} - q_i)^T \ W_i \end{bmatrix} = 0$$

$$\pi^T \times \begin{bmatrix} h_{1i}^T \times (x_{1i} - q_i)^T \\ \cdots \\ h_{ni}^T \times (x_{ni} - q_i)^T \\ \cdots \\ h_{Ni}^T \times (x_{Ni} - q_i)^T \end{bmatrix} = 0$$

$$\tilde{\pi}^T \times \begin{bmatrix} (x_{1i} - q_i)^T \\ \cdots \\ (x_{ni} - q_i)^T \\ \cdots \\ (x_{Ni} - q_i)^T \end{bmatrix} = 0$$

where $\tilde{\pi}_{ni} = \pi_n \times h_{ni}$. Note also that we have used the fact that W_i is invertible since $det(W_i) = 1$.

The final solution of q_i is

$$q_i^{T^*} = \frac{\tilde{\pi}^T X_i}{\sum_{n=1}^{N} \tilde{\pi}_{ni}}. \tag{9.38}$$

Optimal Solution for W_i. Before we explore how to find the optimal solution of W_i, we note that the constraint $det(W_i) = 1$ can be rewritten as

$$\sum_{r=1}^{K_i} (-1)^{r+s} w_{i_{rs}} det(W_{i_{rs}}) = 1 \qquad r, s = 1, \ldots, K_i$$

where $det(W_{i_{rs}})$ is the rsth minor of W_i, and $W_i = [w_{i_{rs}}]$.

This equation can be further rewritten as [12]

$$\sum_{r=1}^{K_i} \sum_{s=1}^{K_i} (-1)^{r+s} w_{i_{rs}} det(W_{i_{rs}}) = K_i. \tag{9.39}$$

To obtain the optimal solution of W_i, we take partial derivative of L with respect to $w_{i_{rs}}, r, s = 1, \ldots, K_i$:

$$\frac{\partial L}{\partial w_{i_{rs}}} = \pi^T \times \begin{bmatrix} 2g_1^T \ U \ \frac{\partial g_1}{\partial w_{i_{rs}}} \\ \cdots \\ 2g_n^T \ U \ \frac{\partial g_n}{\partial w_{i_{rs}}} \\ \cdots \\ 2g_N^T \ U \ \frac{\partial g_N}{\partial w_{i_{rs}}} \end{bmatrix} - \lambda_i \ (-1)^{r+s} det(W_{i_{rs}})$$

$$= \pi^T \times \begin{bmatrix} 2g_1^T \ U \ \times \begin{bmatrix} 0 \\ \cdots \\ \frac{\partial g_{1i}}{\partial w_{i_{rs}}} \\ \cdots \\ 0 \end{bmatrix} \\ \cdots \\ 2g_n^T \ U \ \times \begin{bmatrix} 0 \\ \cdots \\ \frac{\partial g_{ni}}{\partial w_{i_{rs}}} \\ \cdots \\ 0 \end{bmatrix} \\ \cdots \\ 2g_N^T \ U \ \times \begin{bmatrix} 0 \\ \cdots \\ \frac{\partial g_{Ni}}{\partial w_{i_{rs}}} \\ \cdots \\ 0 \end{bmatrix} \end{bmatrix} - \lambda_i \ (-1)^{r+s} det(W_{i_{rs}})$$

$$= \pi^T \times \begin{bmatrix} 2h_{1i}^T \times (x_{1ir} - q_{ir})(x_{1is} - q_{is}) \\ \cdots \\ 2h_{ni}^T \times (x_{nir} - q_{ir})(x_{1is} - q_{is}) \\ \cdots \\ 2h_{Ni}^T \times (x_{Nir} - q_{ir})(x_{1is} - q_{is}) \end{bmatrix} - \lambda_i \ (-1)^{r+s} det(W_{i_{rs}})$$

$$= 2 \sum_{n=1}^{N} \tilde{\pi}_n \ (x_{nir} - q_{ir})(x_{1is} - q_{is}) - \lambda_i \ (-1)^{r+s} det(W_{i_{rs}}).$$

Setting the above equation to zero, we will get

$$2 \sum_{n=1}^{N} \tilde{\pi}_n \ (x_{nir} - q_{ir})(x_{1is} - q_{is}) - \lambda_i \ (-1)^{r+s} det(W_{i_{rs}}) = 0,$$

$$det(W_{i_{rs}}) = \frac{2 \sum_{n=1}^{N} \tilde{\pi}_n \ (x_{nir} - q_{ir})(x_{1is} - q_{is})}{\lambda_i \ (-1)^{r+s}}.$$

Define $w_{i_{rs}}^{-1}$ to be the rsth element of matrix W_i^{-1}. We then have

$$w_{i_{rs}}^{-1} = \frac{(-1)^{r+s} \ det(W_{i_{rs}})}{det(W_i)} = (-1)^{r+s} \ det(W_{i_{rs}})$$

$$= (-1)^{r+s} \ \frac{2 \ \sum_{n=1}^{N} \tilde{\pi}_n \ (x_{nir} - q_{ir})(x_{1is} - q_{is})}{\lambda_i \ (-1)^{r+s}}$$

$$= \frac{2 \ \sum_{n=1}^{N} \tilde{\pi}_n \ (x_{nir} - q_{ir})(x_{1is} - q_{is})}{\lambda_i}$$

$$= 2 \sum_{n=1}^{N} \tilde{\pi}_n \frac{\sum_{n=1}^{N} \tilde{\pi}_n \ (x_{nir} - q_{ir})(x_{1is} - q_{is})}{\lambda_i \sum_{n=1}^{N} \tilde{\pi}_n}$$

$$= \frac{1}{\lambda_i'} \frac{\sum_{n=1}^{N} \tilde{\pi}_n \ (x_{nir} - q_{ir})(x_{1is} - q_{is})}{\sum_{n=1}^{N} \tilde{\pi}_n}$$

where $\lambda_i' = \lambda_i/2 \sum_{n=1}^{N} \tilde{\pi}_n$.

Knowing q_i is just the weighted average of x_{ni}'s, it is easy to observe that $\frac{\sum_{n=1}^{N} \tilde{\pi}_n (x_{nir} - q_{ir})(x_{1is} - q_{is})}{\sum_{n=1}^{N} \tilde{\pi}_n}$ is nothing but the rsth element of the weighted covariance matrix (C_i) of X_i. That is, $C_i = \lambda_i' \ W_i^{-1}$. Take determinant of both sides, we have

$$\lambda_i'^{K_i} det(W_i^{-1}) = det(C_i),$$

$$\lambda_i'^{K_i} = det(C_i,)$$

$$\lambda_i' = det(C_i))^{\frac{1}{K_i}}.$$

Therefore the optimal solution for W_i is

$$W_i^* = (det(C_i))^{\frac{1}{K_i}} \ C_i^{-1}. \tag{9.40}$$

Optimal Solution for U.

$$\frac{\partial L}{\partial u_{rs}} = \pi^T \times \begin{bmatrix} g_{1r} \ g_{1s} \\ \cdots \\ g_{nr} \ g_{ns} \\ \cdots \\ g_{Nr} \ g_{Ns} \end{bmatrix} - \lambda \ (-1)^{r+s} det(U_{rs})$$

$$= \sum_{n=1}^{N} \pi_n \ g_{nr} \ g_{ns} - \lambda(-1)^{r+s} det(U_{rs}).$$

where $det(U_{rs})$ is the rsth minor of U.

Set the above equation to zero, we will get

$$\sum_{n=1}^{N} \pi_n \ g_{nr} \ g_{ns} - \lambda(-1)^{r+s} det(U_{rs}) = 0,$$

$$\sum_{n=1}^{N} \pi_n \ g_{nr} \ g_{ns} = \lambda(-1)^{r+s} det(U_{rs}).$$

This equation has a very similar form to Eq. 9.40. We know its solution is

$$U^* = (det(R))^{\frac{1}{I}} R^{-1} \tag{9.41}$$

where R is the weighted correlation matrix of $g_n, n = 1, \ldots, N$:

$$R_{rs} = \sum_{n=1}^{N} \pi_n \, g_{nr} \, g_{ns} \qquad r, s = 1, \ldots, I.$$

Convergence and Performance. Even though we have obtained explicit optimal solutions

$$q_i^{T*} = \frac{\tilde{\pi}^T X_i}{\sum_{n=1}^{N} \tilde{\pi}_i}, \tag{9.42}$$

$$W_i^* = \left(2 \sum_{n=1}^{N} \tilde{\pi}_n det(C_i) \right)^{\frac{1}{K_i}} C_i^{-1}, \tag{9.43}$$

$$U^* = (det(R))^{\frac{1}{I}} R^{-1}, \tag{9.44}$$

they actually depend on

$$g_{ni} = (x_{ni} - q_i)^T W_i (x_{ni} - q_i), \tag{9.45}$$

which in turn depends on q_i and W_i.

To prove its convergence is not easy, since J is highly non-linear in q_i and W_i. If indeed we can prove its convergence, it is at most an iterative algorithm. In theory this works fine, but for a particular application such as image retrieval, this is not desirable. Fast response time is always one of the most important requirements in image retrieval. The above discussion shows that if both $\Phi()$ and $\Psi_i()$ take quadratic form, the algorithm will not work in practice. We then need to explore other suitable forms for $\Phi()$ and $\Psi_i()$.

9.5.2 Linear in $\Phi()$ and Quadratic in $\Psi_i()$

Because there is no explicit solution for q_i, W_i, and U when both $\Phi()$ and $\Psi()$ are quadratic, we further simplify $\Phi()$ to be a linear function:

$$\min J = \pi^T \times d, \tag{9.46}$$

$$d_n = u^T g_n, \tag{9.47}$$

$$g_n = [g_{n1}, \ldots, g_{ni}, \ldots, g_{nI}], \tag{9.48}$$

$$g_{ni} = (x_{ni} - q_i)^T W_i (x_{ni} - q_i), \tag{9.49}$$

$$s.t. \quad \sum_{i=1}^{I} \frac{1}{u_i} = 1, \tag{9.50}$$

$$det(W_i) = 1, \tag{9.51}$$

$$n = 1, \ldots, N, \tag{9.52}$$

$$i = 1, \ldots, I, \tag{9.53}$$

where the matrix U now reduces to a vector u and its constraint is changed accordingly as shown in Eq. 9.50.

Again, we use Lagrange multipliers to solve this constrained optimization problem:

$$L = \pi^T \times d - \lambda \left(\sum_{i=1}^{I} \frac{1}{u_i} - 1 \right) - \sum_{i=1}^{I} \lambda_i (det(W_i) - 1). \tag{9.54}$$

Optimal Solution for q_i.

$$\frac{\partial L}{\partial q_i} = \pi^T \times \begin{bmatrix} \frac{\partial d_1}{\partial q_i} \\ \cdots \\ \frac{\partial d_n}{\partial q_i} \\ \cdots \\ \frac{\partial d_N}{\partial q_i} \end{bmatrix} = \pi^T \times \begin{bmatrix} u^T \frac{\partial g_1}{\partial q_i} \\ \cdots \\ u^T \frac{\partial g_n}{\partial q_i} \\ \cdots \\ u^T \frac{\partial g_N}{\partial q_i} \end{bmatrix} = \pi^T \times \begin{bmatrix} u^T \times \begin{bmatrix} \frac{\partial g_{11}}{\partial q_i} \\ \cdots \\ \frac{\partial g_{1i}}{\partial q_i} \\ \cdots \\ \frac{\partial g_{1I}}{\partial q_i} \end{bmatrix} \\ \cdots \\ u^T \times \begin{bmatrix} \frac{\partial g_{n1}}{\partial q_i} \\ \cdots \\ \frac{\partial g_{ni}}{\partial q_i} \\ \cdots \\ \frac{\partial g_{nI}}{\partial q_i} \end{bmatrix} \\ \cdots \\ u^T \times \begin{bmatrix} \frac{\partial g_{N1}}{\partial q_i} \\ \cdots \\ \frac{\partial g_{Ni}}{\partial q_i} \\ \cdots \\ \frac{\partial g_{NI}}{\partial q_i} \end{bmatrix} \end{bmatrix}$$

$$= \pi^T \times \begin{bmatrix} -2\, u^T \times \begin{bmatrix} 0 \\ \cdots \\ (x_{1i} - q_i)^T\, W_i \\ \cdots \\ 0 \end{bmatrix} \\ \cdots \\ -2\, u^T \times \begin{bmatrix} 0 \\ \cdots \\ (x_{ni} - q_i)^T\, W_i \\ \cdots \\ 0 \end{bmatrix} \\ \cdots \\ -2\, u^T \times \begin{bmatrix} 0 \\ \cdots \\ (x_{Ni} - q_i)^T\, W_i \\ \cdots \\ 0 \end{bmatrix} \end{bmatrix} = \pi^T \times \begin{bmatrix} -2\, u_i\, (x_{1i} - q_i)^T\, W_i \\ \cdots \\ -2\, u_i\, (x_{ni} - q_i)^T\, W_i \\ \cdots \\ -2\, u_i\, (x_{Ni} - q_i)^T\, W_i \end{bmatrix}.$$

By setting the above equation to zero, we will have

$$\boldsymbol{\pi}^T \times \begin{bmatrix} -2 \, u_i \, (\boldsymbol{x}_{1i} - \boldsymbol{q}_i)^T \, W_i \\ \cdots \\ -2 \, u_i \, (\boldsymbol{x}_{ni} - \boldsymbol{q}_i)^T \, W_i \\ \cdots \\ -2 \, u_i \, (\boldsymbol{x}_{Ni} - \boldsymbol{q}_i)^T \, W_i \end{bmatrix} = 0 \Rightarrow \boldsymbol{\pi}^T \times \begin{bmatrix} (\boldsymbol{x}_{1i} - \boldsymbol{q}_i)^T \\ \cdots \\ (\boldsymbol{x}_{ni} - \boldsymbol{q}_i)^T \\ \cdots \\ (\boldsymbol{x}_{Ni} - \boldsymbol{q}_i)^T \end{bmatrix} = 0.$$

The final solution of $\boldsymbol{q}_i$ is

$$\boldsymbol{q}_i^{T^*} = \frac{\boldsymbol{\pi}^T \, X_i}{\sum_{n=1}^N \pi_n}. \tag{9.55}$$

This solution closely matches our intuition. That is, $\boldsymbol{q}_i^{T^*}$ is nothing but the weighted average of the training samples. It is also exactly the same as the optimal solutions obtained in MindReader (Eq. 9.17) and MLE (Eq. 9.21).

Optimal Solution for W_i.

$$\frac{\partial L}{\partial w_{i_{rs}}} = \boldsymbol{\pi}^T \times \begin{bmatrix} \boldsymbol{u}^T \, \frac{\partial \boldsymbol{g}_1}{\partial w_{i_{rs}}} \\ \cdots \\ \boldsymbol{u}^T \, \frac{\partial \boldsymbol{g}_n}{\partial w_{i_{rs}}} \\ \cdots \\ \boldsymbol{u}^T \, \frac{\partial \boldsymbol{g}_N}{\partial w_{i_{rs}}} \end{bmatrix} - \lambda_i \, (-1)^{r+s} det(W_{i_{rs}})$$

$$= \boldsymbol{\pi}^T \times \begin{bmatrix} \boldsymbol{u}^T \times \begin{bmatrix} 0 \\ \cdots \\ \frac{\partial g_{1i}}{\partial w_{i_{rs}}} \\ \cdots \\ 0 \end{bmatrix} \\ \cdots \\ \boldsymbol{u}^T \times \begin{bmatrix} 0 \\ \cdots \\ \frac{\partial g_{ni}}{\partial w_{i_{rs}}} \\ \cdots \\ 0 \end{bmatrix} \\ \cdots \\ \boldsymbol{u}^T \times \begin{bmatrix} 0 \\ \cdots \\ \frac{\partial g_{Ni}}{\partial w_{i_{rs}}} \\ \cdots \\ 0 \end{bmatrix} \end{bmatrix} - \lambda_i \, (-1)^{r+s} det(W_{i_{rs}})$$

$$= \pi^T \times \begin{bmatrix} u_i \ (x_{1ir} - q_{ir})(x_{1is} - q_{is}) \\ \cdots \\ u_i \ (x_{nir} - q_{ir})(x_{1is} - q_{is}) \\ \cdots \\ u_i \ (x_{Nir} - q_{ir})(x_{1is} - q_{is}) \end{bmatrix} - \lambda_i \ (-1)^{r+s} det(W_{i_{rs}})$$

$$= \sum_{n=1}^{N} \pi_n \ (x_{nir} - q_{ir})(x_{1is} - q_{is}) - \lambda_i \ (-1)^{r+s} det(W_{i_{rs}}).$$

After setting the above equation to zero and going through procedure similar to that in Section 9.5.1, we get

$$W_i^* = (det(C_i))^{\frac{1}{K_i}} \ C_i^{-1} \tag{9.56}$$

where $C_{i_{rs}} = \sum_{n=1}^{N} \pi_n \ (x_{nir} - q_{ir})(x_{1is} - q_{is}), r, s = 1, \ldots, K_i$.

This solution also matches our intuition nicely, since W_i is inversely proportional to the covariance matrix C_i.

Optimal Solution for u. To obtain u_i, set the partial derivative to zero. We then have

$$\frac{\partial L}{\partial u_i} = \sum_{n=1}^{N} \pi_n \ g_{ni} + \lambda \ u_i^{-2} = 0, \ \forall i. \tag{9.57}$$

Multiply both sides by u_i and summarize over i. We then have

$$\sum_{i=1}^{I} u_i \left(\sum_{n=1}^{N} \pi_n \ g_{ni} \right) + \lambda \left(\sum_{i=1}^{I} \frac{1}{u_i} \right) = 0. \tag{9.58}$$

Since $\sum_{i=1}^{I} \frac{1}{u_i} = 1$, the optimal λ is

$$\lambda^* = - \sum_{i=1}^{I} u_i f_i \tag{9.59}$$

where $f_i = \sum_{n=1}^{N} \pi_n \ g_{ni}$.

After substituting Eq. 9.59 into Eq. 9.57, we have

$$u_i^2 \ f_i = \sum_{j=1}^{I} u_j \ f_j. \tag{9.60}$$

From Eq. 9.57, we also know that

$$\lambda = u_1^2 f_1 = \cdots = u_i^2 f_i = \cdots = u_j^2 f_j = \cdots = u_I^2 f_I. \tag{9.61}$$

That is,

$$u_j = u_i \sqrt{\frac{f_i}{f_j}}, \ \forall j. \tag{9.62}$$

After substituting this equation into Eq. 9.60, we obtain the optimal solution for u_i:

$$f_i \, u_i^2 - \left(\sum_{j=1}^{I} \sqrt{f_i \, f_j} \right) u_i = 0, \tag{9.63}$$

$$u_i^* = \sum_{j=1}^{I} \sqrt{\frac{f_j}{f_i}} \tag{9.64}$$

where we discard the solution that $u_i = 0$, because we know $u_i > 0$. Recall that $f_i = \sum_{n=1}^{N} \pi_n \, g_{ni}$; this solution also closely matches our intuition. That is, if the total distance (f_i) of representation i is small (meaning it is close to the ideal query), this representation should receive high weight.

The optimal solutions are summarized in the following equations:

$$q_i^{T^*} = \frac{\pi^T \, X_i}{\sum_{n=1}^{N} \pi_i}, \tag{9.65}$$

$$W_i^* = (det(C_i))^{\frac{1}{K_i}} \, C_i^{-1}, \tag{9.66}$$

$$u_i^* = \sum_{j=1}^{I} \sqrt{\frac{f_j}{f_i}} \tag{9.67}$$

where

$$C_{i_{rs}} = \pi_n \, (x_{nir} - q_{ir})(x_{1is} - q_{is}), \tag{9.68}$$

$$f_i = \sum_{n=1}^{N} \pi_n \, g_{ni}, \tag{9.69}$$

$$g_{ni} = (x_{ni} - q_i)^T W_i (x_{ni} - q_i), \tag{9.70}$$

$$i = 1, \ldots, I, \tag{9.71}$$

$$r, s = 1, \ldots, K_i. \tag{9.72}$$

If we compare this set of equations with Eqs 9.42–9.44, we can see that, unlike Eqs 9.42–9.44, solutions in this approach are de-coupled from each other. No iteration is needed, and they can be obtained by a single step, which is very desirable for image retrieval applications.

Algorithm Descriptions and Computation Complexity Analysis. In this subsection, we will give complete algorithms for both parameter updating and ranked retrieval, and we will examine their computation complexity.

Based on Eqs 9.65–9.70, the algorithm can be summarized as follows:

1. Input: N training samples x_{ni} and their corresponding degree-of-relevance vector π.
2. Output: optimal solutions as defined in Eqs 9.65–9.67.
3. Procedure:
   (a) Compute q_i as defined in Eq. 9.65.
   (b) Compute C_i as defined in Eq. 9.68.
   (c) Compute W_i as defined in Eq. 9.66.
   (d) Compute g_{ni} as defined in Eq. 9.70.
   (e) Compute f_i as defined in Eq. 9.69.
   (f) Compute u_i as defined in Eq. 9.67.

Let a/s and m/d denote the number for addition/subtraction and multiplication/division, respectively. The computation complexity of the above procedure is summarized as follows:

- For q_i, $\forall i$, $a/s = O(N\,(K_i+1))$ and $m/d = O(N\,K_i)$.
- For C_i, $\forall i$, $a/s = O(2N\,K_i^2)$ and $m/d = O(2N\,K_i^2)$.
- For W_i, $\forall i$, $a/s = O(K_i^3 + 2N\,K_i^2)$ and $m/d = O(K_i^3 + 2N\,K_i^2)$.
- For g_{ni}, $\forall n$,$\forall i$, $a/s = O(K_i^2 + 2K_i)$ and $m/d = O(K_i^2 + K_i)$.
- For f_i, $\forall i$, $a/s = O(N(K_i^2 + 2K_i))$ and $m/d = O(N(K_i^2 + 2K_i))$.
- For u_i, $\forall i$, $a/s = O(N(K_i^2 + 2K_i))$ and $m/d = O(N(K_i^2 + 2K_i))$.

Therefore, for a single representation i, the total computation complexity is $a/s = O(K_i^3 + 3NK_i^2 + 3NK_i + N)$ and $m/d = O(K_i^3 + 3NK_i^2 + 3NK_i)$. For all the representations, $a/s = \sum_{i=1}^{I} O(K_i^3 + 3NK_i^2 + 3NK_i + N)$ and $m/d = \sum_{i=1}^{I} O(K_i^3 + 3NK_i^2 + 3NK_i)$. Therefore, the computation complexity is polynomial in all I, N, and K_i. To give readers a more concrete feeling for the complexity, we will plug in some real-world values. Let $I = 5$, $K_i = 10$, and $N = 10$. We will need only 4310 additions/subtractions and 4300 multiplications/divisions to estimate the optimal parameters.

Once we have found the optimal solutions for q_i, W_i, and u, we can then perform the ranked retrieval by computing the distance between an image and the query:

$$d_m = u^T g_m, \tag{9.73}$$

$$g_m = [g_{m1}, \ldots, g_{mi}, \ldots, g_{mI}], \tag{9.74}$$

$$g_{mi} = (r_{mi} - q_i)^T W_i (r_{mi} - q_i), \tag{9.75}$$

$$m = 1, \ldots, M, \tag{9.76}$$

$$i = 1, \ldots, I \tag{9.77}$$

where M is the total number of images in the database and r_{mi} the ith representation vector for the mth image in the database. The algorithm can then be summarized as follows:

1. Input: image representation vectors r_{mi}, query vectors q_i, and weights W_i and u.
2. Output: distance values for all the images in the database.

3. Procedure:
   (a) Compute g_{mi} as defined in Eq. 9.75.
   (b) Compute the overall distance d_m according to Eq. 9.73.

The computation complexity of the above procedure can be summarized as follows:

- For g_{mi}, $\forall m$, $\forall i$, $a/s = O(K_i^2 + 2K_i)$ and $m/d = O(K_i^2 + K_i)$.
- For d_m, $\forall m$, $a/s = O(K_i^2 + 2K_i + I)$ and $m/d = O(K_i^2 + K_i + I)$.

Therefore, for all the images in the database, we need $a/s = \sum_{m=1}^{M} O(K_i^2 + 2K_i + I)$ and $m/d = \sum_{m=1}^{M} O(K_i^2 + K_i + I)$. Therefore, the computation complexity is polynomial in all I, N, and K_i. Again, to give readers a more concrete feeling for the complexity, we will plug in some real-world values. Let $I = 5$, $K_i = 10$, and $M = 10,000$. We will need 1,250,000 additions/subtractions and the same number of multiplications/divisions.

As we can see, a very large portion of computation complexity is not from parameter estimation but from distance computing. This suggests two things. One is that the proposed parameter-updating approach is very efficient. The other is that in order to search quickly, a linear scan of the whole database is not acceptable. We need to develop effective multi-dimensional indexing techniques to improve the query responding time.

9.6 Case Studies

9.6.1 Data Sets

In the experiments reported in this section, the proposed algorithms are tested on one or more of the following data sets.

MESL data set. This image collection was provided by the Fowler Museum of Cultural History at the University of California–Los Angeles. It contains 286 ancient African and Peruvian artifacts and is part of the Museum Educational Site Licensing Project (MESL), sponsored by the Getty Information Institute.

Corel data set. The second image collection was obtained from Corel Corporation. It contains more than 70,000 images covering a wide range of more than 500 categories. The 120 × 80 resolution images are available at `corel.digitalriver.com/commerce/photostudio/catalog.htm`.

Vistex data set A. This data set consists of 384 texture images. The original 24 512 × 512 texture images were obtained from MIT Media Lab at `ftp://whitechapel.media.mit.edu/pub /VisTex/`. Each image is then cut into 16 128 × 128 nonoverlapping small images. The 16 images from the same big image are considered to be relevant images.

Vistex data set B. The same as Vistex data set A, but has more images. The 832 small images are obtained from 52 big images.

MPEG-7 data set. This data set is for MPEG-7 proposal evaluation. We have chosen 300 images to test our algorithms.

9.6.2 Experiments for Algorithms in Sections 9.3.1 and 9.3.2

These experiments are to test the heuristic-based approaches to updating w_i and q_i. The test data set is Vistex data set A. The representations used are co-occurrence matrix texture representation and wavelet texture representation [23]. Each of the 384 images is selected as the query image, and the 15 best matches are returned. Typical retrieval results are shown in Fig. 9.4. The average retrieval precision of 384 query images is summarized in Table 9.1, where the precision is defined as

$$precision = \frac{relevant\ \ images}{returned\ \ images} \times 100\%. \tag{9.78}$$

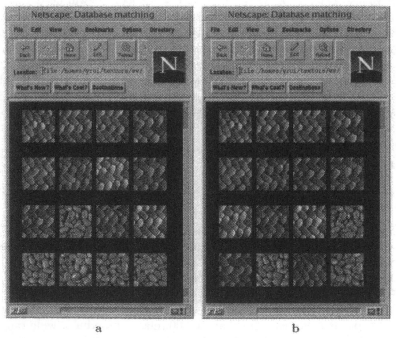

a   b

Fig. 9.4 a Before relevance feedback. **b** After relevance feedback.

There are four columns in the table. The term *0 rf* stands for no relevance feedback; *1 rf* corresponds to one iteration of relevance feedback; and so forth.

The purpose of the experiments is not to compare one texture representation with another, but rather to compare the retrieval performance with relevance feedback versus the performance without relevance feedback. Three observations can be made:

1. The retrieval precision is considerably more improved in the feedback case than in the nonfeedback case.

Table 9.1 Retrieval precision (wv: wavelet based; co: co-occurrence matrix based)

	0 rf	1 rf	2 rf	3 rf
wv(q_i)	77.27	82.33	85.13	85.53
wv(w_i)	77.67	80.27	80.47	80.53
co(q_i)	57.80	63.47	65.13	66.40
co(q_i)	44.33	48.53	48.80	48.80

2. The precision increase in the first iteration of feedback is the largest. Subsequent feedbacks will only achieve minor improvement in precision. This is a very desirable property, because this will guarantee that an acceptable retrieval result is achieved within a limited number of feedback cycles.
3. Updating both the query vectors (q_i) and the weights (w_i) can improve the retrieval performance.

9.6.3 Experiments for Algorithm in Section 9.3.3

These experiments test the heuristic-based parameter updating algorithm for u. The data sets are MESL and Corel. We have chosen these two test sets because they complement each other. The size of the MESL test set is relatively small, but it allows us to meaningfully explore all the color, texture, and shape features simultaneously. On the other hand, although the heterogeneity of the Corel test set makes extracting some features (such as shape) difficult, it has the advantages of large size and wide coverage. We believe that testing our proposed approach on both sets will provide a fair evaluation of the performance of our method.

For the MESL test set, the visual features used are color, texture, and shape of the objects in the image. That is,

$$F = \{f_i\} = \{\text{color, texture, shape}\}.$$

The representations used are color histogram and color moments [33] for the color feature; Tamura [8, 34] and co-occurrence matrix [10, 21] texture representations for the texture feature, and Fourier descriptor and Chamfer shape descriptor [24] for the shape feature.

$$R = \{r_{ij}\} = \{r_1, r_2, r_3, r_4, r_5, r_6\}$$
$$= \{\text{color histogram, color moments, Tamura, co-occurrence}$$
$$\text{matrix, Fourier descriptor, Chamfer shape descriptor}\}.$$

For the Corel test set, the visual features used are color and texture. That is,

$$F = \{f_i\} = \{\text{color, texture}\}.$$

The representations used are color histogram and color moments [33] for color feature; and co-occurrence matrix [10, 21] texture representation for texture feature:

$$R = \{r_{ij}\} = \{r_1, r_2, r_3\}$$
$$= \{\text{color histogram, color moments, co-occurrence matrix}\}.$$

Our proposed relevance feedback architecture is an *open* retrieval architecture. Other visual features or feature representations can be easily incorporated, if needed. The similarity measures used for the corresponding representations are the following. Color histogram intersection [33] is used for the color histogram representation; weighted Euclidean is used for the color moments, Tamura texture, co-occurrence matrix, and Fourier shape descriptor [24] representations; and Chamfer matching [24] is used for the Chamfer shape representation.

There are two sets of experiments reported here. The first set is on the efficiency of the retrieval algorithm, i.e., how fast the retrieval results converge to the true results (objective test). The second set of experiments is on the effectiveness of the retrieval algorithm, i.e., how good the retrieval results are subjectively (subjective test).

Efficiency of the Algorithm. The only assumption that we make in the experiments is that the user is consistent when doing relevance feedback. That is, the user does not change his or her information need during the feedback process. With this assumption, the feedback process can be simulated by a computer.

As we have shown in Fig. 9.1, the image object is modeled by the combinations of representations with their corresponding weights. If we fix the representations, then a query can be completely characterized by the set of weights embedded in the query object Q. Let set s_1 be the *highly relevant* set, set s_2 the *relevant* set, set s_3 the *no-opinion* set, set s_4 the *nonrelevant* set, and set s_5 the *highly nonrelevant* set. The testing procedure is described as follows:

1. Retrieval results of the ideal case: let u^* be the set of weights associated with the query object Q. The retrieval results based on u^* are the ideal case and serve as the baseline for comparing other non-ideal cases.
   (a) Specify a set of weights, u^*, to the query object.
   (b) Set $u = u^*$.
   (c) Invoke the retrieval algorithm (see Section 9.2).
   (d) Obtain the best N_{RT} returns, RT^*.
   (e) From RT^*, find the sizes of sets s_i and n_i, $i = 1, \ldots, 5$. The s_i's are marked by human for testing purpose.
   (f) Calculate the ideal weighted relevant count as

   $$count^* = 3 \times n_1 + 1 \times n_2. \tag{9.79}$$

Note that 3 and 1 are the scores of the *highly relevant* and *relevant* sets, respectively (see Section 9.3.3). Therefore, *count** is the maximal achievable weighted relevant count and serves as the baseline for comparing other non-ideal case.

2. Retrieval results of relevance feedback case: in the real retrieval situation, neither the user nor the computer knows the specified weights u^*. However, the proposed retrieval algorithm will move the initial weights u^0 to the ideal weights u^* via relevance feedback.

   (a) Set $u = u^0$.

   (b) Set the maximum number of iterations of relevance feedback, P_{fd}.

   (c) Initialize the iteration counter, $p_{fd} = 0$.

   (d) Invoke the retrieval algorithm and get back the best N_{RT} returns, $RT(p_{fd})$ (see Section 9.2).

   (e) Compute the weighted relevant count for the current iteration:

   $$count(p_{fd}) = 3 \times n_1(p_{fd}) + 1 \times n_2(p_{fd}) \tag{9.80}$$

   where $n_1(p_{fd})$ and $n_2(p_{fd})$ are the number of *highly relevant* and *relevant* objects in $RT(p_{fd})$. These two numbers can be determined by comparing $RT(p_{fd})$ against RT^*.

   (f) Compute the convergence ratio $CR(p_{fd})$ for the current iteration:

   $$CR(p_{fd}) = \frac{count(p_{fd})}{count^*} \times 100\% \tag{9.81}$$

   (g) Set $p_{fd} = p_{fd} + 1$. If $p_{fd} \geq P_{fd}$, quit; otherwise continue.

   (h) Feed back the current five sets $s_i, i = 1, \ldots, 5$, to the retrieval system.

   (i) Update the weights u according to Eqs 9.10–9.12. Go to step 2(d).

There are three parameters that affect the behavior of the retrieval algorithm: number of feedbacks P_{fd}, number of returns N_{RT}, and specified query weights u^*. For P_{fd}, the more relevance feedback iterations, the better the retrieval performance. However, we cannot expect the user to do relevance feedback forever. In the experiments reported here, we set $P_{fd} = 3$ to study the convergence behavior of the first three iterations. The experiments show that the greatest CR increase occurs in the first iteration of feedback, which is a very desirable property.

In all the experiments reported here, for both the MESL and the Corel test sets, 100 randomly selected images are used as the query images, and the values of CR listed in the tables are the averages of the 100 cases.

CR as a function of u^*. In the MESL test set, there are six representations as described at the beginning of this section. Therefore, both u^* and u^0 have six elements. In addition,

$$u^{0^T} = \begin{bmatrix} \dfrac{1}{6} & \dfrac{1}{6} & \dfrac{1}{6} & \dfrac{1}{6} & \dfrac{1}{6} & \dfrac{1}{6} \end{bmatrix} \tag{9.82}$$

where each entry in the vector u^0 is the weight for its corresponding representation.

In the Corel test set, there are three representations as described at the beginning of this section. Therefore, both u^* and u^0 have three components. In addition,

$$u^{0^T} = \begin{bmatrix} \dfrac{1}{3} & \dfrac{1}{3} & \dfrac{1}{3} \end{bmatrix} \tag{9.83}$$

where each entry in the vector u^0 is the weight for its corresponding representation.

Obviously, the retrieval performance is affected by the offset of the specified weights u^* from the initial weights u^0. We classify u^* into two categories (moderate offset and significant offset) by considering how far away they are from the initial weights u^0.

For the MESL test set, the six moderate offset testing weights are

$$u_1^{*^T} = [0.5\ 0.1\ 0.1\ 0.1\ 0.1\ 0.1],$$
$$u_2^{*^T} = [0.1\ 0.5\ 0.1\ 0.1\ 0.1\ 0.1],$$
$$u_3^{*^T} = [0.1\ 0.1\ 0.5\ 0.1\ 0.1\ 0.1],$$
$$u_4^{*^T} = [0.1\ 0.1\ 0.1\ 0.5\ 0.1\ 0.1],$$
$$u_5^{*^T} = [0.1\ 0.1\ 0.1\ 0.1\ 0.5\ 0.1],$$
$$u_6^{*^T} = [0.1\ 0.1\ 0.1\ 0.1\ 0.1\ 0.5].$$

The six significant offset testing weights are

$$u_7^{*^T} = [0.75\ 0.05\ 0.05\ 0.05\ 0.05\ 0.05],$$
$$u_8^{*^T} = [0.05\ 0.75\ 0.05\ 0.05\ 0.05\ 0.05],$$
$$u_9^{*^T} = [0.05\ 0.05\ 0.75\ 0.05\ 0.05\ 0.05],$$
$$u_{10}^{*^T} = [0.05\ 0.05\ 0.05\ 0.75\ 0.05\ 0.05],$$
$$u_{11}^{*^T} = [0.05\ 0.05\ 0.05\ 0.05\ 0.75\ 0.05],$$
$$u_{12}^{*^T} = [0.05\ 0.05\ 0.05\ 0.05\ 0.05\ 0.75].$$

For the Corel test set, the three moderate offset testing weights are

$$u_1^{*^T} = [0.6\ 0.2\ 0.2],$$
$$u_2^{*^T} = [0.2\ 0.6\ 0.2],$$
$$u_3^{*^T} = [0.2\ 0.2\ 0.6].$$

The three significant offset testing weights are

$$u_4^{*^T} = [0.8\ 0.1\ 0.1],$$
$$u_5^{*^T} = [0.1\ 0.8\ 0.1],$$
$$u_6^{*^T} = [0.1\ 0.1\ 0.8].$$

The experimental results for these cases are summarized in Tables 9.2–9.5.

Table 9.2 MESL moderate offset convergence ratio with $N_{RT} = 12$

Weights	0 feedback	1 feedback	2 feedbacks	3 feedbacks
u_1^*	88.4	99.7	99.7	99.8
u_2^*	64.0	98.4	98.5	98.0
u_3^*	80.7	95.4	93.2	95.8
u_4^*	56.5	94.9	94.9	94.0
u_5^*	66.9	87.9	88.1	88.0
u_6^*	83.5	95.4	97.4	97.5
Avg	73.3	95.3	95.3	95.5

Table 9.3 MESL significant offset convergence ratio with $N_{RT} = 12$

Weights	0 feedback	1 feedback	2 feedbacks	3 feedbacks
u_7^*	40.8	89.6	97.4	98.5
u_8^*	38.9	95.7	98.8	99.0
u_9^*	39.0	77.7	74.2	77.3
u_{10}^*	34.8	91.9	94.6	94.1
u_{11}^*	62.9	85.7	87.1	87.1
u_{12}^*	64.1	87.6	94.4	95.3
Avg	46.8	88.0	91.1	91.9

Table 9.4 Corel moderate offset convergence ratio with $N_{RT} = 1000$

Weights	0 feedback	1 feedback	2 feedbacks	3 feedbacks
u_1^*	57.1	71.7	74.7	75.2
u_2^*	55.2	54.7	54.6	54.6
u_3^*	59.1	71.4	74.0	74.5
Avg	57.1	65.9	67.8	68.1

Table 9.5 Corel significant offset convergence ratio with $N_{RT} = 1000$

Weights	0 feedback	1 feedback	2 feedbacks	3 feedbacks
u_4^*	40.2	63.4	68.8	70.0
u_5^*	31.1	34.4	35.1	35.3
u_6^*	41.8	62.0	66.7	67.6
Avg	37.7	53.3	56.9	57.6

To better represent the process of convergence, we redraw the average CR of the MESL test set and the Corel test set in Fig. 9.5.

Based on the tables and figures, some observations can be made:

– In all the cases, CR increases the most in the first iteration. Later iterations result in only a minor increase in CR. This is a very desirable property, which ensures that the user gets reasonable results after only one iteration of feedback. No further feedbacks are needed, if time is a concern.

- CR is affected by the degree of offset. The lower the offset, the higher the final absolute CR. However, the higher the offset, the higher the relative increase of CR.
- Although the finial absolute CR is higher for the MESL test set than for the Corel test set, the final relative increase of CR is comparable for both test sets (around 10%–20%). The convergence process is more challenging for the Corel test set, because of its larger size and fewer number of feature representations.

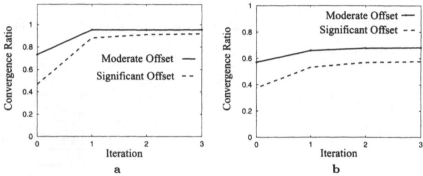

a   b

Fig. 9.5 Convergence ratio curves: **a** MESL test set, **b** Corel test set.

CR **as a function of** N_{RT}. N_{RT} is related to the size of the test data set. Normally only 2%–5% of the whole data set is needed. For the MESL test set, we test $N_{RT} = 10, \ldots, 20$. For the Corel test set, we test $N_{RT} = 850, \ldots, 1100$ The experimental results are listed in Tables 9.6 and 9.7.

Table 9.6 Convergence ratio for MESL test set with u_7^*

N_{NT}	0 feedback	1 feedback	2 feedbacks	3 feedbacks
10	40.4	89.7	97.4	98.6
12	40.9	89.6	97.4	98.5
14	40.8	90.2	97.9	98.4
16	40.2	91.3	98.0	98.2
18	40.4	93.2	97.8	98.0
20	40.5	95.9	97.8	97.9

Some observations can be made based on the experiments:

- The first iteration's CR increases the most when N_{RT} is large. This is because the larger the number of returns, the greater the feedback and thus the better the retrieval performance.

Table 9.7 Convergence ratio for Corel test set with u_4^*

N_{NT}	0 feedback	1 feedback	2 feedbacks	3 feedbacks
850	40.6	63.4	68.5	69.6
900	40.9	63.8	69.0	70.1
950	41.2	64.2	69.6	70.6
1000	41.4	64.6	70.0	71.1
1050	41.6	65.1	70.6	71.7
1100	41.9	65.5	70.9	72.0

− In the second and third iterations, CR is almost independent of different N_{RT}'s. This is because, after first iteration's feedback, most of the desired objects have been found, and later performance is almost independent of N_{RT}.

Effectiveness of the Algorithm. The previous subsection's experiments focused on the convergence of the algorithm. This subsection will focus on how good the returns are *subjectively*. The only way to perform subjective tests is to ask the user to evaluate the retrieval system subjectively. Extensive experiments have been carried out. Users from various disciplines (such as computer vision, art, library science), as well as users from industry, were invited to compare the retrieval performance between the proposed *interactive* approach and the *isolated* approach. All users rated the proposed approach much higher than the *computer-centric* approach in terms of capturing their perception subjectivity and information need.

A typical retrieval process on the MESL test set is given in Figs 9.6 and 9.7.

The user can browse through the image database. Once he or she finds an image of interest, that image is submitted as a query. As an alternative to this query-by-example mode, the user can also submit images outside the database as queries. In Fig. 9.6, the query image is displayed at the upper-left corner and the best 11 retrieved images, with $u = u^0$, are displayed in order from top to bottom and from left to right. The retrieved results are obtained based on their overall similarities to the query image, which are computed from all the features and all the representations. Some retrieved images are similar to the query image in terms of shape feature, while others are similar to the query image in terms of color or texture feature.

Assume the user's true information need is to "retrieve similar images based on their shapes." In the proposed retrieval approach, the user is no longer required to explicitly map an information need to low-level features; rather, the user can express an intended information need by marking the relevance scores of the returned images. In this example, images 247, 218, 228 and 164 are marked *highly relevant*. Images 191, 168, 165, and 78 are marked *highly nonrelevant*. Images 154, 152, and 273 are marked *no opinion*.

Fig. 9.6 The initial retrieval results.

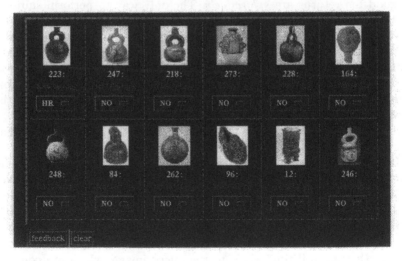

Fig. 9.7 The retrieval results after the relevance feedback.

Based on the information fed back by the user, the system *dynamically* adjusts the weights, putting more emphasis on the *shape feature*, possibly even more emphasis to one of the two shape representations which matches user's perception subjectivity of shape. The improved retrieval results are displayed in Fig. 9.7. Note that our shape representations are invariant to translation, rotation, and scaling. Therefore, images 164 and 96 are relevant to the query image.

Unlike the *isolated* approach, where the user has to precisely decompose an information need into different features and representations and precisely specify all the weights associated with them, the proposed *interactive* approach allows the user to submit a coarse initial query and continuously refine the information need via relevance feedback. This approach greatly reduces the user's effort of composing a query and captures the user's information need more precisely.

9.6.4 Experiments for Algorithms in Section 9.5.2

We have constructed a CBIR system based on the optimization algorithm developed in Section 9.5.2. A user has many options to configure the system and to choose a particular data set. The system can be configured based on the following settings (Fig. 9.8):

- Move query vector versus do not move query vector
- Use full covariance matrix versus use diagonal matrix
- Use any combination of the feature representations

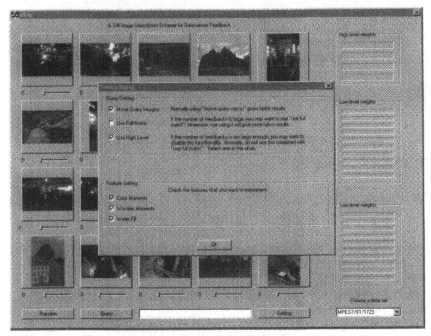

Fig. 9.8 The configuration of the demo system.

Clicking on the "data set" choice box will give a user the choice of selecting different data sets. (Currently, Vistex data set B, MPEG-7 data set, and Corel data set have been populated into the database.) The progress controls on the right-hand side of the interface display the weights (w_i and u) for a particular query. Typical retrieval results before and after (once and twice) relevance feedback are illustrated in Figs 9.9 to 9.11.

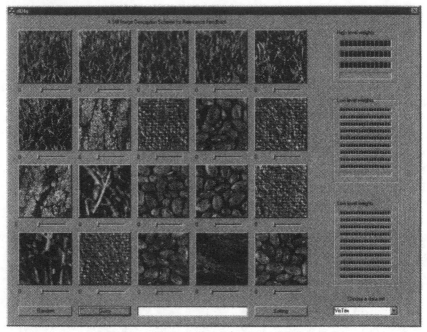

Fig. 9.9 Before the relevance feedback.

9.7 Conclusions

CBIR has emerged as one of the most active research areas in the past few years. Most of the early research effort focussed on finding the "best" image features or representations. Retrieval was performed by summing the distances of individual feature representation with fixed weights. Although this *isolated* approach establishes the basis of CBIR, the usefulness of such systems was limited.

In this chapter, we have introduced a human–computer *interactive* approach to CBIR based on relevance feedback. Unlike the *isolated* approach, where the user has to precisely decompose the information need into precise weights, the *interactive* approach allows the user to submit a coarse initial query and continuously refine his information need via relevance feedback. This approach greatly reduces the user's effort of composing a query and captures the user's information need more precisely.

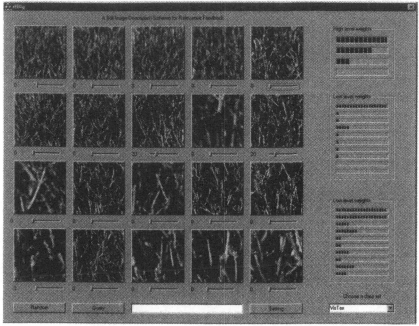

Fig. 9.10 After one iteration of relevance feedback.

Based on the description of the relevance feedback algorithm in Sections 9.2–9.5, we can observe its following properties.

Multimodality. The proposed image object model, and therefore the retrieval model, supports multiple features and multiple representations. In contrast to the *isolated* approach's attempt to find the single "best" universal feature representation, the proposed approach concentrates on how to organize the multiple feature representations, such that appropriate feature representations are invoked (emphasized) at the right place and time. The *multimodality* approach allows the system to better model the user's perception subjectivity.

Interactivity. The proposed approach is *interactive*. The interactivity allows the system to make use of the ability from both the computer and the human.

Dynamic. In contrast to an *isolated* approach's fixed query weights, the proposed approach *dynamically* updates the query weights via relevance feedback. The advantages are for both the user and the system designer. They are no longer required to specify a precise set of weights. Instead, the user interacts with the system, indicating which returns he or she thinks are relevant. Based on the user's feedback, query weights are dynamically updated.

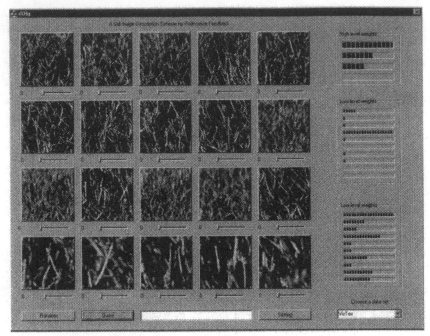

Fig. 9.11 After two iterations of relevance feedback.

In this chapter, we have explored two general approaches to parameter updating via relevance feedback. One is heuristic based while the other is optimization based. The heuristic-based approach is easy to understand and implement but the accuracy is not very high. We have further shown that for certain objective functions, we can obtain explicit optimal solutions, which makes the optimization-based approach very attractive.

9.8 Acknowledgment

The research reported in this chapter was supported in part by NSF/DARPA/NASA DLI Program under Cooperative Agreement 94-11318, in part by ARL Cooperative Agreement No. DAAL01-96-2-0003, and in part by CSE Fellowship, UIUC.

One image collection used in the chapter is from the Museum Educational Site Licensing (MESL) project, sponsored by the Getty Information Institute. Another set of images was obtained from the Corel collection and used in accordance with their copyright statement.

References

1. Buckley C and Salton, G, "Optimization of Relevance Feedback Weights," SIGIR'95, 1995.
2. Callan, JP, Bruce Croft, W, and Harding, SM, "The Inquery Retrieval System," 3rd Int. Conf. on Database and Expert System Application, September, 1992.
3. Chang T and Jay Kuo, CC, "Texture Analysis and Classification with Tree-Structured Wavelet Transform," IEEE Trans Image Proc, 2(4), pp. 429–441, October, 1993.
4. Cheng, B, "Approaches to Image Retrieval Based on Compressed Data for Multimedia Database Systems," PhD dissertation, University of New York at Buffalo, 1996.
5. Cox, IJ, Miller, ML, Omohundro, SM, and Yianilos, PN, "Target Testing and the PicHunter Bayesian Multimedia Retrieval Dystem," Advanced Digital Libraries Forum, Washington D.C., May, 1996.
6. Daneels, D, Campenhout, D, Niblack, W, Equitz, W, Barber, R, Bellon, E, and Fierens, F, "Interactive Outlining: An Improved Approach Using Active Contours," SPIE Storage and Retrieval for Image and Video Databases, 1993.
7. Duda RO and Hart, PE, Pattern Classification and Scene Analysis, John Wiley and Sons, Inc, New York, chapter 6, pp. 211–249, 1973.
8. Equitz, W and Niblack, W, "Retrieving Images from a Database Using Texture - Algorithms from the QBIC System," Technical Report RJ 9805, Computer Science, IBM Research Report, May, 1994.
9. Fagin, R and Wimmers, EL, "Incorporating User Preferences in Multimedia Queries," Int. Conf. on Database Theory, 1997.
10. Haralick, RM, Shanmugam, K, and Dinstein Y, "Texture Features for Image Classification," IEEE Trans on Syst Man Cybern, SMC-3(6), pp. 610–621, 1973.
11. Huang, TS and Rui, Y, "Image Retrieval: Past, Present, and Future," Int. Symposium on Multimedia Information Processing, December, 1997.
12. Ishikawa, Y, Subramanya, R, and Faloutsos, C, "Mindreader: Query Databases Through Multiple Examples," 24th VLDB Conference, New York, 1998.
13. Liu, F and Picard, RW, "Periodicity, Directionality, and Randomness: Wold Features for Image Modeling and Retrieval," IEEE Trans Patt Recog Mach Intell, 18(7), pp. 722–733, July, 1996.
14. Ma, WY and Manjunath, BS, "Texture Features and Learning Similarity," IEEE Conf. on Computer Vision and Pattern Recognition, pp. 425–430, 1996.
15. Ma, WY and Manjunath, BS, "Netra: A Toolbox for Navigating Large Image Databases," IEEE Int. Conf. on Image Processing, 1997.
16. Mandal, MK, Aboulnasr, T, and Panchanathan, S, "Image Indexing Using Moments and Wavelets," IEEE Trans Consum Electron, 42(3), pp. 557–565, August, 1996.
17. Manjunath, BS and Ma, WY, "Image Indexing Using a Texture Dictionary," SPIE Conference on Image Storage and Archiving System, Vol. 2606, 1995.
18. Manjunath, BS and Ma, WY, "Texture Features for Browsing and Retrieval of Image Data," IEEE Trans Patt Recog Mach Intell, 18(8), pp. 837–842, November, 1996.
19. Mao J and Jain, AK, "Texture Classification and Segmentation Using Multiresolution Simultaneous Autoregressive Models," Patt Recog, 25(2), pp. 173–188, 1992.
20. Minka, TP and Picard, RW, "Interactive Learning Using a Society of Models," IEEE Conf. on Computer Vision and Pattern Recognition, 1996.

21. Ohanian P and Dubes, RC, "Performance Evaluation for Four Classes of Texture Features," Patt Recogn, 25(8), pp. 819–833, 1992.
22. Picard, RW and Minka, TP, "Vision Texture for Annotation," ACM Multimed Syst, 3(1), pp. 1–11, 1995.
23. Rui, Y, Huang, TS, and Mehrotra, S, "Content-Based Image Retrieval with Relevance Feedback in MARS," IEEE Int. Conf. on Image Proc., 1997.
24. Rui, Y, Huang, TS, Mehrotra, S, and Ortega, M, "Automatic Matching Tool Selection Using Relevance Feedback in MARS," 2nd Int. Conf. on Visual Information Systems, 1997.
25. Rui, Y, Huang, TS, and Mehrotra, S, "Relevance Feedback Techniques in Interactive Content-Based Image Retrieval," IIS&T SPIE Storage and Retrieval of Images/Video Databases VI, EI'98, 1998.
26. Rui, Y, Huang, TS, Ortega, M, and Mehrotra, S, "Relevance Feedback: A Power Tool in Interactive Content-Based Image Retrieval," IEEE Trans Circ Syst Video Technol, 8(5), pp. 644–655, September, 1998.
27. Salton G and McGill, MJ, Introduction to Modern Information Retrieval, McGraw-Hill Book Company, New York, 1982.
28. Sclaroff, S, Taycher, L, and La Cascia, M, "ImageRover: A Content-Based Image Browser for the World Wide Web," IEEE Workshop on Content-Based Access of Image and Video Libraries, 1997.
29. Shaw, WM, "Term-Relevance Computations and Perfect Retrieval Performance," Inform Process Manag, 31(4), pp. 491–498, 1995.
30. Smith JR and Chang, SF, "Transform Features for Texture Classification and Discrimination in Large Image Databases," IEEE Int. Conf. on Image Processing, 1994.
31. Smith JR and Chang, SF, "Visually Searching the Web for Content," IEEE Multimed Mag, 4(3), pp. 12–20, Summer, 1997.
32. Smith JR and Chang, SF, "An Image and Video Search Engine for the World-Wide Web," SPIE Storage and Retrieval for Image and Video Databases, 1997.
33. Swain M and Ballard, D, "Color Indexing," Int J Comput Vis, 7(1), pp. 11–32, 1991.
34. Tamura, H, Mori, S, and Yamawaki, T, "Texture Features Corresponding to Visual Perception," IEEE Trans on Syst Man Cybern, SMC-8(6), pp. 460-473, 1978.

10. Mix and Match Features in the ImageRover Search Engine

Stan Sclaroff, Marco La Cascia, Saratendu Sethi, and Leonid Taycher

10.1 Introduction

Ideally, an image search engine should support a broad range of possible queries, topics, users, and similarity measures. To date, the general approach in most image search engines has been to deploy an arsenal of modules that compute a broad variety of image decompositions and discriminants: color histograms, edge orientation histograms, texture measures, shape invariants, eigen decompositions, wavelet coefficients, text cues extracted from the text surrounding the image in the WWW document, etc. These features are pre-computed during a WWW crawl, and stored for subsequent queries.

Once all of these features have been computed and stored, the critical issue then becomes how to combine various features in such a way that users find the images they want on the WWW. The search engine must retrieve images that are similar in content, theme, mood, and/or visual appearance to those that each user has in mind. It is well known that human image similarity judgments will vary depending on the context and purpose of a query. For instance, color similarity may be of critical importance in one search, while irrelevant in another. Image search engines must somehow provide the flexibility to allow users to *steer* the relative importance of various features employed in each search. There is a need for algorithms and user interfaces that determine how image feature modules are combined.

The interface used in steering search should be as intuitive and natural as possible. Directly prompting users for feature weightings is problematic, since users may not grasp the technical details of the underlying representation, nor do they have the patience to tweak feature weightings. One way around this is to allow the users to provide example images; this keeps nettlesome image content parameters hidden from the user. This alternative approach is known as the *query by example* (QBE) paradigm [9, 17, 28]. In QBE the user guides the search by selecting or sketching images that demonstrate what she/he is looking for. The user typically makes queries like "find more things like this." Based on the QBE input, the system can infer the appropriate feature weights, and retrieve those images in the index that are closest in the weighted feature space.

Typically the initial set of images retrieved does not fully satisfy the user's search criteria. In order to gain a more satisfying query, the user can iteratively refine a search via *relevance feedback*. Relevance feedback enables the user to iteratively refine a query via the specification of relevant items[33]; i.e., the user marks those images which are most relevant from the images retrieved so far. A WWW image retrieval system can use this feedback to recalculate the feature weightings employed, and retrieve a new set of images. If desired, the user can iteratively steer and refine the query via the relevance feedback mechanism.

In this chapter, we will describe a basic approach to relevance feedback within the context of content-based image retrieval. To obtain a fully integrated system, our discussion must focus on a number of challenges:

Unified feature representation. There is a problem in combining information about images that is obtained in different modalities (e.g., text vs. various image-based features like color, texture, etc.). One promising solution is to employ latent semantic indexing (LSI) for text retrieval combined with vector space approach for image feature representations. Each image is represented by its combined LSI and image feature vector. Subvectors of the feature correspond to the output of various textual and visual indexing modules.

Mixed units. The simplicity of the combined feature space seems appealing, until we consider that each subvector will have different units. Distance metrics in the combined vector space may be invalid or difficult to interpret. One solution is to define similarity based on units-free metrics, or converting to a units-free feature representation.

Feature weights and distance metrics. Each search is different; feature weights and metrics must be selected on the fly, based on relevance feedback. Algorithms for determination of the appropriate weighting of modalities, and selection from among available distance metrics for each subvector will be described. The proposed approach will allow iterative refinement via QBE.

Interactive retrieval. Users would prefer to search an index of millions of images in approximately a second. Adequate retrieval speed combined with relevance feedback can provide a powerful method for *data exploration* or browsing. Interactivity is essential or the user loses interest. Significant reduction in computation can be achieved via a dimensionality reduction of feature vectors used in image indexing.

The proposed approaches have been deployed and tested in the ImageRover search engine. The system details, and example searches will be used to illustrate the basic solution to the aforementioned challenges. As will be seen, these challenges dictate certain design choices to make the components fit together in an efficient and useful system.

10.2   Relevance Feedback: Background

Relevance feedback enables the user to refine a query by specifying examples of relevant items. In some formulations, the user can provide both positive and negative examples. Typically, the system returns a set of possible matches, and the user gives feedback by marking items as relevant or not relevant. This process can be iterated to further refine the query. By including the user in the loop, better search performance can be achieved.

The basic ideas of query by example and relevance feedback have their roots in the text retrieval community [33]. One common approach is to compute a word frequency histogram for each document in the index. Taken together, these histograms form a document vector space. The dot product between histograms corresponding with two documents can be used to estimate the degree of similarity between documents. Salton [33] proposes a number of methods for relevance feedback, based on this vector space representation of documents. Given a user-specified subset of relevant documents, it is possible to compute appropriate term weightings for the user's query.

Cox et al. [5] proposed a Bayesian relevance feedback framework for image retrieval. Basically they compute the probability of all the images in the database of being a relevant image given the complete history of the query (i.e., the images displayed and the corresponding user actions). Experimental results for this technique are given on a small database and the high number of iterations required to find an image make this technique inadequate for the use on a very large database.

Rui et al. [32] formulate image queries in the same framework used for text retrieval. In practice each component of the image feature is considered like the term weight in the classical information retrieval formulation. They report quantitative evaluation on a small texture database and show that on that data set relevance feedback improves retrieval performance.

In Nastar et al. [24] relevance feedback is formulated as probability density estimation. Based on the user feedback (positive and negative) they present an algorithm to estimate feature densities. In this way, they infer in which features the user is mainly interested. Their technique has been evaluated on Columbia [23] and VisTex databases and has been shown to perform slightly better than the standard relevance feedback approach adapted from information retrieval.

Despite sophisticated relevance feedback methods, image statistics are often insufficient for helping users find desired images in a vast WWW index. Users naively expect to search for a specific object or person, but find they can only search for images with a similar distribution of image properties. As suggested elsewhere in this book, perhaps the next best solution is to use text cues as hints at what may be in the image (or at least the topic context in which the image appears) [11, 37]. The operating assumption is that the content of image(s) has something to do with the text in the document that contains them. The problem is then how to combine both textual and visual cues in relevance-feedback and similarity-based retrieval.

10.3 Unified Feature Representation

The solution to the problem of combining textual and visual cues in relevance feedback is best understood through a case study of the ImageRover system [20, 34, 35]. In the ImageRover system, text statistics are captured in vector form using latent semantic indexing (LSI) [6]. Textual and visual information can then be used together in a vector space representation of the images without giving any different and arbitrary importance to the different sources of information. By truly unifying textual and visual statistics, one would in fact expect to get better results than either statistic used separately.

10.3.1 Textual Statistics

The context in which an image appears can be abstracted from the containing HTML document using a method known in the information retrieval community as latent semantic indexing (LSI) [6]. LSI works by statistically associating related words to the semantic context of the given document. The idea is to project words in similar documents to an implicit underlying semantic structure. This structure is estimated by a truncated singular value decomposition (SVD). The LSI procedure is as follows.

To begin with, each image's associated HTML document is parsed and a word frequency histogram is computed. The documents in the database are not similar in length and structure. Also all words in the same HTML document may not be equally relevant to the document context. Hence, words appearing with specific HTML tags are given special importance by assigning a higher weight as compared to all other words in the document.

Different weights are assigned to words appearing in the *title, headers,* and in the *alt* fields of the *img* tags along with words emphasized with different fonts like *bold, italics,* etc. (see Table 10.1). These weight values were chosen heuristically according to their approximate likelihood of useful information that may be implied by the text. Selective weighting of words appearing between various HTML tags helps in emphasizing the underlying information of that document. Related weighting schemes are proposed in Refs. [11, 31].

The given weights are applied by just counting a single occurrence of the word as many times as its corresponding weight value to which it qualifies in the table. For example, if the word "satellite" appears between the *title* tags in the HTML document once, then it is counted as 5.00 instead of 1.00. All subsequent occurrences of the word are counted as weight values according to category in Table 10.1.

In addition, words appearing before and after a particular image are also assigned a weight based upon their proximity to the image. The weighting

Table 10.1 Word weights based on HTML tags

HTML tags	Weights
ALT field of IMG	6.00
TITLE	5.00
H1	4.00
H2	3.60
H3	3.35
H4	2.40
H5	2.30
H6	2.20
B	3.00
EM	2.70
I	2.70
STRONG	2.50
< No Tag >	1.00

value is computed as $\rho e^{-2.0 pos/dist}$, where *pos* is the position of the word with respect to the image and *dist* is the maximum number of words considered to apply such weighting. In the current implementation, the *dist* is 10 and 20 for words appearing before and after the image, respectively. The constant $\rho = 5.0$ so that the words nearest to the images get weighted slightly less than and equal to the words appearing in the *alt* field of that image and the *title* of the web page, respectively. Thus, images appearing at different locations in an HTML document will have different LSI indices. Each image in the web page is now associated with its unique context within a document by selectively weighting words based on proximity to the image.

A term × image matrix A is created; the element a_{ij} represents the frequency of term i in the document containing image j with a weight based on its status and position with respect to the image. Retrieval may become biased if a term appears several times or never appears in a document. Hence, further local and global weights may be applied to increase/decrease the importance of a term in and amongst documents. The element a_{ij} is expressed as the product of the local $(L(i,j))$ and the global weight $(G(i))$. Several weighting schemes have been suggested in the literature. Based on the performance reported by Dumais [8], the *log-entropy* scheme was chosen. According to this weighting scheme, the local and the global weights are given as below:

$$a'_{ij} = L(i,j) \times G(i), \tag{10.1}$$

$$L(i,j) = \log(a_{ij} + 1), \tag{10.2}$$

$$G(i) = 1 - \sum_k \frac{p_{ik} \log(p_{ik})}{\log(ndocs)}, \tag{10.3}$$

$$p_{ik} = tf_{ik} / \sum_k tf_{ik}, \tag{10.4}$$

where tf_{ik} is the pure term frequency for term i in HTML document k not weighted according to any scheme, and $ndocs$ is the number of documents used in the training set.

The matrix $A' = [a'_{ij}]$ is then factored into U, Σ, V matrices using the singular value decomposition:

$$A' = U\Sigma V^T \tag{10.5}$$

where $U^T U = V^T V = I$, $\Sigma = diag(\sigma_1, \ldots, \sigma_n)$, and $\sigma_i > 0$ for $1 \le i \le r$, $\sigma_j = 0$ for $j \ge r + 1$.

The columns of U and V are referred to as the left and right singular vectors respectively, and the diagonal elements of Σ are the singular values of A. The first r columns of the orthogonal matrices U and V define the orthonormal eigenvectors associated with the r non-zero eigenvalues of AA^T and $A^T A$ respectively. For further details about the SVD and the information conveyed by the matrices, readers are directed to Berry and Dumais [3].

The SVD decomposes the original term–image relationships into a set of linearly independent vectors. The dimension of the problem is reduced by choosing k most significant dimensions from the factor space which are then used for estimating the original index vectors. Thus, SVD derives a set of uncorrelated indexing factors, whereby each image is represented as a vector in the k-space:

$$\bar{\mathbf{x}}_{LSI} = q^T U_k \Sigma_k^{-1} \tag{10.6}$$

where q is the word frequency histogram for the image with appropriate weight applied. The resulting LSI vector $\bar{\mathbf{x}}_{LSI}$ provides the context associated with an image and is combined with its computed visual feature vectors and stored in the database index.

Using this technique the text associated with an image, as it appears in the HTML document, is represented by low-dimensional vectors that can be matched against user queries in the LSI "semantic" space. This reduction to semantic space implicitly addresses problems with synonyms, word sense, lexical matching, and term omission.

10.3.2 Visual Statistics

The use of visual statistics for image retrieval has been widely explored in recent years. Particular importance has been given to color and texture representations, as they seem to provide a fair amount of useful visual information and can be computed automatically from image data. For the sake of demonstrating our approach, we therefore focus on these color and texture measures.

Several color spaces, histogram techniques, and similarity metrics have been proposed (see for example Refs. [25, 27, 36, 38]). Due to the lack of a common testbed and the high degree of subjectivity involved in image retrieval task, it is still unclear which technique most closely mimics the human perception [15]. Similarly, the use of texture content [10, 21, 25, 30, 39] for image retrieval has still some unanswered questions. Finally, even though the Brodatz [4] data set has been used extensively in comparing different techniques for texture indexing, it is not clear that one can draw any definitive conclusion about their efficacy for indexing general imagery on the WWW.

The visual statistics we use to describe an image are the color histogram and dominant orientation histogram. The statistics are computed over the whole image and five overlapping image sub-regions [35]. The technique we used provides a reasonable compromise between computational complexity and descriptive power. In any case, the choice of the optimal image statistics is beyond the current scope of the ImageRover system. The ImageRover architecture is general enough to allow inclusion of any image descriptor that can be expressed in vector form. This allows easy extension of the system when future, improved representations become available.

Color. Color distributions are calculated as follows. Image color histograms are computed in the CIE $L^*u^*v^*$ color space [18], which has been shown to correspond closely to the human perception of color by Gargu and Kasturi [14]. To transform a point from RGB to $L^*u^*v^*$ color space, it is first transformed into CIE XYZ space. In our implementation we used the conversion matrix for CIE Illuminant C (overcast sky at noon). The $L^*u^*v^*$ values are then calculated as:

$$L^* = \begin{cases} 25 * (100 * \frac{Y}{Y_0})^{\frac{1}{3}} - 16 & \text{if } \frac{Y}{Y_0} \geq 0.008856, \\ 903.3\frac{Y}{Y_0} & \text{otherwise,} \end{cases} \tag{10.7}$$

$$u^* = 13L^*(u' - u_0'), \tag{10.8}$$

$$v^* = 13L^*(v' - v_0'), \tag{10.9}$$

$$u' = \frac{4X}{X + 15Y + 3Z}, \tag{10.10}$$

$$v' = \frac{9Y}{X + 15Y + 3Z} \tag{10.11}$$

where the reference values are $(X_0, Y_0, Z_0) = (0.981, 1.000, 1.182)$, and $(u_0', v_0') = (0.2010, 0.4609)$, for white under CIE Illuminant C.

Color histograms are computed over N subimages. In the current implementation, $N = 6$; separate histograms are calculated over the whole image and over five image subregions: center, upper right, upper left, lower right, lower left. Each histogram quantizes the color space into 64 (four for each axis) bins. Each histogram is normalized to have unit sum and then blurred.

Texture Orientation. The texture orientation distribution is calculated using steerable pyramids [12, 16]. For this application, a steerable pyramid of four levels was found to be sufficient. If the input image is color, then it is first converted to gray-scale before pyramid computation. At each pyramid level, texture direction and strength at each pixel is calculated using the outputs of seven X–Y separable, steerable quadrature pair basis filters.

The separable basis set and interpolation functions for the second derivative of a Gaussian were implemented directly using the nine-tap formulation provided in Appendix H (Tables IV and VI) of Freeman and Adelson [12]. The resulting basis comprises three G_2 filters to steer the second derivative of a Gaussian, and four H_2 filters to steer the Hilbert transform of the second derivative of a Gaussian.

At each level in the pyramid, the output of these filters is combined to obtain a first order approximation to the Fourier series for oriented energy $E_{G_2 H_2}$ as a function of angle θ:

$$E_{G_2 H_2} = C_1 + C_2 \cos(2\theta) + C_3 \sin(2\theta) \qquad (10.12)$$

where the terms C_1, C_2, C_3 are as prescribed in Freeman and Adelson [12], Appendix I.

Dominant orientation angle θ_d and the orientation strength m at a given pixel are calculated via the following formulae:

$$\theta_d = \frac{1}{2} arg\left[C_2, C_3\right], \qquad (10.13)$$

$$S_{\theta_d} = \sqrt{C_2^2 + C_3^2}. \qquad (10.14)$$

Orientation histograms are computed for each level in the pyramid. Each orientation histogram is quantized over $[-\frac{\pi}{2}, \frac{\pi}{2}]$. As was the case with color, separate histograms are calculated over the whole image and over five image subregions: center, upper right, upper left, lower right, lower left. In the current implementation, there are 16 histogram bins, thus the number of bins allocated for direction information stored per subimage is 64 (four levels × 16 bins/level). Each histogram is then normalized to have unit sum. Once computed, the histogram must be circularly blurred to obviate aliasing effects and to allow for "fuzzy" matching of histograms during image search [29].

In practice, there must be a lower bound placed on the accepted orientation strength allowed to contribute to the distribution. For the ImageRover implementation, all the points with the strength magnitude less than 0.005 were discarded and not counted in the overall direction histogram.

The orientation measure employed in ImageRover differs from that proposed by Gorkani and Picard [16]. While both systems utilize steerable pyramids to determine orientation strengths at multiple scales, there is a difference in how histograms are compared. In the system of Gorkani and Picard, histogram peaks are first extracted and then image similarity is computed in terms of peak-to-peak distances. In practice, histogram peaks can be difficult to extract and match reliably. In our system, histograms are compared directly via histogram distance, thereby avoiding problems with direct peak extraction and matching.

10.3.3 Combined Vector Space

The LSI vector and visual statistics vector are then combined into a unified vector that can be used for content-based search of the resulting image database. By using an integrated approach, the system can take advantage of possible statistical couplings between the content of the document (latent semantic content) and the contents of images (image statistics).

For a given image, each image analysis module and the LSI module contribute subvectors to an aggregate image index vector $\mathbf{X}$. Given M image analysis modules and N subimages, the aggregate image index vector will contain $n = M \times N + 1$ subvectors:

$$\mathbf{X} = \begin{bmatrix} \mathbf{X}_{vis} \\ \bar{\mathbf{x}}_{LSI} \end{bmatrix} = \begin{bmatrix} \mathbf{x}_1 \\ \mathbf{x}_2 \\ \vdots \\ \mathbf{x}_n \\ \bar{\mathbf{x}}_{LSI} \end{bmatrix}. \tag{10.15}$$

Thus, we obtain an image index that combines visual and textual features in a unified vector representation.

In practice, given the potential number of images to index, it is of fundamental importance that the dimension of the visual feature vector $\mathbf{X}_{vis}$ is as small as possible. As pointed out by White and Jain [40], the visual data has an *intrinsic dimension* that is significantly smaller than the original dimension. The use of principal components analysis (PCA) [13] for each visual feature subvector space (color and directionality) allows us to dramatically reduce the dimension of the visual features with a small loss of information [35].

10.3.4 Dimensionality Reduction

In the ImageRover system, it is assumed that the probability distribution of the visual statistics subvectors in $\mathbf{X}_{vis}$ is Gaussian, and that the n subvectors are independent. This may only provide a fair approximation to the true distribution of image statistics; however, the assumption allows us to treat various features independently in relevance feedback. It also simplifies computation considerably and can alleviate the *empty space* problem [7]. Under such assumptions, the covariance matrix for the visual feature vector Σ is block diagonal and the diagonal elements Σ_i are the covariances of the subvectors $\mathbf{x}_i$. Similarly the mean μ_x is composed by the means $\mu_{\mathbf{x}_i}$ of the subvectors $\mathbf{x}_i$:

$$\mu_x = \begin{bmatrix} \mu_{x1} \\ \mu_{x2} \\ \vdots \\ \mu_{xn} \end{bmatrix}, \quad \Sigma = \begin{bmatrix} \Sigma_1 & & & 0 \\ & \Sigma_2 & & \\ & & \ddots & \\ 0 & & & \Sigma_n \end{bmatrix}. \tag{10.16}$$

Note that the Σ_i and $\mu_{\mathbf{x}_i}$ can be precomputed on a large training set of images selected at random from the full database. In practice, we have found that a training set of 50,000 images taken at random from the full image database is sufficient to characterize the distribution. Assuming a diverse training set, there is no need to recompute them when new images are added to the system.

Given Σ_i and $\mu_{\mathbf{x}_i}$, it is then possible to reduce the dimensionality for each subspace via principal components analysis (PCA) [7,13]. For the ith subspace, we compute the eigenvalues and eigenvectors of the subspace co-variance matrix Σ_i. The resulting eigenvalues and eigenvectors are sorted in decreasing order by eigenvalue. The eigenvectors ϕ_j describe the principal axes of the distribution, and are stored as columns in the PCA transform matrix Φ_i. The associated eigenvalues λ_j describe the variances along each principal axes, and are stored in the diagonal matrix Λ_i.

The transform matrix Φ_i decouples the degrees of freedom in the sub-space. Once decoupled, it is possible to compute a truncated feature space that accounts for most of the covariance of the distribution. Although all eigenvectors are required to represent the distribution exactly, only a small number of vectors is generally needed to encode samples in the distribution within a specified tolerance. In practice, the first k eigenvectors are used, such that k is chosen to represent the variance in the data set within some error threshold τ. In our experiments setting $\tau = 0.1$ (10% error), resulted in a dimensionality reduction of over 85%.

Once eigenvalues and eigenvectors are computed, it is possible to compute a truncated basis that accounts for most of the covariance of the distribution and then project each vector $\mathbf{x}_i$ in the new basis to obtain $\bar{\mathbf{x}}_i$:

$$\bar{\mathbf{x}}_i = \Lambda_i^{-\frac{1}{2}} \Phi_i^T (\mathbf{x}_i - \mu_{\mathbf{x}_i}). \qquad (10.17)$$

The transformation (10.17) is computed for each visual feature vector, and the result is used in the index as a much lower dimensional vector:

$$\bar{\mathbf{X}}_{vis} = \begin{bmatrix} \bar{\mathbf{x}}_1 \\ \bar{\mathbf{x}}_2 \\ \vdots \\ \bar{\mathbf{x}}_n \end{bmatrix}. \qquad (10.18)$$

Finally, a dimensionality reduced aggregate index vector for each image:

$$\bar{\mathbf{X}} = \begin{bmatrix} \bar{\mathbf{X}}_{vis} \\ \bar{\mathbf{x}}_{LSI} \end{bmatrix}. \qquad (10.19)$$

10.4 Relevance Feedback

To start a query the user inserts a few words related to the images she/he is looking for. The system projects the query document (i.e., the words given by the user) in the LSI space to obtain a vector representation of the query and compute the k-nearest neighbors in the LSI subspace of the complete feature space. In general, most of the images returned by the system are related by context to what the user is looking for. The user can then steer the system toward the desired images by simply selecting the most relevant images and iterating the query. These refinements of the query are performed in the complete feature space and the different sources of information are mixed through our relevance feedback framework. Dissimilarity is expressed as a weighted combination of distances between subvectors. The main idea is to give more weight to the features that are consistent across the example set chosen as relevant by the user.

10.4.1 Mixing Features and Distance Metrics

In ImageRover, the aggregate feature vector $\bar{\mathbf{X}}$ represents the content of the image. It is the composite of several subvectors: dimensionally reduced color histograms and orientation histograms for each of six overlapping image regions and the LSI descriptor as described above. Different techniques have been proposed to combine different features in a global similarity measure [1, 5, 19, 22, 24, 26, 32]. In some of these techniques the user has to state explicitly the relevance of each feature, in others the system infers the feature weights based on relevance feedback information.

ImageRover employs a linear combination of normalized Minkowski distance metrics. One reason for choosing this approach is its inherent compatibility with optimizations to spatial indexing techniques [2, 35]. Another reason is that a weighted combination of Minkowski distances satisfies the axioms for a metric space. The weights of the linear combination and the order of each Minkowski metric are inferred by the system via the relevance feedback technique described in next section.

As the global distance is a linear combination of distances between the subvectors, normalization is needed to make these distances comparable. To analyze the statistical properties of these distances we randomly selected 50,000 pairs of images from our test database. For all pairs of images, the distance between their index vectors $(\mathbf{x}, \mathbf{y})$ was computed using each of the Minkowski distance measures. We considered separately the distances between subvectors $(\mathbf{x}_i, \mathbf{y}_i)$. We then examined the distributions of these distances for each metric, for each subvector. As these distance distributions appeared to be approximately normally distributed we decided to normalize using a Gaussian model, along the lines of Refs. [19, 26, 32].

In practice for each feature i and for each Minkowski distance metric L_m we computed the mean $\mu_m^{(i)}$ and the variance $\sigma_m^{(i)}$ on the 50,000 random couple of vectors randomly selected from our database. Let $\mathbf{X}$ and $\mathbf{Y}$ denote image index vectors in a database and $\mathbf{x}_i$ and $\mathbf{y}_i$ denote subvectors corresponding to a particular feature. We define the normalized L_m distance between two subvectors:

$$\tilde{L}_m(\mathbf{x}_i, \mathbf{y}_i) = \max\left(0, \frac{L_m(\mathbf{x}_i, \mathbf{y}_i) - \mu_m^{(i)}}{\sigma_m^{(i)}} + \Delta\right) \tag{10.20}$$

where, as stated before, the means and the variances are computed based on the probability distribution of the images contained in the database,

$$\mu_m^{(i)} = E[L_m(\mathbf{x}_i, \mathbf{y}_i)], \quad \sigma_m^{(i)} = Var[L_m(\mathbf{x}_i, \mathbf{y}_i)], \tag{10.21}$$

and Δ is the shift we give to the normalized Gaussian to mimic the original distribution. In our experiment we set $\Delta = 3$.

Note that the expected value $\mu_m^{(i)}$ can be computed off-line over an entire database or a statistically significant subset of it. Moreover, if the database is reasonably large, we do not need to recompute this factor when new images are added to the archive.

10.4.2 Relevance Feedback Algorithm

Our system employs a relevance feedback algorithm that selects appropriate L_m Minkowski distance metrics on the fly. The algorithm determines the relative weightings of the individual features based on feedback images. This weighting thus varies depending upon the particular selections of the user.

Assume that the user has specified a set S of *relevant* images. The appropriate value of m for the ith subvector should minimize the mean distance between the relevant images. The order of the distance metric is determined as follows:

$$m_i = \arg\min_m \eta_m^{(i)} \tag{10.22}$$

where

$$\eta_m^{(i)} = E[\tilde{L}_m(\mathbf{p}_i, \mathbf{q}_i)], \quad \mathbf{P}, \mathbf{Q} \in S, \tag{10.23}$$

and $\tilde{L}_m(\mathbf{x}, \mathbf{y})$ is the normalized Minkowski metric as defined in previous section.

Queries by multiple examples are implemented in the following way. First, the average query vector is computed for S. A k-nearest-neighbor search of the image index then utilizes the following weighted distance metric:

$$\delta(\mathbf{X}, \mathbf{Y}) = \sum_{i=1}^{n} w_i \tilde{L}_{m_i}(\mathbf{x}_i, \mathbf{y}_i) \tag{10.24}$$

where the w_i are *relevance weights*

$$w_i = \frac{1}{\epsilon + \eta_m^{(i)}}. \tag{10.25}$$

The constant ϵ is included to prevent a particular characteristic or a particular region from giving too strong a bias to the query.

10.5 System Implementation

We implemented a fully functional system to test our approach. At the point of this writing, the robots have collected more than one million unique and valid images. Two images are considered unique if they have different URLs. An image is valid if both its width and height are greater than 64 pixels; images not satisfying this heuristic are discarded. In practice, some of the documents and the images are duplicated as sometimes the same document or the same image appears with a different URL due to name aliasing.

To have a significant sampling of the images present on the WWW a list of links related to more diverse topics is needed. To gather a diverse image index, the robots followed all the site links in Yahoo's WWW listing, found at www.yahoo.com. This run resulted in a test collection of over one million unique images. A demo version of the resulting ImageRover index is available on-line at http://www.cs.bu.edu/groups/ivc/ImageRover, and the reader is encouraged to try it. The demo system contains a subset of approximately 350,000 images.

The current implementation of the system is not optimized for speed. The query server and the web server run on an SGI Origin 200 with four R10000 180 MHz processors and 1 GB of RAM. As all the data can be kept in memory a brute force search of the k nearest neighbors takes around 3 s. In the case of the page zero we have to compute the LSI vector corresponding to the keywords provided by the user. This is a simple vector by matrix product, and, even though the dimension is high, it takes less than 1 s.

10.5.1 User Interface

The user interacts with the system through a web browser. The user specifies a set of keywords; this set of keywords can be considered as a text document. An LSI index is computed for the keywords and used to match nearest neighbors in the subspace of all LSI vectors in our image database. This yields *page zero*.

The query can then be further refined via relevance feedback. Once the user finds and marks one or more relevant images to guide the search, the user can initiate a query with a click on the search button. Similar images (the number of returned images is a user chosen value) are then retrieved and shown to the user in decreasing similarity order. The user can then select other relevant images to guide next search and/or deselect one or more of the query images and iterate the query. There is no limit either on the number of feedback iterations or on the number of example images employed.

10.5.2 Example Search

Figures 10.1–10.3 show an example search in our system. Figure 10.1 shows the first 15 matches in response to the query "mountain bike race." This is

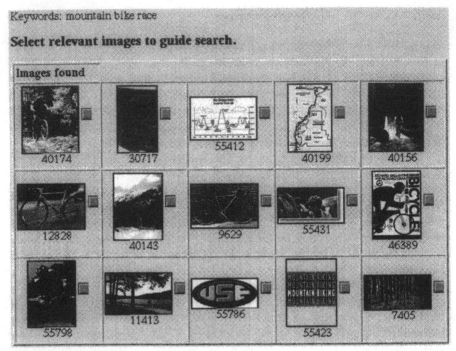

Fig. 10.1 Example of *page zero* search. Response to the keywords "mountain bike race." The numbers below images represent the indexes of the images in the database.

page zero. It is evident that there are several kinds of images that are related to the same keywords. By providing relevance feedback, the user can now narrow the search to images that share not only the same keywords but also similar visual properties.

Fig. 10.2 is a set of images found by the system using the relevance feedback; the top two images are the images selected by the user: images of cyclists racing together. The next three rows contain the retrieved 15 nearest neighbors. Images are displayed in similarity rank order, right to left, top to bottom. In this particular example, ImageRover ranked other racing photographs as closest to the user-provided examples. The other returned images not only share similar text cues, but also share similar color and texture distributions.

Similarly, Fig. 10.3 shows the output of the system given the same page zero, but different relevance feedback from the user. In this example, the user selected images of mountain bikes. The system then retrieved images that were relevant to the user's query. As can be seen, the use of visual features and relevance feedback is very useful in removing the ambiguity that is almost always present when using keywords to retrieve images.

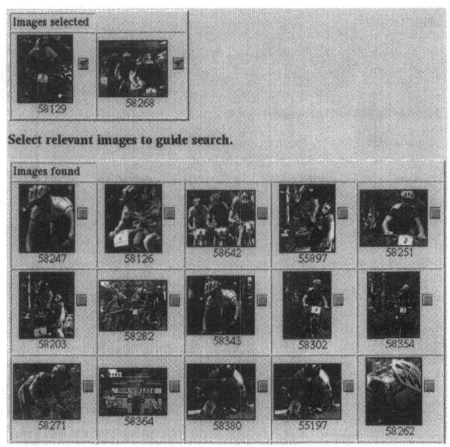

Fig. 10.2 Example of search using relevance feedback. The top two images are the *relevant* images selected by the user. The others are the response of the system.

10.6 Summary

Searching for images involves the mixture of different features, both visual and textual. Due to the broad range of users and possible types of queries, the relative importance of different features cannot be known ahead of time. Instead, the mixture must be obtained via relevance feedback given by the user during each query. The various textual and visual features cannot be combined blindly, because the features have different units.

To mix textual and visual features, we introduced a common vector space approach. Textual statistics were captured in vector form using latent semantic indexing (LSI) based on text in the containing HTML document. Visual statistics are captured in a vector using standard histogram techniques. The

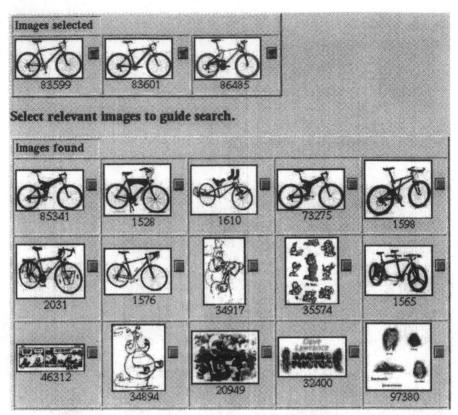

Fig. 10.3 Example of search using relevance feedback. The top three images are the *relevant* images selected by the user. The others are the response of the system.

system employs dimensionality reduction via a PCA on the original higher-dimensional vector space and then stores the result in the index. The combined approach allows improved performance in conducting content-based search as demonstrated by experiments reported in Sclaroff et al. [34].

A method for units-free comparison was proposed based on distance distributions. As these distance distributions appeared to be approximately normally distributed we decided to normalize using a Gaussian model, along the lines of Refs. [19, 26, 32].

We described a novel relevance feedback algorithm that selects a weighted combination of L_m distance metrics appropriate for a particular query. The user can implicitly specify what image properties are appropriate by selecting example images. Thus, naive users are shielded from the need to understand the particular image measures and statistics employed in the index.

The resulting relevance feedback mechanism allows the user to perform queries by example based on more than one sample image. The user can collect the images she/he finds during the search, refining the result at each iteration. The main idea consists of giving more importance to the elements of the feature vectors with the lowest variances. These elements very likely represent the main features the user is interested in. Experimental results have confirmed this behavior [34].

Another advantage of our formulation is that it is not dependent on the particular features employed in representing the visual content of images. We can reasonably expect that the system will remain efficient as more image analysis modules are included, because the computational complexity will scale linearly with the number of features employed.

Acknowledgments

This work was sponsored in part by through grants from the United States National Science Foundation (Faculty Early Career Award #IRI-9624168, and CISE Research Infrastructure Awards #CDA-9623865 and #CDA-9529403). Marco La Cascia was sponsored in part by a doctoral scholarship from the Italian Ministry for University and Scientific Research.

References

1. Ardizzone E and La Cascia, M, "Automatic Video Database Indexing and Retrieval," Multimed Tools Applic, 4(1), pp. 29–56, 1997.
2. Arya, S, Mount, DM, Netanyahu, NS, Silverman, R, and Wu, AY, "An Optimal Algorithm for Approximate Nearest Neighbour Searching in Fixed Dimensions," ACM SIAM Symposium on Discrete Algorithms, pp. 573–582, 1994.
3. Berry, MW and Dumais, ST, "Using Linear Algebra for Intelligent Information Retrieval," Technical Report UT-CS-94-270, University of Tennessee, Computer Science Department, December, 1994.
4. Brodatz, P, Textures: a Photographic Album for Artists and Designers, Dover Publications Inc, New York, NY, 1966.
5. Cox, IJ, Miller, ML, Omohundro, SM, and Yanilos, PN, "Pichunter: Bayesian Relevance Feedback for Image Retrieval," Int. Conf. on Pattern Recognition, Wien, Austria, 1996.
6. Deerwester, S, Dumais, ST, Landauer, TK, Furnas, GW, and Harshman, RA, "Indexing by Latent Semantic Analysis," J Soc Inform Sci, 41(6), pp. 391–407, 1990.
7. Duda, RO and Hart, PE, Pattern Classification and Scene Analysis, John Wiley & Sons, 1973.
8. Dumais, ST, "Improving the Retrieval of Information from External Sources," Behav Res Meth Instrum Computers, 23(2), pp. 229–236, 1991.

9. Flicker, M, Sawhney, H, Niblack, W, Ashley, J, Huang, Q, Dom, B, Gorkani, M, Hafner, J, Lee, D, Petkovic, D, Steele, D, and Yanker, P, "Query by Image and Video Content: The QBIC System," IEEE Computer, 28(9), pp. 23–32, 1995.

10. Francos, JM, Meiri, A, and Porat, B, "A Unified Texture Model Based on a 2-D Wold Like Decomposition," IEEE Trans Sig Process, pp. 2665–2678, August, 1993.

11. Frankel, C, Swain, M, and Athitsos, V, "Webseer: An Image Search Engine for the World Wide Web," Technical Report 96-14, University of Chicago, 1996.

12. Freeman, W and Adelson, EH, "The Design and Use of Steerable Filters. IEEE Trans Patt Anal Mach Intell, 13(9), pp. 891–906, September, 1991.

13. Fukunaga, K and Koontz, W, "Application of the Karhunen-Loeve Expansion to Feature Selection and Ordering," IEEE Trans Commun, 19(4), 1970.

14. Gargi, U and Kasturi, R, "An Evaluation of Color Histogram Based Methods in Video Indexing," International Workshop on Image Databases and Multimedia Search, August, 1996.

15. Gevers T and Smeulders, A, "A Comparative Study of Several Color Models for Color Image Invariant Retrieval," Image Databases and Multimedia Search Workshop, pp. 17–26, August, 1996.

16. Gorkani, MM and Picard, RW, "Texture Orientation for Sorting Photos at a Glance," IEEE Int. Conf. on Pattern Recognition, Jerusalem, 1994.

17. Gupta, A, "Visual Information Retrieval Technology: A Virage Perspective," Technical report, Virage Inc, 177 Bovet Rd, Suite 540, San Mateo CA, 1996.

18. Hunt, R, Measuring Color, John Wiley & Sons, 1989.

19. Jain, AK and Vailaya, A, "Shape-Based Retrieval: A Case Study With Trademark Image Databases," Technical Report MSU-CPS-97-31, Michigan State University, November, 1997.

20. La Cascia, M, Sethi, S, and Sclaroff, S, "Combining Textual and Visual Clues for Content-Based Image Retrieval on the World Wide Web," IEEE Int. Workshop on Content-based access of Image and Video Libraries, June, 1998.

21. Liu, F and Picard, RW, "Periodocity, Directionality and Randomness: Wold features for Image Modeling and Retrieval," IEEE Trans Patt Anal Mach Intell, 18(7), pp. 722–733, 1996.

22. Minka, TP and Picard, RW, "Interactive Learning Using a Society of Models," Patt Recogn, 30(4), 1997.

23. Murase, H and Nayar, S, "Visual Learning and Recognition of 3-D Objects from Appearance," Int J Computer Vision, 14(1), pp. 5–24, January, 1995.

24. Nastar, C, Mitschke, M, and Meilhac, C, "Efficient Query Refinement for Image Retrieval," IEEE Conf. Computer Vision and Pattern Recognition, Santa Barbara, CA, 1998.

25. Niblack, W, Barber, R, Equitz, W, Flickner, M, Glasman, E, Petkovic, D, Yanker, P, and Faloutsos, C, "The QBIC Project: Querying Images by Content Using Color, Texture and Shape," IS&T/SPIE Symposium on Electronic Imaging: Science and Technology – Storage Retrieval for Image and Video Databases I, San Jose, CA, 1993.

26. Ortega, M, Rui, Y, Chakrabarti, K, Mehrotra, S, and Huang, TS, "Supporting Similarity Queries in MARS," ACM Multimedia, 1997.

27. Pass, G, Zabih, R, and Miller, J, "Comparing Images using Color Coherence Vectors," ACM Multimedia, 1996.

28. Pentland, A, Picard, RW, and Sclaroff, S, "Photobook: Tools for Content-Based Manipulation of Image Databases," SPIE: Storage and Retrieval Image and Video Database II, 1994.
29. Picard, RW and Minka, TP, "Vision Texture for Annotation," Multimed Syst, 3(3), pp. 3–14, 1995.
30. Rao, AR and Lohse, GL, "Towards a Texture Naming System: Identifying Relevant Dimensions of Texture," IEEE Conf. on Visualization, San Jose, CA, October, 1993.
31. Rowe, NC and Frew, B, "Finding Photograph Captions Multimodally on the World Wide Web," AAAI Spring Symposium on Intelligent Integration and Use of Text, Image, Video, and Audio Corpora, March, 1997.
32. Rui, Y, Huang, TS, and Mehrotra, S, "Content-Based Image Retrieval with Relevance Feedback in MARS," IEEE Int. Conf. on Image Processing, Santa Barbara CA, 1997.
33. Salton, G and McGill, MJ, Introduction to Modern Information Retrieval, McGraw-Hill, 1989.
34. Sclaroff, S, La Cascia, M, Taycher, L, and Sethi, S, "Unifying Textual and Visual Cues for Content-Based Image Retrieval on the World Wide Web," Computer Vision Image Understand, 75(1/2), pp. 86–98, July/August, 1999.
35. Sclaroff, S, Taycher, L, and La Cascia, M, "ImageRover: A Content-Based Image Browser for the World Wide Web," IEEE Int. Workshop on Content-based access of Image and Video Libraries, June, 1997.
36. Smith, JR and Chang, SF, "Tools and Techniques for Color Image Retrieval," IS&T SPIE: Storage and Retrieval Image and Video Database IV, San Jose, CA, 1996.
37. Smith, JR and Chang, SF, "Visually Searching the Web for Content," IEEE Multimed, 4(3), pp. 12-20, July, 1997.
38. Swain, M and Ballard, D, "Color Indexing," Int J Computer Vision, 7(11), 1991.
39. Tamura, H, Mori, S, and Yamawaki, T, "Textural Features Corresponding to Visual Perception," IEEE Trans Syst Man Cybern, 8(6), 1978.
40. White, DA and Jain, R, "Algorithms and Strategies for Similarity Retrieval," Technical Report VCL-96-101, University of California, San Diego, July, 1996.

11. Integrating Analysis of Context and Image Content

Vassilis Athitsos, Charles Frankel, and Michael J. Swain

11.1 Introduction

In some situations of considerable interest, images are found embedded within documents. For example HTML, Word, Powerpoint, Framemaker, LATEX, and other document layout languages all permit the inclusion of images. Therefore, the World Wide Web, and other archives of documents in these formats, often contain images within the context of text relevant to the content of the images. Since magazines and newspapers typically contain many photographs and other images, archives of the images in these publications may preserve the associated text along with the photographs.

Retrieval of images that are embedded within documents can take advantage of at least three sources of information, including relevant text from the documents, header information from the images, and the image content. Hyperlinked documents, such as documents on the World Wide Web, contain further information in the connectivity of the hyperlinks, that can be of value for indexing, especially in ranking responses. We distinguish the image content from the other types of information, which will be referred to as the image context.

This chapter will describe heuristics for determining which text in a document is relevant to the images the document contains. It will then discuss the ways in which content analysis can complement the information available from the text. One way it can do so is to categorize the images into classes of interest to the user. There are many types of classification that may be of interest to the user, though only some can be done reliably given the state of the art in image processing and computer vision. Examples of the latter include distinguishing color from gray-scale images, distinguishing photographs from non-photographs (for instance, computer-generated graphics), and distinguishing portraits, or images of people, from non-portraits. On the World Wide Web, another classification of considerable interest to search engines is to separate the "family-friendly" images from the "adult," or pornographic, content. Note that, to make all of these classifications it is typically valuable to consider both the image content and its context.

11.2 Extracting Relevant Text

Conceivably, one could use sophisticated natural language understanding techniques to discover which text of the document was relevant to each image within the document. As natural language understanding remains largely an open problem, simpler techniques, based on the analysis of the structure of the document, have been employed with considerable success, at least in the case of HTML documents found on the World Wide Web.

11.2.1   Relevant Text in HTML Documents

Sources of text relevant to a given image include: text from the filename, alternate text, text determined to be a caption, and text in hyperlinks to the image. If relevant text from different sources is weighted according to the likelihood of the text being relevant, the sum of the weights of the matching terms can be used to rank the matches to a query. A second weighting scheme, similar to the standard *tf.idf* weighting from information retrieval, should be used to weight the matches of multiple queries according to their frequency of occurrence in the total document set. This second weighting scheme is described in more detail below.

Filenames. Filenames can be extremely valuable in providing cues to the content of the image. Caveats are that multiple words may be combined in a filename, and that words may be abbreviated for convenience, or out of necessity because of restrictions imposed by the operating system. Standard separators such as "_", ".", and "-" can be assumed to separate words in filenames.

Text other than the filename from the full URL of an image could be valuable for indexing it. For instance, the directory name could contain the name of the subject of the image, as in "/marilyn/mm-trans.gif," in a reference to an image of Marilyn Monroe. However, indexing text from the URL other than the filename has a possibly unwanted side effect, in that images from the same site are more likely to be returned as the result of a query. Since users can browse individual sites themselves, they appreciate *variety* in the responses, that is, results from different websites are preferred over results from one website, all other things being equal.

Hyperlink Text. In most cases, images are embedded within HTML documents, using an `<img>` tag. However, images can also be referenced through hyperlinks. In fact, there can be more than one hyperlink, typically on different web pages, that point to the same image. The text in these hyperlinks is a good indication of the content of the image. Text adjacent to a hyperlink, in the same sentence or phrase (e.g., within an item of a list) can also be relevant. An example is "Marilyn *jumping for joy*," where the italicized text is the text of the hyperlink.

Alternate Text. Images embedded in HTML documents often have *alternate text* associated with them, indicated by an `alt=""` tag. This text is displayed when images have been turned off by the user, and as a *tool tip* when the mouse pointer is located over the image. This text is often relevant to the content of the image.

Image Captions. Text can be determined to be an image caption by being within the same `<center>` tag as the image, within the same table cell, or within the same paragraph. However, since tables are often used for layout purposes in HTML, large portions of a document may be located within one table cell. The text in such table cells can be excluded by placing restrictions on the table cell, such as it must not contain more than a threshold number of words, another table, more than one image, or multiple paragraphs. The same restrictions were placed on paragraphs interpreted to be captions.

The relative weights of these sources of relevant text were assigned in an *ad-hoc* manner in WebSeer [4], with filenames, alternate text, and anchor text receiving the highest weights, because they were the most likely to contain relevant text.

When the amount of text used from a page to index an image is greater, the likelihood of higher recall is increased. However, when more irrelevant text is included, the precision may decrease.* WebSeer was only tested on a modest-sized collection, about three million images, compared to the total number of images on the web, which most likely exceeds 100 million unique images. In very large collections on the scale of the entire set of images on the Web, using a smaller amount of highly relevant text may generate adequate recall while maintaining high precision and reducing the size of the index. Lycos's image search engine appears to have taken this approach, using only hyperlink text and filenames.

Like WebSeer, the Marie-3 system used heuristics to extract relevant text from Web pages for indexing purposes [10]. The Marie-3 system used some basic linguistic analysis to aid in determining relevant text.

Since matching a query on the relevant text has similar characteristics to matching whole HTML pages, standard information retrieval techniques can be applied [9]. Recently, ranking techniques based on the hyperlinked structure of the World Wide Web, and the popularity of user selection of response selection, have proven to be valuable additions to the text-based ranking described above. The hyperlinked structure can be used to estimate the *quality* of a page, where high quality pages are considered to be those that are pointed to by a large number of high quality pages. An iterative algorithm, known as the Kleinberg algorithm, can be used to compute quality given this circular definition [2, 8]. Can these ranking techniques be applied to image search? Although they had not been employed at the time of writing, the answers seem clear. The popularity-based approach can be applied directly, without modification. Image *quality* could be defined to be the sum of the quality of the pages that embed or link to the image. Of course, we don't know if a page is considered high quality because of the images it includes, or because of the text. Image search engines typically provide access to the web pages that reference the images, so users will benefit from the text as

---

* Images retrieved using such text and the weighting scheme described above will be ranked down from the top responses, so the precision of the most-viewed responses is unlikely to decrease, given a large enough collection.

well. In fact, the text could provide high quality descriptions or background information for the image the user was searching for.

11.2.2 Other Document Formats

Some document layout languages, such as LATEX, contain explicit caption tags. When such tags exist it is straightforward to determine which text is the caption for each image. The sentences that refer to the document may also be of value; in LATEX these can usually be found because of the \ref tags that will be found embedded in the sentences. The \ref tags include an argument that is the LATEX label of the image being referenced. Some other document layout languages, such as Word, contain less structural information about the document, and may be less readily mined for relevant text.

Some types of images, typically graphs and charts, are encoded in such a way that preserves the text represented as characters. The Macintosh PICT format is able to do so, for instance. In such cases, if the language used to create the image is known, the text can be mined from the image itself. Detecting text in images represented as pixels is a more difficult problem, though e.g., Wu et al. [15] have shown some success in doing so.

11.3 Image Classification: Photographs vs. Graphics

Image classification allows the user to search a subset of the total collection that is more relevant to the user's search. An important distinction on the World Wide Web is between photographs and non-photographs - typically computer-generated graphics. Computer-generated graphics, including buttons, separators, most image maps, and so on, make up a large fraction of the total number of images on the Web. Yet users of image search engines typically want to search for photographs. On the other hand, there are some, such as those making their own web pages, who wish to search specifically for graphics. Other useful distinctions that can be drawn include color versus black and white images, and photographs that are portraits versus those that are not. These types of classification are discussed in Section 11.3.11.

A few basic statistical observations can be made concerning the differences between computer-generated graphics and photographs that appear on the Web:

– Color transitions from pixel to pixel follow different patterns.
– Some colors appear more frequently in graphics than in photographs. For instance, highly saturated colors are much more common in graphics.
– Graphics tend to have different sizes and shapes than photographs. In particular, they tend to be smaller and more elongated.
– Graphics have fewer colors than photographs.

Based on those observations, several *image metrics* can be derived. The metrics are functions from images to real numbers, which will be termed *metric scores*. A simple metric is the number of colors in an image. The goal is to design metrics in which graphics tend to have scores in different ranges than photographs. This way, the metric scores can be used to decide if an image is a photograph or a graphic.

The scores obtained from individual metrics are rarely definitive. In order to achieve high accuracy rates, it is necessary to combine scores from several metrics to make the final decision. Decision trees can be used for such a role. The trees can be constructed in an automated way, based on the metric scores of large sets of images, randomly chosen to be downloaded from the Web, pre-classified by hand as photographs or graphics.

11.3.1 What are Photographs and Graphics?

For most images a human has no trouble deciding if they are photographs or computer-generated graphics. However, there are some common cases when it is not clear whether an image is a photograph or a graphic, and there are also cases when neither of the two categories is applicable:

Mixed images. A significant fraction of web images have both a photograph and a computer-generated part. Examples are photographs with a frame around them, photographs with text overlaid on them, and images that are half photographs and half graphics.
Hand drawings. Examples could include sketches of people or landscapes.

Images falling into these categories were not used for training or testing in the experiments described below.

11.3.2 Differences Between Photographs and Graphics

By looking at many photographs and graphics, one can easily notice certain basic differences between them that are easy to describe in quantitative terms. These are the differences we used as a starting point in the design of our metrics:

- Color transitions from pixel to pixel follow different patterns in photographs and graphics. Photographs depict objects of the real world, and regions of constant color are not common in the real world, because objects tend to have texture (Fig. 11.1). In addition, photographs of objects always contain a certain amount of noise, which causes even nearby pixels to have different RGB values. On the other hand, graphics tend to have regions of constant color. Figure 11.1 displays a typical example. The graphic has only eight different colors, and most of the pixels have the same color as their neighbors.

Fig. 11.1 An example of a photograph (left). An example of a graphic (right).

Edges in graphics tend to be much sharper. Typically an edge occurs between a region of constant color and another region of constant color, and the transition takes place over one pixel. In photographs, boundaries between objects are often blurred because the camera is not focussed precisely on them. In addition, many color transitions do not correspond to boundaries between objects, but to light variations and shading. Such transitions are much smoother.

– Certain colors are much more likely to appear in graphics than in photographs. For example, graphics often have large regions covered with highly saturated colors. Those colors are much less frequent in photographs.

– Graphics have fewer colors than photographs. This is related to the fact that they tend to have large one-color regions. On the Web in particular, people often prefer to use graphics with a small number of colors, because they compress better.

– Graphics tend to have different shapes than photographs. They are often narrow, much longer in one dimension than in the other. Photographs tend to be more square. In addition, graphics frequently come in small sizes, which is very rare for photographs.

11.3.3 Image Metrics

Based on the general observations we have described in the previous section, we implemented several metrics, which map images to real numbers. Photographs and graphics tend to score in different ranges in those metrics. Because of that, the metric scores are evidence we can use to differentiate between those two types.

In the following discussion, we assume that images are represented by three two-dimensional arrays, each corresponding to their red (R), green (G), and blue (B) color bands. The entries of those arrays are integers from 0 to 255. The color vector of a pixel p is defined to be (r, g, b), where r, g, and b are respectively the red, green, and blue component of the color of the pixel.

The metrics we use are the following:

The number of colors metric. The score of the image in this metric is the number of distinct colors that appear in it. As we mentioned before, graphics tend to have fewer colors than photographs.

The prevalent color metric. We find the most frequently occurring color in the image. The score of the image is the fraction of pixels that have that color. Since graphics tend to have a small number of colors and large one-color regions, they tend to score higher than photographs in this metric.

The farthest neighbor metric. This metric is based on the assumptions we made about color transitions in graphics and photographs. On the one hand, graphics tend to have large regions of constant color. On other hand, they tend to have sharper color transitions.

For two pixels p and p', with color vectors (r, g, b) and (r', g', b') respectively, we define their color distance d as $d = |r - r'| + |g - g'| + |b - b'|$. Since color values range from 0 to 255, d ranges from 0 to 765. Each pixel p_1 (except for the outer pixels) has neighbors up, down, left, and right. A neighbor p_2 of p_1 is considered to be the farthest neighbor of p_1 if the color distance between p_1 and p_2 is larger than the color distance between p_1 and any other of its neighbors. We define the transition value of p_1 to be the distance between p_1 and its farthest neighbor.

In the farthest neighbor metric, we have to specify a parameter P between 0 and 765. The score of the image is the fraction of pixels that have a transition value greater than or equal to P.

Since graphics tend to have large one-color regions, they tend to have many pixels whose transition value is 0. Therefore, if we set P to 1, we expect photographs to score higher than graphics as a rule. On the other hand, color transitions are usually sharper in graphics, so for high values of P graphics tend to score higher than photographs.

We use a second version of the same metric to accentuate the difference in scores between graphics and photographs for high values of P. In the second version, the score of an image is the fraction f_1 of pixels with transition value greater than or equal to P, divided by the fraction f_2 of pixels with transition value greater than 0. Since f_2 tends to be larger for photographs, graphics have even higher scores with respect to photographs than they do in the first version.

The saturation metric. This metric is based on the assumption that highly saturated colors are more common in graphics than in photographs.

For a pixel p, with color vector (r, g, b), let m be the maximum and n be the minimum among r, g, and b. We define the saturation level of p to be $|m - n|$.

The score of the image is the fraction of pixels with saturation levels greater than or equal to a specified parameter P. For high values of P we expect graphics to score higher than photographs, since saturated colors occur more frequently in them.

The color histogram metric. This metric is based on the assumption that certain colors occur more frequently in graphics than in photographs. In contrast to the saturation metric, here we don't assume anything about the nature of those colors. We simply collect statistics from a large number of graphics and photographs and construct histograms which show how often each color occurs in images of each type. The score of an image depends on the correlation of its color histogram to the graphics histogram and the photographs histogram.

A color histogram is a three-dimensional table of size $16 \times 16 \times 16$. Each color (r, g, b) corresponds to the bin indexed by $(\lfloor \frac{r}{16} \rfloor, \lfloor \frac{g}{16} \rfloor, \lfloor \frac{b}{16} \rfloor)$ in the table (where $\lfloor x \rfloor$ is the floor of x). The color histogram of an image contains at each bin the fraction of pixels in that image whose colors correspond to that bin.

The correlation $C(A, B)$ between two histograms A and B is defined as $C(A, B) = \sum_{i=0}^{15} \sum_{j=0}^{15} \sum_{k=0}^{15} (A_{i,j,k} B_{i,j,k})$, where $A_{i,j,k}$ and $B_{i,j,k}$ are respectively the bins in A and B indexed by (i, j, k).

We create a graphics color histogram H_g by picking hundreds or thousands of graphics, and taking the average of their color histograms. We similarly create a photographs color histogram H_p using a large set of photographs.

Suppose that an image I has a color histogram H_i. Let $a = C(H_i, H_g)$ and $b = C(H_i, H_p)$. The score of the image in the color histogram metric is defined as $s = \frac{b}{a+b}$. Clearly, as $C(H_i, H_p)$ increases, s goes up, and as $C(H_i, H_g)$ increases, s goes down. Therefore, we expect photographs to score higher in this metric.

The farthest neighbor histogram metric. This metric is based on the same assumptions as the farthest neighbor metric, but provides a different means of testing an image.

The farthest neighbor histogram of an image is a one-dimensional histogram with 766 bins (as many as the possible transition values for a pixel). The ith bin (starting with 0) contains the fraction of pixels with transition value equal to i. We create a graphics histogram F_g by averaging the farthest neighbor histograms of hundreds or thousands of graphics, and we create a photographs histogram F_p in the same way, using a large set of photographs. We define the correlation $D(A, B)$ between histograms A and B as $D(A, B) = \sum_{i=0}^{765} A_i B_i$, where A_i and B_i are respectively the ith bins of A and B.

Let F_i be the farthest neighbor histogram of the image, $a = D(F_i, F_g)$ and $b = D(F_i, F_p)$. Then, the score s of the image in this metric is defined as $s = \frac{b}{a+b}$. As in the color histogram metric, we expect photographs to score higher than graphics.

The dimension ratio metric. Let w be the width of the image in pixels, h be the height, m be the greatest of w, and h and l be the smallest of w and h. The score of an image is $\frac{m}{l}$. Graphics very often score above 2, whereas photographs rarely do.

The smallest dimension metric. The score of an image is the length of its smallest dimension in pixels. It is much more common for graphics to score below 30 in this metric than it is for photographs.

Table 11.1 gives some indicative results for each metric. We used as a testing set 259 GIF graphics and 223 GIF photographs. The columns have the following meanings:

- Metric is the name of the metric. If the metric uses a parameter, we give that in parentheses.
- T is the threshold we used.
- G_l is the percentage of graphics whose score was less than T.
- P_l is the percentage of photographs whose score was less than T.

Table 11.1 Individual metrics

Metric	T	G_l	P_l
Number of colors	80	73.0	25.1
Prevalent color	0.2	11.6	80.3
Farthest neighbor version 1 (1)	0.83	81.2	14.5
Farthest neighbor version 2 (264)	0.20	16.4	86.7
Saturation (63)	0.71	31.7	96.0
Color histogram	0.39	79.7	3.8
Farthest neighbor histogram	0.32	76.0	5.6
Dimension ratio	2.34	57.9	98.7
Smallest dimension	57	61.4	7.6

Combining the Metric Scores. The individual metric scores are not definitive. To make the final decision, a decision-making module is needed to make the final classification based on those scores. We used multiple decision trees for that task [1].

Each decision tree is a binary tree. Each non-leaf node n has a test field, which contains a metric M_n, a parameter P_n to be used with M_n (if applicable), and a threshold T_n. Each leaf node contains a real number, between 0 and 1. To classify an image using a tree, we do the following:

1. Starting at the root of the tree, while we are at a non-leaf node n, we get the score S_n of the image using the metric M_n with the parameter P_n. If $S_n < T_n$ we move to the left child of n, otherwise we move to the right child of n.
2. When we get to a leaf node, we return the number that is stored in that node. The higher the number is, the higher the probability is that the image is a photograph.

At the end, we find the mean A of the results that we get from all decision trees. If A is less than a given threshold K, the image is considered a graphic, and otherwise it is considered a photograph. The value of K affects the accuracy of the photo detector. As K increases, the accuracy rate for graphics increases and the error rate for photographs also increases.

11.3.4 Constructing the Decision Trees

To construct a decision tree, we have to specify a training set of images S, and a set of metrics M. For each metric $m \in M$ that requires a parameter, we need to also specify the set of parameters P_m that it can use. Images in S have been hand-classified as photographs or graphics. The following is a recursive description of how a decision tree gets constructed.

To construct the tree, we start at the root. If the images in S are all photographs or all graphics, we stop. Otherwise, we pick the optimal test for the root, with respect to the training set. A test for a node is specified by the metric m, parameter p, and threshold t to use at that node. We use the same criteria as Amit and Geman [1] to determine what the optimal test in a given node is. If an optimal test doesn't exist, we stop. Otherwise, we recursively construct the left and right subtree under the root. For the left subtree we use as training set all images in S whose metric score with metric m and parameter p is less than t. We use the rest of the images as a training set for the right subtree.

It is possible that an optimal test doesn't exist for a given node. For example, if the training set consists of just one photograph and one graphic, and the two images score exactly the same in all metrics in M, no test can separate them.

11.3.5 Preparation of Training and Testing Sets

Web images appear in the GIF and JPEG format. Much better results can be obtained by using different decision trees to classify images in each format. Images in the two formats have important differences, that make them score differently in our metrics. For example, JPEG images have thousands of colors regardless of whether they are photographs or graphics, because of the way JPEG compression works. Therefore, the number-of-colors metric is not useful for JPEGs. Furthermore, JPEG compression tends to make sharp color transitions smoother. This affects the scores in the farthest neighbor metric: JPEG graphics, still score higher than JPEG photographs (for high parameters), but JPEGs in general score much lower than GIFs. So, we maintain different training and testing sets for the two formats.

To create the decision trees, we used as a training set 1025 GIF graphics, 362 GIF photographs, 270 JPEG graphics, and 643 JPEG photographs. To construct the mean color histograms and the mean farthest neighbor histograms we used about as many images, which were not included in the training sets for the decision trees.

After we have created tens of different trees, we manually put them together into several sets, which we test in order to pick the set among them that gives the highest accuracy rate.

11.3.6 Reasons for Using Multiple Decision Trees

We can consider as a feature vector of an image the vector that contains the scores of the image in all metrics for all possible parameters. The dimensionality of the space of feature vectors is in the order of thousands. The goal of the decision-making module is to assess the conditional probability that an image is a photograph, given its feature vector.

Single decision trees are a reasonable way to deal with the high dimensionality of the feature space in this domain. Decision trees can combine scores from different tests without making any assumption about how those scores are correlated with each other. They can use the most useful parameters of the feature vector, and ignore other parameters that are highly correlated with the ones already used. This is desirable in this domain, where groups of tests are highly correlated, whereas other groups of tests give relatively independent results. For example, the test scores obtained by applying the saturation metric with parameters ranging from 100 to 110 are highly correlated with each other. On the other hand, those scores are relatively independent of the score of the image in the dimension ratio test.

Multiple decision trees offer several advantages over single decision trees:

- We have so many possible tests (combinations of metrics and parameters) that, given the size of our training sets, we cannot use all the information we get in a single decision tree.
- We can add additional metrics without having to increase the size of the training set, or alter the training and classification algorithms.
- Multiple decision trees offer increased accuracy over single decision trees, even if all trees are built based on the same metrics (as long as the metrics are used in different order in the different trees).

We performed some experiments to verify this last claim. In one experiment, we constructed three decision trees. For each of the trees we used the saturation and the farthest neighbor metrics with parameters between 50 and 200 for the saturation metric and between 50 and 500 for the farthest neighbor metric. The three trees demonstrated error rates 0.170, 0.170, and 0.210 respectively. The group of the three trees together had an error rate of 0.154. For the testing we used approximately 1000 JPEG graphics and 1000 JPEG photographs. The group of trees was more accurate than any single tree both for graphics and for photographs.

The reason the group of trees was more accurate than any individual tree was that it handled better some borderline cases. Several images were classified incorrectly by one of the three trees, but they were classified correctly by the other two trees, and by the group as a whole.

Boosting algorithms such as AdaBoost provide an alternative, principled way of generating and combining results from multiple decision trees [5, 6, 13].

11.3.7 Results

As we mentioned before, after we get the average of the results of all decision trees for an image, we compare it with a threshold K and consider the image to be a graphic if the average is less than K. The choice of K affects directly the accuracy of the system for images of each type. As we increase K, we get a higher error rate for graphics and a lower error rate for photographs. The error rate for images of a given type (photographs or graphics) is defined as the percentage of images of that type that the system classifies incorrectly.

We tested the system on random images we downloaded from the Web and classified by hand. The test images consisted of 7245 GIF graphics, 454 GIF photographs, 2638 JPEG graphics, and 1279 JPEG photographs. None of those images was used in constructing the mean color and farthest neighbor histograms, in constructing the trees, or in experimenting with sets of trees to decide which set we should use. Tables 11.2 and 11.3 give error rates, as percent values for different choices of K, for the two formats.

Table 11.2 Error rates for GIF images

K	GIF graphics	GIF photographs
0.37	8.2	3.5
0.38	6.0	3.9
0.40	5.0	6.9
0.42	4.2	7.4
0.44	3.6	12.1
0.50	2.4	17.8

In WebSeer, we set K to 0.5 both for GIF and JPEG images. GIF graphics are by far the most common image type on the Web. They occur about 15–20 times as often as GIF photographs, and about six times as often as JPEG images of any kind. Our choice of K allows GIF graphics to be classified correctly at a rate of 97.6%. If we had allowed a 5% error rate for GIF

Table 11.3 Error rates for JPEG images

K	JPEG graphics	JPEG photographs
0.40	20.0	3.4
0.44	16.4	4.4
0.47	11.8	6.4
0.50	9.3	8.7
0.55	6.1	15.3
0.59	5.0	17.6

graphics, about 40% of all images classified as photographs in WebSeer would have been GIF graphics.

Table 11.4 gives the number of images the system can classify per second. The measurements were done on an UltraSPARC-1, running at 167 MHz, using the same images that were used for the results in Tables 11.2 and 11.3. The images were read from disk. The performance table also contains the mean file size and the mean number of pixels for those images.

Table 11.4 Performance of the system

Format	Images/s	Mean file size	Mean # of pixels
GIF graphics	4.5	6K	44,521
GIF photos	2.7	35K	50,653
JPEG graphics	2.4	15K	62,185
JPEG photos	1.6	29K	82,070

11.3.8 Improved Learning Techniques

Combining the results of the individual trees by simply averaging them is an ad hoc approach that does not have any theoretical justification. Averaging works best when the results of the different trees are independent. In this case, there is no theoretical reason for the results to be independent, and the evidence suggests that they aren't.

Much larger training sets can be obtained from the Web, with their size restricted only by the amount of effort that can be spent in labeling them. Using a graphical interface written for the task, we were able to hand-classify images at a rate of 2,500 images an hour.

11.3.9 Image Context Metrics

We define as image content all the information that is stored in the color bands of the image. That information is the dimensions of the image and the

RGB values for every pixel. The metrics we have described up to now draw their information from the image content.

Image context is the information we have about an image that does not come from the image content. We already make de facto use of an image context metric: the image format. We use the fact that the image content in GIF images differs significantly from the image content in JPEGs, and the fact that the ratio of graphics to photographs is much higher among GIFs than among JPEGs. Using different sets of trees for the two formats is effectively the same thing as having a format metric at the root node of each tree.

The GIF image format contains additional useful information that our current implementation ignores. GIF files can specify that one color is to be used as a transparent color. Transparent colors occur much more frequently in graphics than in photographs. We can define a binary metric that takes the value 0 if the image does not have a transparent color, and the value 1 if it does. In addition, GIF files contain extensions, which give some extra information about the image, like the application that created it. Extensions are much more common in graphics than in photographs. We could have a metric testing for the presence of such extensions.

Finally, the HTML context of an image is also a useful source of information. If an HTML page has a link to an image but does not display it, that image is usually a photograph. Images with the USEMAP and ISMAP attributes are usually graphics. The URL of an image can give statistically useful information; words like "logo," "banner," "bar," appear much more frequently in URLs of graphics. Images that are grouped together on a page are usually of the same type. Finally, the text around an image can be useful, even if we just scan it for the appearance of specific words, like "photograph."

We plan to combine information from both image-context metrics and image-content metrics to improve the accuracy of our system. Our first results indicate that individual image-context metrics do not separate photographs and graphics as well as our image-content metrics do. On the other hand, they will probably be useful, at least in creating an initial bias for the classification of some images.

11.3.10 Mixed Images

Mixed images are images that contain a photograph, but are not pure photographs. A mixed image can be a collage of photographs, a photograph with text on top, a photograph with a computer-generated frame or background around it, or any other combination of photographic and computer-generated material.

A medium term goal is to extend the system to perform some segmentation, so that it can identify photographic parts in at least some mixed images. For at least some cases, like photographs with one-color backgrounds, it is pretty easy to segment out the computer-generated part, and give the rest to the system to classify. The problem is that segmentation would also be applied to graphics, and might eliminate portions of graphics that would provide evidence that the whole images are graphics. For instance, graphics with a large one-color background score very high in the prevalent color metric. If we segmented out one-color backgrounds we would not be able to use that information to classify the rest of the image as a graphic. As long as we ignore mixed images, we can use any evidence of artificiality in an image as evidence that it is a graphic. If we allow for mixed images, we can't do that anymore. Therefore, some methods that we can use to classify mixed images in a meaningful way make it much harder to classify graphics correctly.

11.3.11 Other Approaches

The WebSeek search engine [14] classified images in those two types by comparing their color histograms to the average color histograms of a training set of photographs and a training set of graphics. The training sets consisted of 200 images. An image is assigned the type of the average histogram that is closest to its own color histogram. The distance between histograms h and m is defined as $|T(h - m)|$, where T is a matrix that gets derived from the training set, in order to achieve maximum separability between the two classes.

The work described in Rowe and Frew [10] has several things in common with the approach described here. The image content features that are used to discriminate between photographs and graphics are the squareness of the image, the number of colors, the fraction of impure colors (colors that are not pure white, black, gray, red, green, or blue), the neighbor variation (fraction of horizontally-adjacent pixels of the same color) and the color dispersion (fractional distance of the mean color in the sorted color histogram). In addition, the filename portion of the image URL is tested for the occurrence of words that are usually associated with only one of the two image types. The existence or not of such words is an additional feature that is considered in the classification. The actual classification is done by a linear-classifier perceptron neuron. The image is considered to be a photograph if a weighted sum of input factors exceeds a certain threshold.

Because these systems have been tested on different test sets, with different definitions of the boundaries of what constitutes a photo and a graphic, it is difficult to compare their accuracy. Smith and Chang [14] claim a recall rate of 0.914 for web photographs and 0.923 for web graphics. However, they don't specify exactly what they consider photographs and graphics. Such definitions have a direct impact on the error rate. Rowe and Frew [10] also don't specify exactly what they consider photographs and graphics.

11.4 Other Types of Image Classification

Three useful types types of image classification on the World Wide Web are detecting portraits, distinguishing color from gray-scale images, and filtering potentially offensive (in this case, pornographic) images. Algorithms for detecting these classes of images rely on statistical observations, combine image content with image context, and make use of the availability of huge amounts of training data from the Web. Examples of additional image types that could be detected using such methods are maps, charts, and cartoons. Detecting such types would be very useful in indexing images in extensive collections of multimedia documents, like the Web, in a meaningful way.

11.4.1 Detecting Portraits

Pictures of people are an important class of images on the World Wide Web. It is estimated that there are people present in more than one-third of the photographs on the Web [7]. In WebSeer, Rowley et al.'s face finder [11] was used to locate the faces in the images. The interface allowed users to select the number of people in the image, and the size of the largest face, indicating a close-up, bust shot, half-body shot, or full-body shot. These portrait categories were distinguished using the size of the largest face in the image. While the system suffered false negatives, usually due to rotated, small, or partially occluded faces, there were few false positives. Rowley's work has since been extended to locate faces rotated about the image axis [12].

11.4.2 Distinguishing Color from Gray-Scale Images

This distinction is not as trivial to detect as it might seem. While checking that all R, G, and B values for each pixel are equal will guarantee that an image is a gray-scale image, many images that most people would consider gray-scale images have pixel values that vary from this precise definition. The most common reason for this is the use of a color scanner to digitize gray-scale photographs. In such a situation, any photograph will produce at least some non-gray pixels. Some photographs will give a more systematic deviation, because they have faded over time, giving them a typically brownish tint. Nonetheless, this is a much easier problem than, e.g., distinguishing photographs and graphics, and it can be approached using standard statistical techniques.

11.4.3 Filtering Pornographic Images

For the task of filtering pornographic images and video, there are a number of different sources of information that can be fruitfully combined. Researchers have shown that adding skin-color detection to an automated text-analysis system significantly increases the detection rate of pornographic images [3, 7]. The skin-color detector described by Jones and Rehg [7] is a learning-based system, trained on large amounts of labeled skin data sampled from the World Wide Web (over one billion pixels). This system could be improved by texture analysis, and face recognition, which can determine if a large patch of skin viewed in the image is explained by being part of a face or not.

11.5 User Interface Issues in Image Search

The user interface to an image search engine will be used by millions of different people, most of whom want quick results and will have not read any instructions on how to use the system. Therefore, the user interface should be simple as possible. Text search boxes are familiar to users of text search engines, and search engines have tended to stick to this familiar interface for multimedia search as well.

An image search engine will bring users to the sites that it indexes, so these sites are usually predisposed to be cooperative in order to maximize their user traffic. However, site operators can become concerned if the users visit their sites without viewing the banner ads from which they receive their revenue. For this reason, some search engines, such as AltaVista's multimedia search, link only to the HTML pages linking to or embedding the images, not the images themselves – even though each image is addressable by its own URL. Others, such as Scour.net, link directly to the images themselves. If sites do not wish to be indexed they can typically easily block the crawler visiting the site. In fact, crawlers almost universally respect a standard protocol known as the robots exclusion standard for protecting content from being indexed.

Some concern has been expressed by copyright holders over reduced-resolution thumbnail summaries of images as being "whole works," and therefore possibly copyright violations in themselves. Search engines contend that their use of thumbnails falls under fair use, as do the summaries supplied by text search engines. Copyright holders can become particularly concerned about indexes of illegal copies of their images, encouraging users of the search engine to view their images at sites other than the sites they sanction and receive royalties from.

References

1. Amit, Y and Geman, D, "Shape Quantization and Recognition with Randomized Trees," Neur Comput, 9, pp. 1545–1588, 1987.
2. Brin, S and Page, L, "The Anatomy of a Large-Scale Hypertextual Web Search Engine," Proc. 7th Annual World Wide Web Conference (WWW7), 1998.

3. Chan, Y, Harvey, R, and Smith, D, "Building Systems to Block Pornography," Proc. 2nd UK Conf. on Image Retrieval, 1999.
4. Frankel, C, Swain, MJ, and Athitsos, V, "WebSeer, an Image Search Engine for the World Wide Web," Technical Report TR-96-14, Department of Computer Science, University of Chicago, 1996.
5. Freund, Y and Schapire, RE, "Experiments with a New Boosting Algorithm," Proc. Thirteenth International Conference on Machine Learning (ICML-98), 1996.
6. Freund, Y and Schapire, RE, "A Decision-Theoretic Generalization of On-Line Learning and an Application to Boosting," J Computer Syst Sci, 55, pp. 119–139, 1997.
7. Jones, MJ and Rehg, JM, "Statistical Color Models with Application to Skin Detection," Proc. IEEE Conf. on Computer Vision and Pattern Recognition (CVPR-99), 1999.
8. Kleinberg, JM, "Authoritative Sources in a Hyperlinked Environment," Proc. 9th ACM-SIAM Symposium on Discrete Algorithms, 1998.
9. Robertson, SE and Sparck Jones, K, "Simple, Proven Approaches to Text Retrieval," Technical Report 356, University of Cambridge Computer Laboratory, 1997.
10. Rowe, NC and Frew, B, "Automatic Caption Localization for Photographs on World Wide Web Pages," Inform Process Manag, 34, pp. 95–107, 1998.
11. Rowley, H, Baluja, S, and Kanade, T, "Neural Network-Based Face Detection," IEEE Trans Patt Anal Mach Intell, 20, pp. 23-38, 1998.
12. Rowley, J, Baluja, S, and Kanade, T, "Rotation Invariant Neural Network-Based Face Detection," Technical Report CMU-CS-97-201, Computer Science Department, Carnegie Mellon University, 1997.
13. Schapire, RE, "A Brief Introduction to Boosting," Proc. Sixteenth International Joint Conference on Artificial Intelligence (IJCAI-99), 1999.
14. Smith, JR and Chang, SF, "Searching for Images and Videos on the World Wide Web," Technical Report 459-96-25, Center for Telecommunications Research, Columbia University, 1996.
15. Wu, Z, Manmatha, R, and Riseman, E, "Finding Text in Images," Proc. 2nd ACM Int. Conf. on Digital Libraries, 1997.

12. Semantic-Based Retrieval of Visual Data

Clement Leung, Simon So, Audrey Tam, Dwi Sutanto,
and Philip Tse

12.1 Introduction

A fundamental problem in any visual information processing system is the ability to search and locate the information that is relevant [1, 5, 9, 10, 15, 19]. Visual information is widely regarded [9, 10] as powerful information bearing entities that will fundamentally affect the way future information systems are built and operate. The key to effective visual information search hinges on the indexing mechanism. Without an effective index, the required information cannot be found, even though it is present in the system.

Search effectiveness in an image database is always a trade-off between the indexing cost and semantic richness. While low-level semantically limited automatic indexing of information content is available, its use has been found to be very limited, and it is necessary to move on to a semantically rich paradigm that mirrors how humans view and search visual information [4, 7, 9]. The aim is to develop mechanisms for the indexing, search and retrieval of semantically rich visual information.

12.2 Two Kinds of Visual Contents

It is essential to make a distinction between two main types of image contents for the purpose of visual information search: primitive contents and complex contents.

12.2.1 Primitive Contents

Primitive image content is typically *semantically limited* and may include:

- A machine's view of the image including the raw elements from which the image is constituted (e.g., color)
- Those features of the image that can be recognized and extracted automatically (e.g., shapes, boundaries, lines); these normally form aspects or parts of an object without defining the complete object.

In addition, primitive contents may be related to aspects of an image not directly visible to the human eye, e.g., ultraviolet and infrared features. Primitive contents are often quantitative in nature from which spatial relationships, positional, geometric, or other quantitative properties may be extracted [2, 4, 7].

12.2.2 Complex Contents

Complex image contents are *semantically rich* and they often correspond to the patterns within an image that are perceived as meaningful by human users. Complex contents are *object based* in the sense that entire visual objects within an image often form the basis of retrieval. Examples of objects are *car*, *child*, and *violin* (Fig. 12.1(a)). In addition, although complex contents are almost invariably built from objects, they need not be confined to isolated objects. Complex contents cannot normally be automatically identified by the computer and are often qualitative in nature, although their extraction may be, in a limited way, assisted by the identification of primitive contents. For example, if one is searching for the object *swimming pool*, then the appropriate use of color and texture information would help to narrow down the search.

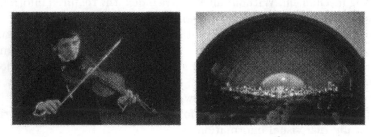

a   b
Fig. 12.1 a Objects: child, violin. b Object group: concert.

12.2.3 Levels of Complex Contents

Very often, *groups of objects* combined in specific ways may constitute higher-level concepts that may form the basis of meaningful retrieval; examples of these include *Christmas party, celebration, riot*, and *concert* (Fig. 12.1(b)). Semantically rich contents may be identified at different levels, depending on the requirement of the underlying applications. At the *primary level*, the representation will be based only on a basic level of common sense knowledge which allows the recognition the objects in an image (Fig. 12.1(a)). Sometimes, a representation may go beyond the primary level to a *secondary level*, which involves an interpretation of the objects seen in the image, and is based on additional knowledge acquired from familiarity with the subject matter. For example, "The Last Supper" (Fig. 12.2) goes beyond the basic representation of different objects within the image and involves significant additional knowledge beyond the basic recognition of factual objects within the image. The primary level offers a more objective factual description, while the secondary level, depending on the background and perception of the viewer, can be highly subjective. The successful implementation of a high-level pictorial database often hinges on adopting a correct level of description for the particular application.

Fig. 12.2 Secondary level representation.

12.3 Multiple Search Space Pruning

Searching images by their contents invariably involves some degree of imprecision. To keep the candidate set of returned images manageable, it is necessary to use an assortment of means to limit the relevant search space. Pruning the search space itself will not deliver the target images and large-scale multiple pruning of the search space represents the first steps in the search process. Once the search space has been reduced to manageable proportions, the more semantically rich indexing and query languages may then be activated to identify more precisely the target images.

This process is shown in Fig. 12.3. The candidate set represents the set of images where the target images may be found, which results from large-scale pruning of the search tree. The candidate set is obtained by pruning away large chunks of the underlying search tree. The double arrow linking the target set with high precision search signifies that a degree of iteration is involved since, unlike a conventional database search, some relevance feedback is necessary. Multiple pruning may be distinguished into:

– Multiple paradigm pruning
– Multiple media pruning

The first one is concerned with using a variety of means of filtering out irrelevant visual information. All of the paradigms operate on a single medium, typically images, but a variety of approaches and querying specification mechanisms is used to reduce the search space. Search pruning may be regarded as a "negative" approach, since it does not directly pick out the required target

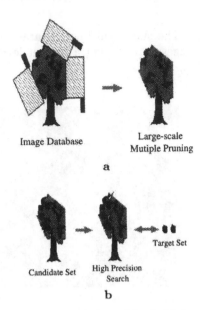

Image Database | Large-scale Mutiple Pruning

a

Target Set

Candidate Set | High Precision Search

b

Fig. 12.3 a Large-scale pruning followed by **b** high-precision search

images, but rather rules out unwanted ones. The multi-paradigm approach will be discussed in greater detail elsewhere in this chapter. For example, content matching and query specification based on both low-level features and high-level semantic content entail the use of multiple paradigms. Values or ranges for metadata can be specified using an SQL-like language or a form with fill-in text fields and drop-down choice boxes. Certain metadata naturally lend themselves to visual specification: for example, clicking or dragging on a map can indicate geographic locations of interest. To retrieve an image with an old man riding a blue bicycle originated after 1997 from a photo database, one might specify:

```
SELECT object, originator
FROM photo-db
WHERE year > 1997
AND content = "old MAN Riding blue BICYCLE"
```

Here, some metadata are included but the essential part is the content specification, which may stand by itself or, more typically, be mixed with metadata specifications. Due to the complexity of various Boolean combinations and the ambiguity of textual description, the content field would point to parsing and searching procedures that would require separate database tables [13].

The interface may also provide mechanisms to facilitate visual query composition, as conventional query languages are unable to capture the pictorial character of visual queries. The user could use a feature palette to create icons of size, shape, color, and texture that they wish to find, with the option of dragging and dropping these icons onto a structure template to specify the layer (foreground, etc.) or position in an image of the objects represented by the icons. Many software packages for creating graphics organize images in layers and some domains have standard names for these layers, e.g., in geographical mapping and architecture. Such structure can be incorporated into the query interface at this level.

In many visual information collections, other media types such as text and sound are typically involved. A further method for pruning the search space is to use multiple search on multiple media. It will follow the scheme as shown in Fig. 12.4, where we have only shown three media types, but this may be generalized to more than three media types. Here, we shall mainly focus on searches carried out on visual data and audio data for the purpose of image identification.

Search Tree: Search Tree: Search Tree:
Medium 1   Medium 2   Medium 3

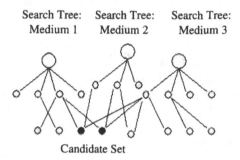

Candidate Set

Fig. 12.4 Multiple media search tree for producing target images.

In many image or video collections, audio information is often present and such information often provides valuable clues about the visual content of the data. In such situations, searching the audio information can produce significant pruning of the search tree, enhancing the overall search efficiency. Parsing of audio content can be done by signal processing and spectrum analysis. Although it is unlikely that automatic extraction is possible due to the multiplexing of different sources of sound in real-life scenarios, some simple sound patterns such as door bells, hand-clapping or engine vibration can be detected using model-based approaches if the background noise can be minimized. Here, we are not interested in the audio information directly as in Wold et al. [21], but only in making use of to assist in locating visual information.

It is now common to use MIDI (Musical Instrument Digital Interface) and digital instruments to generate and encode certain standard audio patterns and use them as accompaniments to visual information. The sound representation using MIDI can achieve a saving of hundreds of times (a compression ratio of hundreds of times over a wave file or MP3 file representation is not

Search Key
GM#123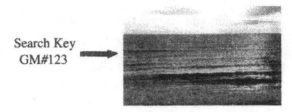

Fig. 12.5 Using General MIDI as a structured search argument for visual information.

uncommon) in storage requirements. The standard GM (General MIDI) has a 128-sound set with a variety of sound and patterns represented. For example, to retrieve an image of "seashore" (Fig. 12.5) when the image contains MIDI sound accompaniments, one can search for GM Program Number 123, which corresponds to the sound of a seashore. This is much more efficient than using complex image processing algorithms to retrieve images of seashores. Similarly, image classification may be automatically done through categorizing the style of music accompanying the image.

However, General MIDI is limited in scope as it does not contain musical style and rhythm data en bloc as a unit rhythm has to be recognized from the many constituent channels and drum patterns, which in itself is of comparable complexity to recognizing patterns and objects from images. A standard set of programs on musical styles would be advantageous for image retrieval, as the musical style tends to have greater scope for expressing the contents of an image and, if judiciously used, can produce substantial reduction of the search space. In such applications of the technique, it fulfills the following purposes.

- It serves to *exclude* irrelevant (rather than include relevant) images; e.g., images of peaceful natural scenery are unlikely to be accompanied by certain music styles such as disco or hard rock. As a result, significant pruning of the search tree may be accomplished.
- It provides a summary of the general character of an image, similar to the use of outline and summary facts in the Ternary Fact Model [12] for visual data search.

The SMF (Standard MIDI File Format) is unable to represent the richness of audio information accompanying the images, since it supports only audio data such as note number, patch, on/off velocity, start time, duration, and pedal information. To represent richer audio data, different manufacturers tend to have their own standards. For example, Cakewalk uses a file format with extension *.wrk (in addition to the MIDI *.mid), Voyetra's Digital Orchestra Plus uses *.orc files, and Technics uses the proprietary file format (Tech). In particular, it is interesting to note that the Technics file format supports a total of five file types and *.seq allows the full rhythm patterns and music styles to be represented en bloc. It is thus useful just to examine the rhythm number (similar to the principle illustrated in Fig. 12.3 (a) in order to determine the *irrelevance* of certain images and therefore, exclude them from the candidate image set. By themselves, the MIDI events or rhythm patterns will not pinpoint the visual information, but they can significantly help to reduce the search space when used in combination with visual search techniques.

Other audio searches (e.g., narrative, if present) are also useful and will provide meaningful image identification. This works by making use of speech recognition software and converts the spoken narrative to text, from which keywords may be extracted for index building as in Wactlar et al. [20]. However, a simple keyword approach by itself suffers from a number of semantic limitations, including the inability to express relationships and properties of visual objects.

12.4 Using Feature Vectors for Primitive Content Search

Most current approaches to image databases tend to concentrate on low-level features that are typically represented as feature vectors [22]. These are designed to provide a quantitative characterization of an image using purely automatic algorithms. Although such algorithms do not directly support the extraction of complex contents, they should play a part in helping to limit the search process in producing images that are similarly characterized. Primitive contents such as colors, textures, and shapes can be numerically computed to form a representation of the image. These may take the form of a set of numerical quantities that may be represented as vectors, arrays, or histograms. The characteristics of all images in the databases may be indexed using these signatures. Ideally, similar images should have a close match in the signature space, while dissimilar ones should exhibit very different signatures.

The extent of search reduction, however, depends on the probability distribution of the image collection over the appropriate vector space used for their characterization. In general, for a k-dimensional space, if we denote the probability density of image distribution over the neighborhood of $x = (x_1, \ldots, x_k)$ by $f(x_1, \ldots, x_k)dx$ and if the search space is characterized by

$$m_1 \leq x_1 \leq M_1,$$
$$m_2 \leq x_2 \leq M_2,$$
$$\ldots$$
$$m_k \leq x_k \leq M_k,$$

then a search with argument x and nearness ϵ will result in retrieving all images lying within the index region

$$R = \{y : \mid x_i - y_i \mid \leq \epsilon\},$$

which gives a conditional search reduction of

$$\rho(\boldsymbol{x}) = \frac{\int_R f(x_1, \ldots, x_k) d\boldsymbol{x}}{\prod_{i=1}^{k} (M_i - m_i)}.$$

The unconditional search reduction is therefore

$$\rho = \int_{\Omega} \rho(\boldsymbol{x}) d\boldsymbol{x}$$

where Ω represents the entire search space. The density $f(x_1, \ldots, x_k) d\boldsymbol{x}$ could be incrementally measured and stored in the image database. The range constants $\{m_i\}$ and $\{M_i\}$ may be estimated from the collection as

$$\tilde{m}_i = \min\{x_i\},$$
$$\tilde{M}_i = \max\{x_i\}$$

where the minimum and maximum are taken over the entire image collection. From [6], it can be shown that these estimates will converge to their true values:

$$E(\tilde{m}_i) \approx \frac{M_i - m_i}{n+1} + m_i \to m_i \text{ as } n \to \infty,$$

$$E(\tilde{M}_i) \approx \frac{nM_i}{n+1} \to M_i \text{ as } n \to \infty$$

where n is the number of images in the database (sample size). The maximum and minimum values may be stored as a special search space boundary record in the image database and would need to be updated as images are inserted and deleted from the database. Although insertion updating may be comparatively straightforward, deletion updating may involve searching the entire index. Such a record consists of $4k$ fields and is shown in Fig. 12.6.

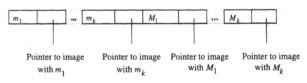

Fig. 12.6 Search space boundary record.

When a new image is inserted, its index vector is computed. Let its values be $z_1, \ldots, z_k$. The boundary record is retrieved and comparisons are made between z_i and m_i as well as M_i for $i = 1, \ldots, k$. If $m_i \leq z_i \leq M_i$ for all i, the boundary record will remain unchanged. If this is not true, then the appropriate m or M value and the associated pointer in the boundary record will need to be updated to refer to the newly inserted image. It is possible that several values of the boundary record may be altered as a result of a single image insertion. In the case where an existing image I is deleted from the database, the process for correctly updating the boundary record is considerably more elaborate. Let the index vector of image I be $w_1, \ldots, w_k$, which has been previously computed when the image was inserted and is available in the database. As before, comparisons are made between w_i and m_i as well as M_i for $i = 1, \ldots, k$. If $m_i \leq w_i \leq M_i$ for all i, then no action is required. If this is not the case, then all the index vectors in the entire database will need to be searched to determine the value of the field to be updated. Suppose $w_i = m_i$ and the corresponding pointer points to I. If the image with the next corresponding smallest value is I' with associated value m_i', then m_i will be replaced by m_i' and its pointer will now point to I' instead. Such an exhaustive search for updated values tends to be inefficient and it causes considerable delay for image deletion. It is much better to use an additional supplementary record that has the same structure as the boundary record and contains all the relevant information relating to the next level of extreme boundary values together with the associated pointers. In this way, the deletion process can be accomplished much more quickly while maintaining the boundary record up-to-date. The supplementary record may be updated at a later stage when there is less activity on the database.

If the estimation of the density f proves to be impractical, then the uniform distribution may be employed to determine the search reduction that, in this case, simplifies to

$$\rho = \frac{(2\epsilon)^k}{\prod_{i=1}^{k}(M_i - m_i)}.$$

For a given search reduction factor ρ, the nearness ϵ may be adjusted to provide the required level of reduction:

$$\epsilon \leq \frac{1}{2} \sqrt[k]{\rho \prod_{i=1}^{k}(M_i - m_i)}.$$

Many techniques may be used to generate signatures and different signatures may be used together for search space reduction. Invariance properties are highly desirable for signatures. In particular, a robust signature scheme should be invariant at least with respect to rotation and scaling. Since all the signatures of the image repository are pre-computed, the computational cost of this search method is confined to numerical comparison.

Although this method is primarily used for low-level contents, it has been shown that it is able to provide an effective index [8]. In many applications that contain only single or well-defined objects, this method is advantageous. However, it does not appear to work well for complex or natural objects, but could be very valuable if used in conjunction with higher-level methods. Although we have used only a single feature vector, it is possible that several feature vectors or arrays may be combined into one and the same calculations will also apply. For example, the indexing scheme proposed by Zheng and Leung [22] consists of two vectors, each of which has six components. Conceptually, these vectors may be amalgamated into a single vector consisting of 12 components, which is designed to provide a quantitative characterization of an image using purely automatic algorithms.

In the next two sections, we shall examine two specific schemes that we have developed to assist search and identification targeting at the low-level contents, and high-level contents, respectively. The first scheme, composite bitplane signatures [16, 17], improves search efficiency by making use of image groups whereby comparisons may be made on a group basis, rather than on an individual image basis. The second scheme focusses on the building of structured high-level content indexes by making use of a semantic data model.

12.5 Inverted Image Indexing Using Composite Bitplane Signature

A computed signature of an image (e.g., feature vectors) serves to provide a useful, if not unique, characterization of the low-level image contents. Here, we use a pre-defined size of bitplane as the "signature canvas." A retrieval method is chosen so that we can translate the image into a set of bit patterns and onto the bitplane. It then serves as the signature of the image. The method must be reasonably unique to avoid too many images hashed onto the same bit patterns. Ideally, we would like the signature scheme to exhibit the following properties:

- Precise retrieval of an image if the corresponding signature is used to query the database.
- The signature scheme must be able to associate similar images with similar signatures.
- The computational cost for comparing two signatures should be very efficient.
- The data structure to organize the signatures should be highly searchable and easily manipulated.

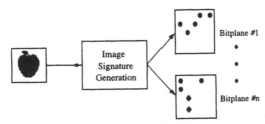

Fig. 12.7 Image hashing.

The selection of progressively better signature schemes is the subject of ongoing research. Basically, we are looking for a good hashing scheme (Fig. 12.7) that can translate an image into a set of signature bitplanes. As long as the scheme generates a reasonably unique and sparse signature bitplane and possesses the above properties, we can use it for the purpose of generating image signatures. In this chapter, we use the technique of retaining the signs of the m largest-magnitude wavelet coefficients for signatures as proposed in Refs. [11,18]. Figure 12.8 outlines the processing steps of our signature scheme. It is our view that the raw data should undergo a preprocessing step before any wavelet decomposition takes place. Not only is the usual conversion of different color spaces included in this step, it should also incorporate measures to achieve high degrees of translational, rotational, and scale invariance. Standard Haar wavelet decomposition for each color channel is then performed and a set of six bitplane signatures is generated as a result of retaining the signs of the m largest-magnitude coefficients.

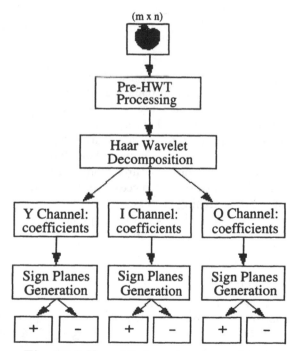

Fig. 12.8 Haar bitplane signature scheme.

Given a set of bitplane signatures with similar patterns, we can overlay the bitplanes and form a composite bitplane signature as in Fig. 12.9. In fact, it is possible to overlay non-similar patterns as long as we do not populate the bitplane excessively. Otherwise, the uniqueness of the composite bitplane signature will be eroded.

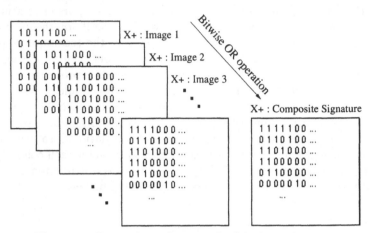

Fig. 12.9 Formation of composite bitplane signature.

To search a particular signature embedded in a composite bitplane signature, we rank all the composite bitplane signatures not by the number of matched bits but by the least difference of 1 bit from the query signature. The logic behind this searching mechanism is that the likelihood of finding the target image in a composite bitplane signature is higher in those composite bitplane signatures having least difference. Therefore, it would be logical to search the individual bitplane signatures from those composite bitplane signatures having a higher chance of success first. Moreover, false matches are taken care of in the searching process.

Expressed mathematically, the composite bitplanes $\{C^+, C^-\}$ for a set of sign planes $\{I_k^+, I_k^-; 1 \leq k \leq N\}$ are defined as

$$C^+(x, y) = \oplus_{k=1}^{N} I_k^{\,+}(x, y),$$
$$C^-(x, y) = \oplus_{k=1}^{N} I_k^{\,-}(x, y)$$

where $\oplus$ is the bitwise inclusive OR operator.

The individual scores between the sign planes of the query image Q and the target image T for all three color channels ($1 \leq i \leq 3$) are computed by the following equations:

$$S_i^+ = \sum_{x=0}^{m} \sum_{y=0}^{n} Q_i^+(x, y) \wedge \sim T_i^+(x, y),$$
$$S_i^- = \sum_{x=0}^{m} \sum_{y=0}^{n} Q_i^-(x, y) \wedge \sim T_i^-(x, y).$$

The overall score for all three color channels is the weighted sum of the individual scores

$$S = \sum_{i=1}^{3} (a_i S_i^+ + b_i S_i^-).$$

With this measure, a higher score means a weaker match between the query image and the target image. We rank the target images by the least scores. Also, we can place different emphases on the color channels by assigning different weights to a_i and b_i. For example, if the color scheme is YIQ, ranking by intensity only (i.e., ignoring colors) can be performed by setting the weights for I and Q scores to zero.

On the surface, in numbers of bits, the size of the bitplanes appears to be prohibitively large. In fact, this is not true: on the contrary, the storage requirement for our bitplane signatures is quite small. Only approximately 150 bytes are needed for a 256×256 sign bitplane (with $m = 128$) in compressed form. Hence, we can store all six sign planes of an image area with just a few hundred bytes.

The composite bitplane signatures eliminate the need to examine individual bitplane signatures in the entire image repository. For any query image, we still have to search through the composite bitplane signatures sequentially. Although it is much better than searching the entire collection of images as many existing retrieval methods do, performance will suffer if the image database contains a large number of composite bitplane signatures. Because of our unique ranking scheme, a hierarchical structure of composite bitplane signatures can be constructed for extremely fast image searching.

12.6 Using Semantic Data Models for Complex Content Search

In order to build a usable high-level content index, it is necessary to use a data model that is able to transform the intrinsic *heterogeneous unstructured* image contents to a set of homogeneous structured data items to support rapid searching, while at the same time providing adequate scope for the semantic richness of the images to be incorporated. Although the use of free text can be employed to assist content identification, as already indicated above, it has intrinsic limitations. A homogeneous structured data approach would be able to capture greater semantic richness and, at the same time, allows greater search efficiency.

The pivotal role played by semantic data modeling is illustrated in Fig. 12.10. Here, the data model allows a meaningful identification of the structural components within an image, from which a standardized logical representation of image content may be derived. In doing so, we are also proceeding from the *appearance value* to the *semantic value* [10] of visual objects. The logical image representation, typically in symbolic form, may be processed and the relevant index entries may be extracted. The extracted index items will be suitably organized and used to build the basic content index. At a later stage, the basic content index may be further processed and manipulated to provide a level of content index that is semantically richer than the basic level.

Fig. 12.10 Semantic indexing.

12.6.1   The Basic Ternary Fact Model

Our data model is principally concerned with complex image contents and, instead of focusing on unstructured patterns, the basic construct of our data model consists of discrete facts. Conceptually, image facts may be distinguished into the following types:

Elementary facts. These are the patterns within the image that are perceived as meaningful to human users but are not generally machine-recognizable. An elementary fact merely states the presence of a particular item in the image. In general, there are a number of elementary facts in an image. Some examples are flower, mountain, boat, and chair.

Modified facts. These are the elementary facts augmented with descriptive properties through the use of modifiers. Examples: *red* flower, *tall* building.

Binary facts. These are the patterns within the image linking together exactly two elementary or modified facts; each represents a specific interaction between these facts in the image that is meaningful to the human user. Such links may correspond to verbs or prepositions. Examples: boy *rides* elephant, girl *sits upon* yellow chair.

Ternary facts. These are the patterns within the image linking together exactly three elementary or modified facts; each represents a particular interaction between these three facts in the image that is meaningful to the human user. Example: woman *hits* cat *with* racket.

We may limit the above components to objectively verifiable, factual contents of an image, so that modifiers such as *large* or *young*, which can be subjective, are not allowed. This will ensure a high degree of precision in any subsequent image identification and retrieval process across a spectrum of applications. In one sense, this is restricting the content to a primary level.

However, this will also considerably limit the semantic richness of the model. In our approach, we do not explicitly prevent subjective facts. In addition, it is expected that the retrieval process will normally involve a number of iterations, so that the potential number of additional iterations introduced by the exclusion of subjective facts is unlikely to make a significant difference to the performance of the system.

12.6.2 Natural Language Indexing

The facts in our data model are expressed in natural language and are entered into the system using voice-input software. All facts are expressed in simple sentences and a parser is then responsible for extracting the different components for indexing (Fig. 12.11). Thus, our indexing language [14] has a very simple and intuitive structure; this will facilitate the indexing process as well as limit the cost of indexing.

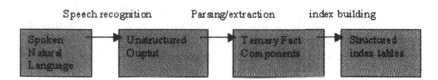

Fig. 12.11 Index building process.

Since the ternary fact model covers only a subset of the natural language, full natural language processing is not needed. An index sentence is broken into several index clauses. Each clause is then processed separately. A sketch of the ternary fact model component extraction procedure is given below:

Step 1. The first noun identified from a clause is considered as an elementary subject fact candidate.

Step 2. If one or more nouns directly follow the elementary subject fact candidate, then the last noun is considered as the actual elementary subject fact.

Step 3. Words (nouns or adjectives) before the elementary subject fact are considered as modifiers.

Step 4. The first noun extracted from the clause after the elementary subject fact will be treated as an elementary object fact candidate. The actual elementary object fact is determined as in step 2.

Step 5. Words between the elementary subject fact phrase and elementary object fact phrase are considered as fact linkers.

The same procedure applies for finding the ternary fact components.

Optionally, the user can choose to have the extracted facts run through the thesaurus.

12.6.3 Logical and Physical Index Structure

The main objective of the index structure is to be able to relate the image content to an image-ID. It is considered that the relational model is better suited for efficiency reasons, although it is also possible to use the object-relational paradigm. In all the relational tables, the image-ID constitutes the unique identifier and is the only candidate key available. However, it is rarely used as the search argument; rather, it would form the answer to the queries with other fields acting as the search arguments. It is possible to use a single "universal relation" to represent all the index information:

R (Modifier11, Modifier12, Fact1, Modifier21, Modifier22, Fact2, Modifier31, Modifier32, Fact3, Link, Image-ID).

This has the advantage that only a single table needs to be searched, with no decision required as to which of several tables is to be examined first. However, this suffers from the limitation of having a fixed number of modifiers for each fact (only two modifiers are shown for each fact, but the number can be increased). A second disadvantage is that, for many of the facts, most fields would have a null value, resulting in a rather sparse table. To overcome these limitations, we adopt the following four basic (relational) index tables as constituting the image content index:

- R_1 (Fact, Image-ID, Weight)
- R_2 (Fact, Modifier, Image-ID)
- R_3 (Fact1, Fact2, Link, Image-ID)
- R_4 (Fact1, Fact2, Fact3, Link, Image-ID)

where R_1 corresponds to the Elementary Fact Table, R_2 corresponds to the Modified Fact Table, R_3 corresponds to the Binary Fact Table and R_4 corresponds to the Ternary Fact Table. To facilitate search, it is possible that a given fact may be stored in more than one table.

In building up the index, a numerical weight will be assigned to each elementary fact in the indexing process. This signifies the visible prominence of a fact within an image in the case of direct fact indexing. For derived facts (see next section), this would provide a mechanism for indicating the likelihood of finding them within an image. Weights are only assigned to facts and not to modifiers.

It needs to be recognized that, although a relational structure is adopted, we are still dealing with a one-to-many relationship: one fact will normally correspond to several images. For rapid search identification for some queries, we build additional main memory search structures to speed up the process (Fig. 12.12).

- **Bit vector index**. This will be used to index facts and is built from the four index tables. All facts will be ordered, with each giving rise to a bit vector containing as many elements as the total number of images.

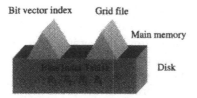

Fig. 12.12 Main memory structures for speeding up search identification.

These vectors would reside in main memory and provide the fastest means of image identification. Rapid identification is possible by looking for the "presence" or 1 bit in each vector, whose positions would indicate the relevant picture numbers.

- **Grid files/bit matrices.** These will be used to index the link between commonly occurring facts. Although the exact nature of the relationship cannot be included in these structures, they can be used to indicate the *presence* or *absence* of a relationship and may, in certain circumstances, allow unique identification. These will also reside in main memory and can be automatically generated from the basic index tables.

12.7 Discussion and Future Work

An index provides a means of identifying an image without involving the full image. An index may be defined as [3] "a systematic guide to items contained in, or concepts derived from, a collection. These items or derived concepts are represented by entries arranged in a known or stated searchable order, such as alphabetical, chronological, or numerical." With feature indexing, the aim therefore is to use some information derived from the image, much smaller than the image itself, to summarize certain features and characteristics of the image for identification purposes.

In a conventional database, part of the record (one or more fields) is designated as the key that is used as a search argument to assist in retrieving the full record from the database. In terms of size, an index should represent a small fraction of a record, regardless of whether the underlying record is highly structured, text based or pictorial; the smaller the index, the more efficient would be the search. In a conventional database, the user typically has knowledge of the key and enters it into the system for querying purposes. In the case of images, entering symbolic information, such as those based on the ternary fact model, will provide a useful means of image identification and retrieval. When applicable, this approach can be quite powerful, involving minimal specification on the part of the user while potentially able to deliver large target sets of images as the query result. For example, entering boat will allow the identification of all images with boats in them, if the image contents are properly modeled and represented. However, it is unlikely that symbolic specifications will suffice since the basic object of retrieval is non-symbolic in character.

Conventional indexing of textual contents requires a matching of the indexing language and the query language. In the case of visual data content, this is not always possible as a variety of aspects of an image must be indexed and these have to be incorporated in different ways. As far as possible, we adopt a systematic approach for structuring the contents and incorporate the same structure within a suitably extended query language.

Retrieval of visual information always requires substantial pruning of the search space and such pruning needs to be achieved by different means. This structure facilitates the indexing of high-level contents and allows semantically rich concepts to be efficiently incorporated. The methods presented form part of an integrated scheme for the effective retrieval of images based on a spectrum of image characteristics and it is intended that such a scheme may be implemented for wider usage and experimentation on the Internet.

In this chapter, we have used a multiple search approach to raise search effectiveness and, at the same time, provide a good degree of semantic richness. By providing an expressive data model that can incorporate a relatively rich and precise representation of semantic content and, coupling that with less labor-intensive semi-automatic indexing algorithms, a good degree of retrieval precision may be attained. However, we find that in order to achieve this level of semantic precision (preliminary indication gives a precision of over 80%), the indexing cost remains high. Even with natural language spoken input, the indexing time is measured in minutes. We have started work on perfecting a scheme that will partially automate this process that we call visual data mining and it works as follows.

It is necessary to distinguish between two types of indexing paradigms for high-level contents: *explicit indexing* and *implicit indexing*. The former requires an explicit entering of every index entry (the feature vectors and facts in the previous sections), while the latter includes a component for implicit deduction of index items for inclusion. Implicit indexing may be viewed as a kind of visual data mining. After the initial indexing of the image database, it should be possible to perform data mining to detect additional meaningful patterns in the content of images.

The value of conventional data mining stems from the fact that it is impossible to explicitly specify or enter all useful knowledge into the database. In the present context, the value of data mining stems from the impossibility of indexing everything explicitly in an image. Here, the association between the presence of different kinds of image contents would be discovered and indexed automatically. We shall focus our attention on the following two types of rules, which will allow semantically richer contents to be identified and indexed:

High-level contents $\rightarrow$ *Higher*-level contents,
High-level contents + Low-level features $\rightarrow$ *Higher*-level contents.

The second type relates to data mining on a combination of metadata, high-level and low-level content. For example, an image containing people buying and selling food and craftwork could be a marketplace if it also contains a specific combination of colors, textures, and shapes that indicate a market environment. However, it is expected that the first type of rules will play a more dominant role in the expansion of the index. Here, we use the ternary fact model by expanding and restructuring facts into a hierarchy. This expansion of the index makes use of rules to automate the generation of a higher-level fact from elementary facts, a higher-level fact from other high-level fact(s) or a combination of these two. In Fig. 12.13 (where a dot represents a fact and a line to the left of the fact relates that particular fact with its components), elementary facts are labeled as lower-level objects, while higher-level facts are labeled as medium- or high-level concepts. Basically, apart from the left-most set, which contains all the elementary facts, the remaining sets contain higher-level facts. Although Fig. 12.13 shows only one medium-level concept, in practice we can have several.

Lower Level   Medium Level   Higher Level

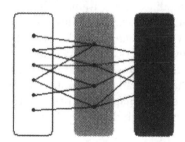

Fig. 12.13 Framework for automatic index expansion.

Indexes are entered manually by humans or extracted automatically by computer from the image in terms of atomic objects. These indexes become "native" indexes. Native indexes are defined as indexes originally entered by the human indexer or extracted by the computer directly from the image. From these native indexes we can obtain derivative indexes using rule deduction or rule induction. Rule deduction produces a single higher-level derivative index from several native indexes. Rule induction, on the other hand, produces several lower-level derivative indexes from a single native index. In terms of object/concept hierarchy, low-level derivative indexes represent primitive objects/concepts (e.g., nose, eyes, lips) while high-level derivative indexes represent complex objects/concepts (e.g., head, person, human).

A rule consists of an IF part that lists all condition(s) that must be satisfied, followed by a THEN part that concludes the rule given that the condition(s) are met. A rule deduction comprises an IF part with several lower-level objects/concepts linked by Boolean AND (defining the necessary conditions) or Boolean OR (defining the optional conditions), followed by a THEN part with a single high-level object/concept. A rule induction, on the other hand, comprises an IF part with a single high-level object/concept, followed by a THEN part with several low-level objects/concepts linked by Boolean AND or OR.

Rule deduction is used to build higher-level indexes from the existing indexes. By validating the rule conditions with existing indexes (which could be elementary facts or higher-level facts) and obtaining higher-level indexes from the hypothesis of the rule, we can create a new index entry automatically. Thus, we build a higher-level index from lower-level indexes that might be directly recognizable from the image. For retrieval, users can take advantage of these high-level indexes to speed up the searching time and narrow down the search space. An example of defining a high-level object "computer" is: IF there exists a monitor AND a CPU AND a keyboard AND a mouse THEN the OBJECT is a computer. In turn, we can treat the object "computer" as an intermediate-level index and then use it as a condition for a higher-level concept description. In this way, the index representation is structured into a hierarchy. Therefore, several native indexes will be able to define intermediate indexes and several intermediate index items will be able to define higher-level indexes and so on. Another benefit of this method is the reusability of the rules. Once a rule that defines an object is created, it can be used to define several other higher-level objects/concepts. This may be compared to the use of *feature groups* in making low-level features more expressive [9]. This indexing mechanism will have the added advantage of avoiding inconsistency and subjective perception of the image concepts. Continuous incremental indexing will also be possible through dynamic incorporation of new rules into the knowledge base.

Rule induction is used to elaborate the existing indexes. By validating the rule condition with existing indexes and obtaining more detailed indexes from the hypothesis of the rule, we can create new index entries automatically. These new indexes will have lower levels of hierarchy that provide details to the existing indexes. Rule induction reduces much of the indexing time and cost. Now indexers can enter high-level description in their indexing process and leave the details of the indexes to be generated automatically by the rule. Similar to the rule deduction above, we can also create a rule hierarchy in rule induction. This mechanism will allow us to go beyond a single-level index expansion and, together with rule deduction, will eventually increase the query recall and precision.

References

1. Adjeroh, D and Nwosu, KC, "Multimedia Database Management - Requirements and Issues," IEEE Multimed, 4(4), pp. 24–33, 1997.
2. Aslandogan, Y and Yu, C, "Techniques and Systems for Image and Video Retrieval," IEEE Trans Knowledge Data Eng, 11(1), pp. 56–63, 1999.

3. Borko, H and Bernier, C, Indexing Concepts and Methods, Academic Press, 1978.

4. Campanai, M, Del Bimbo, A, and Nesi, P, "Using 3D Spatial Relationships for Image Retrieval by Contents," IEEE Workshop on Visual Languages, 1992.

5. Del Bimbo, A, "A Perspective View on Visual Information Retrieval Systems," IEEE International Workshop on Content-based Access of Image and Video Libraries, pp.108–109, 1998.

6. Feller, W, An Introduction to Probability Theory and its Applications, Vol. 2, Wiley, 1971.

7. Flicker, M, Sawhney, H, Niblack, W, Ashley, J, Huang, Q, Dom, B, Gorkani, M, Hafner, J, Lee, D, Petkovic, D, Steele, D, and Yanker, P, "Query by Image and Video Content: The QBIC System," IEEE Computer, 28(9), pp. 23–32, 1995.

8. Gevers, T and Smeulders, A, "Evaluating Colour- and Shape-Invariant Image Indexing for Consumer Photography," International Conference on Visual Information Systems, Melbourne, pp. 293–302, February, 1996.

9. Gupta, A and Jain, R, "Visual Information Retrieval," Comm ACM, 40(5), pp. 70–79, 1997.

10. Gupta, A, Santini, S, and Jain, R, "In Search of Information in Visual Media," Comm ACM, 40(12), pp. 35–42, 1997.

11. Jacobs, CE, Finkelstein, A, and Salesin, DH, "Fast Multiresolution Image Querying," ACM SIGGRAPH, pp. 277–286, 1995.

12. Leung, CHC and Zheng, ZJ, "Image Data Modelling for Efficient Content Indexing," IEEE International Workshop on Multi-media Database Management Systems, New York, pp. 143–150, August, 1995.

13. Pereira, F, "MPEG-7: A Standard for Content-Based Audiovisual Description," 2nd International Conference on Visual Information Systems, San Diego, pp. 1–4, December, 1997.

14. Salton, G and McGill, M, Introduction to Modern Information Retrieval, McGraw-Hill, 1983.

15. So, WWS, Leung, CHC, and Zheng, ZJ, "Analysis and Evaluation of Search Efficiency for Image Databases," in Image Databases and Multi-media Search, Smeulders, A and Jain, R (eds), World Scientific, pp. 253–262, 1997.

16. So, WWS and Leung, CHC, "Inverted Image Indexing and Compression," SPIE Multimedia Storage and Archiving Systems II, Dallas, Texas, 3229, pp. 254–263, November, 1997.

17. So, WWS and Leung, CHC, "A New Paradigm in Image Indexing and Retrieval using Composite Bitplane Signatures," IEEE International Conference on Multimedia Computing and Systems, Florence, 1, pp. 855–859, June, 1999.

18. Stollnitz, EJ, DeRose, TD, and Salesin, DH, Wavelets for Computer Graphics: Theory and Applications, Morgan Kaufmann Publishers, 1996.

19. Tam, AM and Leung, CHC, "A Multiple Media Approach to Visual Information Search," ACM SIGIR International Workshop on Multimedia Indexing and Retrieval, Melbourne, pp. 1–6, August, 1998.

20. Wactlar, HD, Kanade, T, Smith, MA, and Stevens, SM, "Intelligent Access to Digital Video: Information Project," IEEE Computer, 29(5), pp. 46–52, May, 1996.

21. Wold, E, Blum, T, Keislar, D, and Wheaton, J, "Content-Based Classification, Search and Retrieval of Audio," IEEE Multimedia, 3(3), pp. 27–36, 1996.

22. Zheng, ZJ and Leung, CHC, "Quantitative Measurements of Feature Indexing for 2D Binary Images of Hexagonal Grid for Image Retrieval," IS&T/SPIE Symposium on Electronic Imaging: Science and Technology, San Jose, California, 2420, pp. 116–124, 1995.

13. Trademark Image Retrieval

John P. Eakins

13.1 Introduction

Trademark image searching is potentially one of the most important application areas for automated content-based image retrieval (CBIR) techniques. There are many large and growing collections of trademark images in electronic form. The task of maintaining manual indexes to these image collections is becoming increasingly onerous. And there is a paramount need for accurate and reliable searching, since the images can be of major commercial significance.

A trademark can be defined as "a word, phrase, symbol or design which distinguishes the goods or service of one party from those of others."[*] Trademarks form an important part of any company's intellectual property, associating the company's reputation for quality and reliability with its goods and services, and hence providing valuable information to customers and clients. The majority of trademarks in current use are officially registered with national or international patent or trademark offices, to provide their owners with legal protection in the case of any dispute.

Most trademarks consist purely of text words or phrases (often referred to as *word* marks). In some cases, the design of the text characters may itself form part of the trademark (*stylized word* marks). However, an increasing proportion of trademarks (around 40% in the case of the UK Trade Marks Registry) now contain image data, either on their own (*device only* marks) or in combination with text (*device and word* marks). Examples of each type of mark are illustrated in Fig. 13.1.

Fig. 13.1 Examples of different kinds of trademark: **a** word only, **b** stylized word, **c** device only, **d** device and word.

---

[*] Shortened from the definition given in Ref. [1].

13.2 Trademark Search and Registration

13.2.1 General Principles

Although the process of trademark registration differs between one country and another, the principles under which a trademark can be granted are broadly similar. The UK Patent Office, for example, requires candidate trademarks to meet a number of specified criteria before they can be accepted for registration, including the following:

- They must be signs (words or images) that can be represented graphically.
- They must be distinctive (they cannot consist exclusively of words or images in common use in a particular trade).
- They cannot be immoral, illegal, or deceptive.
- They cannot include protected emblems such as coats-of-arms or national flags.
- They cannot be so similar to any existing mark for the same type of goods or services that the public might confuse one with the other.

While some of these criteria can be decided purely by examining the candidate mark itself, the only way to establish the final criterion in the above list is to examine all potentially conflicting marks. Since many national registries now contain hundreds of thousands of marks (the UK Trade Marks registry holds over 120,000 marks containing some form of image data), this represents a significant retrieval problem. Over the years, patent and trademark offices have therefore developed a variety of manual and automated systems for searching their archives.

13.2.2 Searching for Word Marks

The process of establishing that a word mark is sufficiently different from existing marks to avoid confusion is far from straightforward. Since two marks can conflict if they either look or sound alike (for example, the words *Rightmark* and *Writemarque* would be easily confused by most English speakers), examiners have to search the registry both for identical text strings, and for strings containing phonetically similar characters (e.g., substituting "p" for "b," "f" for "v," and almost any vowel for any other). A further problem is that the semantic connotations of word marks can often be crucial (*Aquamatic* and *Watermatic* would be considered sufficiently close to cause confusion, for example). No automatic system yet developed can take account of all these variations, though the knowledge-based systems now in use at some of the larger commercial search services can overcome many of the problems involved. Most current systems aim to provide a variety of exact and approximate string-matching facilities, leaving it to the examiners to exercise their judgment on the most appropriate search strategies to use. An example of such a system is OPTICS, used by the UK Trademark Registry to carry out searching of word marks. This offers its users the ability to search for the presence of a required substring anywhere within a word, with options for wildcard matching, approximate character matching (identifying all words which differ from the search string by just one or two characters), and matching by phonetic similarity.

13.2.3 Searching for Device (Image) Marks

Searching for potentially conflicting marks when the candidate mark consists primarily of image data is, if anything, an even more difficult task. Similarity is in the eye of the beholder, and the examiner has to try to judge whether any feature of the candidate mark (such as shape, spatial layout, or subjective connotation) could reasonably cause a member of the general public to confuse it with an existing mark. The example shown in Fig. 13.2(a) illustrates the importance of shape similarity. Examiners faced with the image in Fig. 13.2(a) would almost certainly identify those shown in Fig. 13.2(b-d) as sufficiently similar to warrant detailed examination to decide whether any of them might be confused with (a). All these images appear to most observers as a triangle enclosing a circle – though the shape of the triangle can vary (b), it does not need to be solid (c), and its presence may need to be inferred from other shape elements (d).

a   b   c   d

Fig. 13.2 a An abstract image; **b-d** other images which show some element of shape similarity.

In the example shown in Fig. 13.3, shape similarity appears to be of less importance than the spatial arrangement of image components. All images in this set consist primarily of arrangements of vertical bars. The shape of these bars varies within quite wide limits – in (b) they have straight rather than round ends, in (c) they are much wider, and in (d) much narrower in relation to their length than those in the query image (a). Similar variations can be tolerated in their number, absolute orientation, and linear separation. The presence of parallelism and equal spacing is enough to suggest potential for confusion.

The most widely adopted methods for trademark image searching in current use are those based on classification systems such as the so-called Vienna Classification [2]. This aims to divide up the entire universe of possible types

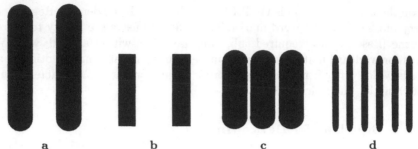

a   b   c   d

Fig. 13.3 An example where similarity of spatial layout appears to be the predominant influence.

of image into 29 major categories, each of which is subdivided into a second and sometimes a third level of sub-categories, permitting quite detailed description of an image. An example of some of these categories is presented:

1. Celestial bodies
   1.1 Stars
   1.3 Sun
   1.7 Moon
   . . .
2. Human beings
3. Animals
   . . .
25. Ornamental motifs
26. Geometrical figures
   26.1 Circles
   26.3 Triangles
      26.3.1 One triangle
      26.3.2 Two triangles, one inside the other
      26.3.3 More than two triangles inside one another
      26.3.4 Several triangles, joined or intersecting
      . . .
   26.4 Quadrilaterals
   26.5 Other polygons
   26.11 Lines or bands
   . . .
27. Forms of writing
   . . .

Vienna Classification codes can be used as the basis for either manual or automated retrieval systems. An example of the latter is TRIMS from the UK Patent Office, which allows searching by Boolean combination of Vienna codes, and subsequent on-screen display of retrieved images. As with all such systems, however, initial assignment of codes remains a purely manual process.

A limitation of any system based on Vienna Classification codes is that in order to be reasonably sure of retrieving all relevant images, it is often necessary to use broad search terms. In the larger national and international trademark registries, this can result in the retrieval of excessively large numbers of images – in some cases running into thousands – all of which need to be examined manually. The example search shown in Fig. 13.2 could be narrowed down to some extent by specifying the presence of both categories 26.1 (circles) and 26.3 (triangles), though this would not distinguish between circles within triangles and triangles within circles. There is no easy way of expressing the concept of parallelism for the search in Fig. 13.3, which would have to be formulated simply as category 26.11 (lines or bands), leaving the examiner to pick out possibly relevant marks by searching sequentially through the output images displayed on the screen.

13.2.4 Other Applications of Trademark Image Matching

Trademark registration is not the only application making use of image matching techniques. Automatic recognition of company trademarks on invoices or letterheads is an increasingly important aspect of automatic document image processing (DIP). Developers in this field normally refer to the process as *logo recognition* [25]. The objectives of this process are significantly different from those of trademark registration. In trademark registration, the aim is to identify all existing images whose similarity to a given candidate image exceeds a certain threshold. The query image is not normally expected to match any previously registered mark. In logo recognition, by contrast, the unknown image is assumed to be a (possibly corrupted) example of an image previously known to the system. Its similarity to other images in the collection is largely irrelevant. While both types of application make use of similar technology for image matching, it is important to appreciate the difference in motivation. Logo recognition is in general a much easier process, as exhaustive retrieval is seldom required, and the matching process can be based on simple physical similarity. For trademark registration, exhaustive retrieval is essential. To achieve this, it is necessary to take a much wider range of factors into account, such as shape features and semantic connotations implicit in an image.

13.2.5 System Requirements for Trademark Image Retrieval

An image retrieval system for trademark registration must have the ability to find all existing trademark images which might be deemed confusingly similar to any new candidate mark, and to display them to a human examiner for a decision on whether registration should be permitted. Such similarity could take a number of different forms. Most experts in the field agree that shape similarity (whether explicit or implied) is the single most important determining factor, as illustrated in Fig. 13.2. However, the examples in Fig. 13.3 show that the picture is not quite as simple as this – higher-level patterns made by individual components also play a part. Not all aspects of image shape are considered significant. For example, borders surrounding an image are generally ignored. Hence the requirements for any image retrieval system used for trademark registration purposes are to be able to identify and match the most salient aspects of a candidate mark's appearance, whether these comprise its overall shape, the shapes of important image components, shapes defined by perceptually significant groupings of components, or key aspects of image structure.

Other aspects of a trademark's appearance are generally considered to be less important. Color is seldom a major issue, at least at present, since the vast majority of trademarks are registered in black an white to give maximum protection (if registered in color, protection extends only to color combinations explicitly claimed in the registration). Texture is again seldom important; in most cases, a textured component is held to be identical to a solid or outlined component of the same shape. Figure 13.4 illustrates some of the foregoing points.

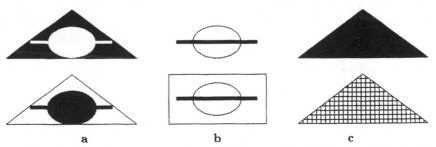

a   b   c

Fig. 13.4 Pairs of images which would be considered identical for trademark registration purposes: **a** figure-ground reversal; **b** presence or absence of borders; **c** presence or absence of texture.

Finally, appearance is not the only aspect that trademark examiners need to consider. The semantic connotations of an image are also important. One trademark prominently featuring (say) a dog, or a crown, might be confused with another, even though their actual shapes were quite different. They might even be confused with an abstract pattern of shapes suggesting an image of a dog or a crown. Ideally, a trademark registration and retrieval system should be able to cope with all the aspects of similarity outlined above. As discussed below, some of the problems involved are much harder for more current technology to solve than others.

13.3   Techniques for Image Retrieval

13.3.1   The Growth of CBIR

Traditional methods for providing subject access to large collections of images have centered around the use of manually assigned keywords or classification codes. However, the labour-intensive and subjective nature of this task has led to an upsurge of interest in methods for automatic image analysis and retrieval. Research in this area has grown enormously in volume over the last 10 years, and there is now a bewildering variety of alternative techniques open to system designers. Several comprehensive reviews of the subject have been published, including Idris and Panchanathan [31], and Rui et al. [57]. The majority of techniques are based on comparing primitive image features such as color, texture, and shape, though there now is a growing body of research into image retrieval by semantic content.

13.3.2   Techniques for Shape Retrieval

Shape matching is by common consent the single most important technique for trademark image retrieval. The aim of all techniques is the same – to derive a quantitative measure $D(q, s)$ of the difference between a query object q and some stored object s of similar type such that D depends purely on the differences in shape between q and s. This raises the question of how an object's shape is defined, and what constitutes shape similarity. For most 2-D image processing applications, the shape of an object within an image can be defined as the contour traced by its boundary. No valid definition of shape similarity has yet been formulated, principally because for most purposes (including trademark registration) the ultimate judge of similarity is a (subjective) human observer. Uncovering the rules that govern human shape similarity judgment is no easy task. Such rules do exist, however, and have been the subject of investigation by a number of psychologists, most notably Goldmeier [27].

Goldmeier's experiments set out to understand the underlying rules by which humans judge which of two shapes A or B is more similar to a third shape C. His experiments demonstrated that a number of seemingly plausible hypotheses about similarity failed to account for all observed results. Instead, his results suggested that in many cases similarity judgments were based on comparing perceptually significant groups of image elements, as predicted by Gestalt theory [71]. Perhaps the most significant aspect of his work relates to his experiments on *Prägnanz*, a term which he translates as *singularity*. Certain aspects of a shape – the presence of straight lines, right-angles, parallelism, components of equal size or spacing – appear to play a special part in its perception. Hence small changes to a shape which disturb such aspects can have a disproportionately large impact on similarity judgments. One might thus expect matching techniques which take account of these principles to be more successful than those which do not.

As well as the ability to match human similarity judgments, the ideal shape matching technique needs to fulfill several other criteria. Robustness to noise or small deformations in an image, and invariance to translation, rotation and scaling are frequently quoted as desirable criteria – though there can be cases where the ability to detect differences in scale or orientation is important. Low computational complexity is also desirable, though perhaps less important than in the past, in view of the steady increase in power of each new generation of processors.

A wide variety of techniques meeting at least some of these criteria has been described in the literature. One important class of methods is based on direct matching of complete (information-preserving) representations of object shape, such as chain-codes or splines. Such methods can have high discriminating power, at least when matching highly similar shapes, but are often computationally very expensive. A second class of methods is based on the extraction and comparison of features such as edge direction histograms or moment invariants, which may capture important aspects of an object's appearance, but which cannot be used to reconstitute its entire shape. These methods often have lower discriminating power, but tend to scale up better to large image collections. Further distinctions can be drawn between measures computed purely from a region boundary, and those computed from every pixel within the region. Some can be derived directly from raw pixel data; others require the region boundary to be converted to vector format (lines, arcs, or splines) before they can be computed. Some of the more important techniques are briefly described below. For full details, the reader is referred to the original literature.

Techniques which involve direct matching of information-preserving representations of shape boundaries include:

String-matching of chains of boundary pixels. Given a set of boundary pixels represented as Freeman chain codes [26], it is possible to define a number of measures representing the distance between the two chains. Cortelazzo et al. [13] examined this problem in detail, showing that three different methods based on summation of substring differences or string rewriting rules were all capable of giving intuitively sensible distance measures between contours of simple trademark images. Given suitable normalization, these distance measures can be rendered invariant to translation, rotation, scaling, and choice of starting point for string matching – though this is not a trivial problem.

Measurement of turning angle. For any given shape, it is possible to represent its boundary by the turning function $\Theta(s)$, measuring the angle of the tangent to the boundary as a function of s, the normalized distance along the boundary from a given reference point. The difference in shape between two objects a and b can thus be computed [59] as:

$$\int\limits_0^1 |\Theta_a(s) - \Theta_b(s)| ds. \tag{13.1}$$

Such measures are inherently invariant to translation or scaling, and can be rendered invariant to rotation given an appropriate choice of starting point.

Elastic deformation of templates. A potentially powerful, though computationally expensive technique for matching unknown and query shapes is to deform the boundary of the query shape until it matches a given stored shape, and then to measure some function Φ which gives an indication of the cost of the deformation process. The earliest reported use of this technique was in the Photobook system from MIT [50], which computed the energy required to deform one shape into another from their finite element representations. Further variations of the technique have been reported by del Bimbo and Pala [14], who used elastic deformation energies together with correlations between curvature functions of query and stored shapes to train a back-propagation network to match human similarity judgments, and by Jain et al. [32], whose method applies displacement functions to the query template in order to compute its goodness of fit with a given image region. Jain's method appears to have better invariance to rotation than del Bimbo's.

Matching using non-information-preserving features normally involves calculating and storing a feature vector characterizing selected aspects of the shape of each image in the database, and then calculating the similarity between the feature vector computed from the query with each image in the database, using some measure such as Euclidean distance $L_2 = ||v_i - v_j||$, where v_i and v_j represent the feature vectors of images i and j. Commonly used types of feature include:

Simple global features. Several computationally simple measures of global shape (characteristic of the shape as a whole rather than any individual part) have been proposed over the years (e.g., Refs [40, 68]). They may be calculated either from region boundaries or from the complete set of pixels making up a region. Some of the more frequently used global features are aspect ratio (L_1/L_2), circularity ($4\pi A/P^2$), and transparency (A/H)* , all of which are invariant to translation, rotation, and scaling.

Local features. Local features can represent shape characteristics of small regions of an image, and can thus act as a useful complement to global measures. Most are calculated from region boundaries. Examples include the line-angle line triplet features used by Eakins for the SAFARI [17] system, and the longer segment sequences used by Gary and Mehrotra [46]. When normalized, they are effectively invariant to translation,

---

* where L_1 is the length of the perpendicular to the longest chord in the shape, L_2 the length of its longest chord, A the area, P the perimeter, and H the area of its convex hull.

rotation, and scaling. The advantage of these features is that they can recognize similarity between partly occluded shapes, and between regions of articulated objects such as scissors.

Edge direction histograms. Another indirect measure of shape within an image is to compute a histogram of edge directions. Jain and Vailaya's technique [33] uses the Canny edge detector [7] to identify edge pixels and compute edge directions, and then accumulates these into bins at 5° intervals. Histograms of this kind can give an indication of the predominant edge directions within the image, capturing at least some aspects of its shape. They are translation-independent, can be rendered scale-invariant to some extent by normalization, but are rotation-independent only to a limited extent.

Fourier descriptors. It is possible to represent the cumulative curvature around the boundary of any region as a function of curve length, and expand this function as a Fourier series [75]:

$$\theta(t) = \mu_0 + \sum A_k \cos(kt - a_k) \tag{13.2}$$

in which the coefficients A_k and a_k, the kth harmonic amplitude and phase angle respectively, are known as the Fourier descriptors of the curve. Strictly speaking, this form of representation is information-preserving, but only if an infinite Fourier expansion is used. In practice, a relatively small number of Fourier descriptors can provide a description of the curve which appears to reflect its overall shape fairly consistently, and is invariant to rigid transformations.

Moment invariants. For any digital image $I(x, y)$, it is possible to compute a series of central moments μ_{pq}, defined as:

$$\mu_{pq} = \sum_x \sum_y (x - \bar{x})^p (y - \bar{y})^q I(x, y) \tag{13.3}$$

from which a series of *moment invariants* $\varphi(n)$ can be derived which characterize shape in a manner which is invariant to scaling, rotation and translation [30]. The first three invariants in the series are defined as $\varphi_1 = (\eta_{20} + \eta_{02})$, $\varphi_2 = (\eta_{20} + \eta_{02})^2 + 4\eta_{11}^2$, and $\varphi_3 = (\eta_{30} - 3\eta_{12})^2 + (3\eta_{21} - \eta_{03})^2$, where:

$$\eta_{pq} = \frac{\mu_{pq}}{\mu_{00}^\gamma}, \text{ and } \gamma = \frac{p + q}{2} + 1. \tag{13.4}$$

Moment invariants have been widely used in image analysis for many years (e.g., Ref. [16]).

Zernike moments. Another set of moments with the useful property of orthogonality was proposed by Teague [65]. The Zernike moment of order n with repetition m for image $I(r, \theta)$ is defined as:

$$A_{nm} = \frac{n+1}{\pi} \sum_{\rho} \sum_{\theta} \left(R_{nm}(r)e^{im\theta} \right)^* I(r, \theta) \tag{13.5}$$

where $R_{nm}(r)$ are the set of radial polynomials defined by Zernike [66]. Zernike moments are defined only within the unit circle ($r \leq 1$), so images need to be transformed to lie within this circle before computation begins. Zernike moments computed from images transformed in this way are effectively invariant to rotation, translation, and scaling.

13.3.3   Other Types of Image Retrieval Techniques

A wide range of techniques for retrieval by other types of primitive image feature has been developed. Some of the best-known of these include:

Color. Image matching by color similarity is widely used in CBIR, using techniques such as comparison of color histograms [63]. This involves computing histograms of pixel distributions in some suitable color space for each image in the database, and then calculating the similarity between histograms from query and stored images using an appropriate distance measure. Several useful refinements to the technique have been proposed, including region-based color matching [8].

Texture. Retrieval by similarity of texture has proved useful in a variety of contexts. Favored measures, generally derived from second-order statistics of pixel intensity distributions, include *contrast, coarseness, directionality,* and *regularity* [64], and *periodicity, directionality,* and *randomness* [43]. Gabor filters have also been successfully used for texture retrieval [45]. Texture similarity between query and stored images is normally computed from the distance between feature vectors comprising such measures. An alternative approach is to derive a *texture thesaurus* of codewords representing important classes of texture within a collection [44], and to retrieve images on the basis of similarity between their textured regions and codes stored in the thesaurus.

Spatial location. Several techniques have been proposed for allowing users to search for images containing *objects* in defined spatial relationships with each other. One well-established technique for rapid matching of images on the basis of similarity of spatial layout is 2-D iconic indexing, introduced by S.K. Chang et al. [11]. This generates a string representation of the partial ordering of objects within an image along both x- and y-axes, which can readily be used as the basis for similarity matching. Its lack of rotational invariance can be a problem in some contexts. An alternative method described by Gudivada et al. relies on computing edge graphs between the centroids of every significant object in the image [28]. Query and stored images can then be matched by comparing the relative orientation of corresponding edges. Unlike Chang's method, the technique is insensitive to rotation – though it does require all image objects to be labeled before matching can begin, which could be a problem in trademark retrieval applications.

Another technique, developed by Hou et al. [29], might prove more useful for trademark retrieval. It involves partitioning the image into closed regions, calculating the centroid positions of each region and of the image as a whole, and then storing these positions as a list of 3-tuples (w_i, d_i, θ_i), where w_i, is the weight (typically its relative area) associated with region i, d_i, the distance between c_i, the centroid of region i, and c_0, the image centroid, and θ_i, the angle between $c_0 c_i$ and $c_0 c_{i+1}$. Spatial similarity between query and stored images can then be measured by computing a suitable cost function for transforming the query pattern into the stored pattern.

Spatial indexing has also proved effective in combination with other cues such as color [60].

Multiresolution transformations. Promising retrieval results have been reported using coefficients derived from multiresolution transformations of pixel intensities. These include the *wavelet transform*, widely used for a variety of image processing applications including image retrieval [41], and *retrieval by appearance*, based on multi-scale Gaussian filtering. Two versions of this latter method have been developed, one for whole-image matching and one for matching selected parts of an image. The part-image technique involves filtering the image with Gaussian derivatives at multiple scales [52], and then computing differential invariants; the whole-image technique uses distributions of local curvature and phase [53].

Techniques for retrieval by image semantics are much less well developed, as this constitutes a much more difficult problem. Despite some success in developing techniques for automatic scene classification [48], object identification within limited domains [24], and learning semantic associations from users [10, 39], progress in this area remains slow.

13.3.4 Retrieval Efficiency

A significant limitation of current CBIR technology is the problem of efficiently retrieving the set of stored images most similar to a given query. In contrast with text retrieval, where efficient searching is normally accomplished by the use of inverted-file indexes allowing direct access to all documents containing a given keyword, many CBIR systems rely on sequential matching of the query with every stored image, since it is not feasible to construct this type of index for the real-valued feature data characteristic of most image retrieval applications.

Finding index structures which allow efficient searching of an image database is still an unsolved problem [23]. The most well-researched approach to date is multidimensional indexing, using structures such as the R^*-tree [5], TV-tree [42], or SS^+-tree [38]. Most of these techniques were designed for applications where access to spatial or volume data were required. They work well in only two or three dimensions, but lose efficiency rapidly as the number of dimensions increases (the so-called "curse of dimensionality"). The considerable overheads of using such complex index structures for image retrieval (where the number of dimensions is governed by the size of the feature set, and can often be 20 or more) have limited their use so far to a few experimental systems. More recent approaches, which seem to offer better prospects of success, are the use of similarity clustering of images, allowing hierarchical access for retrieval as well as providing a means of database browsing [69], and *vantage objects* [70]. For this latter technique, a set of mutually dissimilar objects is selected from the database, and similarities (in terms of distances between their feature vectors) calculated between these vantage objects and every other object in the database. Query matching is performed by measuring the distances between the query and each vantage object, and then retrieving stored objects with the most similar sets of distance measures to the query.

13.3.5 Effectiveness of CBIR Techniques

A crucial question for designers and potential system users alike concerns the effectiveness of current techniques for image retrieval. Is it possible to identify any one set of techniques as consistently superior to any other? Some attempts have been made to answer this question, but the task is fraught with difficulties. Different sets of researchers, even within the restricted world of trademark retrieval, have tested out their systems on different collections of images, using different measures of effectiveness, different sources of query images, and different ways of judging the correctness of system output. Since in the last analysis, image retrieval systems have to model subjective human judgments, reliable ground truth is hard to come by. This situation is in marked contrast to the text information retrieval field, where the problems of judging the relevance of a given document to a given query have been extensively researched, even if they remain far from solved [22], where there are long-established performance measures such as *precision* (number of retrieved documents which are relevant to the query) and *recall* (number of relevant documents that are retrieved) [12], and where the annual series of TREC retrieval experiments provides a common test-bed for comparative trials of system effectiveness [61].

Recognition is growing in the image retrieval field that system developers need to base any claims for success on some quantitative measure of retrieval effectiveness. Several researchers use established measures derived from the information retrieval field, such as precision and recall, though others use a variety of *ad hoc* indicators. But there are few widely available collections of images available for comparative studies, let alone sets of standard queries and relevance judgments to provide essential ground truth for comparative studies. And there is still remarkably little awareness of the need to obtain independent relevance judgments for evaluating system effectiveness. Judgments made by members of the development team are inherently flawed, as shown clearly by the experiments of Squire and Pun [62] comparing human and machine performance in partitioning images into similar groups. Overall, they found only moderate agreement between human and machine judgments. However, the judgments made by the paper's author (who had also developed much of the software) correlated far better with the machine's results than any of the independent observers.

For this reason, the results of the comparison of alternative shape measures performed by Mehtre et al. [47] cannot be relied upon. This is unfortunate, because their study was in other respects well-planned. It compared the retrieval effectiveness of three boundary-based shape methods (Cortelazzo's string-matching technique, and two varieties of Fourier descriptors), three region-based methods (moment invariants, Zernike moments, and pseudo-Zernike moments), and a combination of moment invariants with the two kinds of Fourier descriptor. Fifteen queries were put to a database of 500 trademarks, and retrieval effectiveness computed using a measure known as *retrieval efficiency*, which is equivalent to the precision when the number of images presented to the user is less than the number of relevant images in the data, and recall otherwise. All shape measures seemed to perform well, with moment invariants the best single technique, and moment invariants plus UNL Fourier descriptors [51] giving the highest overall scores. It would be interesting to see how these techniques compare in experiments using independent relevance judgments.

One comparative trial of different shape retrieval techniques which did make use of independent human judgments was that of Scassellati et al. [59]. Their experiments were designed to assess the relative effectiveness of a number of different methods for automated shape similarity assessment, including algebraic moments, parametric curve distance, turning angle, and modified Hausdorff distance. Forty human subjects were asked to scan a database of 1415 simple shapes to identify all shapes they considered similar to each of 20 query shapes. Machine rankings of similarity were then obtained for each of the 20 queries, with each similarity assessment technique. The effectiveness of each technique was calculated from the total number of times each of the top 20 retrieved images had been selected by human judges. In contrast to the results from Mehtre's experiments, no technique performed particularly well (scores were typically only 20–30% of perfect performance), though turning angle appeared to give better overall performance than any other method. It is clearly premature to draw any firm conclusions about either the absolute or relative effectiveness of different shape retrieval techniques.

13.4 Trademark Image Retrieval Systems

13.4.1 System Design Issues

At least 10 separate groups of researchers have reported investigations into some aspect of trademark image matching over the last five years, using a variety of different approaches. The majority have proposed or exploited techniques derived from mainstream image processing, typically involving the extraction and matching of feature vectors held to represent key aspects of each image's appearance. There are, however, significant differences in approach between different groups of investigators. For example, some have attempted to develop novel image representation methods specifically for trademarks, while others have adapted standard CBIR techniques for use with trademark images. Some have proposed very specific techniques to solve one aspect of the problem, while others have attempted to build and test complete systems.

Some of the most important issues for designers of trademark image retrieval systems to address are the following:

Systems scope.
- What range of trademarks is the system designed to handle – all types of trademark containing image data (including stylized word and device-and-word), just device-only marks, or just geometric shapes?
- What kinds of similarity matching are to be supported – shape, structure, semantics, or other types of feature? How are they to be combined?

Image representation.
- Feature set. What types of image feature are selected for use in similarity matching?
- Granularity. Should the image be represented as a single entity, or as discrete components? If the latter, how are the components defined?

Processing issues.
- What image segmentation (if relevant) and feature extraction techniques are to be used at the input stage?
- What feature matching techniques and similarity measures will be employed? What indexing techniques (if any) are needed to improve search efficiency?

Interface issues.
- Input. What degree of human intervention is needed when images are first added to the system?
- Output. How are queries to be formulated, and results returned to the searcher? Are facilities required for modifying search strategy in the light of search output?

An indication of the variety of ways in which these issues have been tackled by different research teams can be found in the detailed description of each system in the following sections.

13.4.2   Manually Based Techniques for Similarity Retrieval

One group has chosen to base its approach on manually assigned image features, rather than adopting methods derived from image processing technology [72]. Their technique was originally derived for retrieval of fine art images, but has also been applied to trademark images. It involves deriving a set of descriptors from features used by human subjects to discriminate between pairs of distinct images, supplemented in the case of trademark retrieval by terms from the Vienna Classification. After considerable refinement, a feature set of 150 terms was derived, which could then be used to index trade mark images manually. A pilot database was set up which offered automated similarity matching of query and stored images on the basis of the number of descriptors they had in common. Some preliminary evaluation experiments were carried out using this database – though since it contained only 115 images, and effectiveness was assessed largely on the basis of how well it retrieved a single target image, it is difficult to draw any firm conclusions about system effectiveness. The authors presented no convincing evidence that their system represented any advance over operational systems based on the Vienna Classification, such as the UK Patent Office's TRIMS.

13.4.3   CBIR Techniques for Similarity Retrieval

TRADEMARK (Kato, 1992). The first image retrieval system described in the literature which specifically addressed the needs of trademark images was Kato's TRADEMARK* [35]. The system formed part of a comprehensive research programme into CBIR, based on the principles that effective image retrieval required visual interaction with the user, automatic interpretation of image content, similarity measurements based on human perception, and the use where appropriate of textual cues. His TRADEMARK system, influenced by studies of human perception of graphic symbols, represented each trademark image as a series of graphic features (collectively referred to as a *GF-vector*) intended to convey essential aspects of its visual appearance in compact form. The principal features making up the GF-vector of an image were *Gray8* and *Edge8*, the number of black pixels and edge pixels respectively appearing in each cell of an 8×8 grid laid over the image, *RunB/W* and *RunW'*, frequencies of different run-lengths of black, white, and any type of pixel in four vertical and horizontal divisions of the image, and *Corr4* and *Cont4*, local measures of similarity and contrast in average pixel values between each cell in a 4×4 grid. The system as described in the article made no attempt to segment images into component regions – all image analysis and similarity matching was performed at the whole-image level.

Queries could be submitted either as example images for matching, or as hand-drawn sketches. In either case, a GF-vector was calculated for the query

---

* **TRA**demark and **DE**sign database with **M**ultimedia **A**bstracted image **R**epresentation on **K**nowledge base.

image, and a distance measure similar to the city-block metric computed between this vector and the GF-vector from each stored image in the database. All stored images were then ranked in order of similarity to the query, and those deemed most similar displayed on the screen. The system was tested using a series of whole-image and sketch queries put to a database of 2000 graphic symbols. The target image was retrieved within the first 10 images displayed in 100% of whole-image queries, and 95% of sketch queries.

Kato's paper is noteworthy in several respects. It was the first to describe features specifically designed for use with trademark images, and was among the first to emphasize the importance of human perception in retrieval system design. It also contained the earliest published use of the term *content-based image retrieval*. However, the effectiveness of the TRADEMARK system itself is harder to judge. The feature set chosen appears to offer a robust method of shape discrimination, though it is not clear how well it would recognize the types of similarity illustrated in Fig. 13.2 and 13.3. The fact that images are treated as a whole, not segmented into components, limits the system's ability to recognize images with similar structure or with similar-shaped components. GF-vector representations appear to provide rotational invariance only in steps of 90°. The rather limited evaluation experiments described in the paper give little indication of its effectiveness in retrieving different but similar images; they simply test the system's ability to identify identical or near-identical images (the logo recognition problem described earlier).

STAR (Wu et al., 1996). A much more sophisticated system for trademark registration is STAR,* developed by Jian-Kang Wu and colleagues from the National University of Singapore [74]. This remains one of the most comprehensive trademark retrieval systems described to date, though it relies heavily on manual intervention at key stages. It is potentially capable of dealing with word, device-and-word or device-only marks, though this review covers only its image handling capabilities. The system works on the premise that trademark images can be viewed both as a whole, and as a series of individual components. Both the shapes of these components and their spatial layout (structure) can be important in the way the image is perceived. Recognition of perceptually significant components is considered too difficult a task to automate, so all monochrome images are segmented into regions at input by human operators (automatic segmentation of color images is apparently possible). A limited set of relationships between components is also coded at this stage, including outline/background, parallelism, intersection, and formation of regular patterns such as squares or triangles.

STAR treats trademark image similarity matching as a multidimensional process which can involve shape, structure, and semantics (subjective meaning which observers might read into the image). The shape of each component is represented in three different ways: as Fourier descriptors [75], moment invariants [30], and as projections (average pixel densities for every value of x

---

* System for Trademark Archival and Registration.

or y in the image). The first two of these are invariant to translation, rotation, and scaling. The third, at least as described in the paper, is not. Image structure is represented by the component relationships described above. Semantics are represented by added keywords or Vienna Classification codes. When matching a new candidate trademark with a set of stored marks, individual distance measures for shape, structure, and semantics are calculated both at the image and individual component level. These are then combined additively according to the following formula:

$$d_j = \frac{\sum_k (w_m d_{mkj} + w_p d_{pkj} + w_g d_{gkj})}{N} \tag{13.6}$$

where d_j is the overall distance between a query image and stored image j, N is the number of components in the query image, d_{mkj}, d_{pkj}, and d_{gkj} are individual distances between the meaning, structure, and shape aspects respectively of component k of the query image and a component of stored image j, and w_m, w_p, and w_g are corresponding weighting factors. The process of adjusting the values of these weights for optimal performance is not straightforward. A four-layer back-propagation network was therefore used to train the system with examples of desired output. Weights for the summation process above could be taken directly from the trained network.

Evaluation of STAR on a pilot database of 500 trademarks generated results which were considered satisfactory by trademark office representatives. Unfortunately no quantitative evaluation results appear to have been published. In the absence of such measurements of retrieval performance, the effectiveness of the system cannot be objectively judged. However, it is clear that the designers of STAR have tried to address all the major questions surrounding trademark image retrieval for registration purposes. Their system is comprehensive enough to cover all types of trademark currently in use, and is capable of handling queries involving shape, structure or semantics. The need for substantial human effort at the input stage – in segmenting images into components and assigning keywords or Vienna codes – is clearly a drawback. The system as described also has a number of limitations, particularly the lack of facilities for partial image matching, or for varying the weights given to various aspects of image similarity in the light of user feedback. Although inverted files are used to improve the efficiency of searching for word marks, it appears that image matching is performed sequentially. Matching times for the 500-image pilot database on a Unix workstation averaged 5 s.

ARTISAN (Eakins et al., 1998). The original objectives of the ARTISAN[*] project at the University of Northumbria [20] were twofold: firstly, to develop a workable system for content-based retrieval of at least one class of trademark images, and secondly, to establish the extent to which techniques developed for earlier systems such as SAFARI [17] could be adapted

---

[*]   Automatic Retrieval of Trademark Images by Shape ANalysis.

for trademark retrieval. The underlying philosophy behind ARTISAN, developed in collaboration with the UK Patent Office, is in many ways similar to that behind STAR, though there are important differences. ARTISAN's scope is deliberately more restricted than that of STAR: it makes no attempt to capture textual cues within a trademark, or any aspect of semantic image similarity, being designed purely for use with device-only trademarks consisting of abstract geometric designs. This decision was taken for purely pragmatic reasons: abstract geometric shapes were the ones causing trademark examiners most difficulty, and appeared most likely to be suitable for CBIR techniques. Like STAR, ARTISAN aims to view trademark images at multiple levels. Three different levels are distinguished: the whole image, perceptually significant regions within the image, and individual components within each region. A further important distinction between STAR and ARTISAN is that the latter aims to identify perceptually significant regions automatically, rather than asking human users to do this at the input stage. Although it seems unlikely that this process can ever be completely automated, it was felt to be important to investigate the extent to which automation was in fact feasible.

To ensure that the similarity judgments made by ARTISAN reflected those of human trademark examiners as well as possible, discussions were held with a number of trademark examiners, followed by observation of several examiners in action. This suggested that examiners identify and keep in mind the most distinctive feature of the query image (which might be a single large shape within the image, or a group of objects making up a recognizable pattern), and then try to identify stored images containing components matching that feature. These investigations resulted in ARTISAN's main element of novelty – its technique for grouping image components into perceptually significant regions or *families*. Individual image components are first identified by standard edge-detection techniques, and their boundaries approximated by sequences of straight lines and circular arcs. This form of approximation then allows individual components to be clustered into families on the basis of *non-accidental properties* [73] such as proximity, parallelism, co-curvilinearity, and shape similarity between the line and arc segments making up their boundaries [18]. An example of family formation is shown in Fig. 13.5.

Two sets of shape feature are extracted from the image, a *boundary shape vector* (aspect ratio, circularity, transparency, and relative area computed from the envelope of each individual component and family in the image) and a *family characteristics vector* (right-angleness, sharpness, complexity, directedness and straightness averaged over all individual components of each family). Some of these properties, notably right-angleness (the fraction of well-defined corners approximating to a right angle), and straightness (the fraction of the region boundary more readily approximated by straight lines than curved arcs), implement aspects of the singularity principle noted by

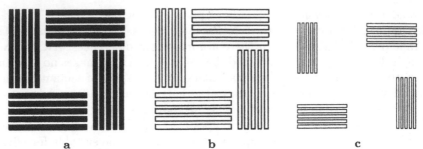

Fig. 13.5 An example of boundary family creation in ARTISAN: **a** original image, **b** line approximation of component boundaries, **c** grouping of components into four families on the basis of proximity and parallelism.

Goldmeier [27]. The set of shape vectors from each family and component making up the query image can then be matched with the set of shape vectors from each stored image. The query module permits matching at the level of the family, the individual component, or both. It also provides a number of alternative ways of computing similarity scores in cases where query and stored images have different numbers of components. The most successful matching paradigm overall was found to be *asymmetric simple*, which averages the best matching scores between each component in the query image and any component of the stored image, irrespective of the number of components in each image.

Evaluation of ARTISAN was carried out on a sample of just over 10,000 images from the UK Trade Marks Registry. A selection of past queries put to the database by trademark examiners in the course of their work was put to this database, and results of human and machine matching compared. Three statistics were used for evaluation, normalized recall R_n, normalized precision P_n, and last-place ranking L_n [19]. Overall performance figures for the set of 24 queries chosen by the UK Patent Office were $R_n = 0.90 \pm 0.12$, $P_n = 0.63 \pm 0.24$, and $L_n = 0.56 \pm 0.31$. Search times were quite long (around 3–4 min to search the entire database on a 120 MHz Pentium machine), since all images were searched sequentially. While encouraging, these figures showed that the initial prototype was insufficiently developed for operational use. Analysis of cases of retrieval failure showed that the commonest causes were:

- Failure to recognize implied shape features (Fig. 13.2, above)
- Inappropriate segmentation of query or stored images, particularly in the presence of noise (Fig. 13.6)
- Failure of shape measures alone to capture perceived image similarity (Fig. 13.3, above)

– Miscellaneous problems, including failure to cope with textured regions, inconsistent handling of borders surrounding query images, and unhelpful grouping of components into families

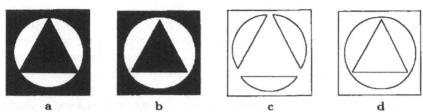

a   b   c   d

Fig. 13.6 Problems caused by inappropriate image segmentation. Although most observers would agree that images **a** and **b** are very similar, ARTISAN segments the inner features of the former image into three sectors of a circle **c**, but corresponding areas of the latter image into a triangle within a circle **d**. Hence it fails to recognize their similarity.

Work is now under way on the development of an improved version of the ARTISAN prototype, reflecting some – though not all – of the lessons learnt from failure analysis of version 1. Important changes being incorporated into version 2 include the use of multiresolution analysis to remove texture and improve the system's ability to cope with noisy images, new ways of grouping low-level components into higher-level regions, and a wider range of shape and structural features – especially local features of the type used for SA-FARI [17]. A parallel investigation is also under way into the perception of visually significant regions within trademark images, and how these might be modeled in future image retrieval systems [55]. Human observers appear to base decisions on how to segment an image on a relatively small number of rules; it is hoped that this rule set can be implemented within an extended version of ARTISAN's family creation framework.

Jain and Vailaya (1996). Jain and Vailaya have used databases of trademark images to test a number of techniques for image matching and retrieval. Their initial experiments [33] investigated the usefulness of color matching using histograms of RGB color distribution and shape matching using histograms of edge direction. Color similarity matching was found to be more effective than shape similarity matching in retrieving scaled, rotated, and noisy versions of a given reference image – though a weighted average of the two performed better than either measure alone. However, no attempt was made to assess the effectiveness of these techniques for similarity matching. Hence they would appear more relevant to the problem of logo recognition than retrieval for trademark registration.

A later paper by Jain and Vailaya [34] does address the question of similarity retrieval. The authors propose a two-stage system for trademark retrieval, based on rapid screening using a combination of edge direction histograms and moment invariants [30], followed by a more detailed matching stage using deformable templates [32]. They report results from a prototype system which implements the first of these two steps. Five hand-drawn queries were matched against a database of 1100 trademarks, and the results compared with those of human similarity judgments. On average, around 40% of relevant images were retrieved within the top 20 positions – though this increased to over 70% when all closed regions within the image were filled in with solid black before feature extraction or matching. However, these figures must be viewed with caution. They relate to only three queries, and in only one of these cases (a stylized bull's head, where judgments may well have been influenced by the shape's semantic connotations) were similarity judgments obtained from independent observers. Future extensions envisaged for the system include segmentation of images into components along similar lines to STAR or ARTISAN, the use of local features for similarity matching, the provision of some relevance feedback facility, and automatic clustering of images to improve retrieval efficiency.

Peng and Chen (1997). While not a complete system, the approach to trademark shape matching proposed by Peng and Chen [49] has a number of interesting features. Like the majority of system developers in the field, they regard trademark images as consisting of a number of individual components. They go one step further than this, however, regarding each component as the union of the set of all possible closed contours which can be derived from component boundaries. Each contour is reduced to an angle-code string representing the sequence of vertex angles from its polygonal approximation (Fig. 13.7).

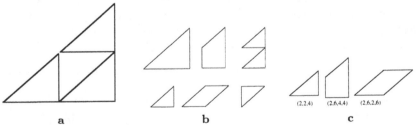

a   b   c

Fig. 13.7 An example of Peng and Chen's contour redrawing process. **a** component boundary extracted from original image, **b** contours derived from component boundary, **c** examples of angle-code strings for three of these contours. (**a** and **b** reproduced from Ref. [49] by permission of Elsevier Science.)

Matching is then performed at three levels: (i) contour–contour, (ii) component–component, and (iii) image–image. The similarity between two contours is computed from the maximum degree of overlap between their angle-code strings; similarity measures derived at one level are then propagated to the next level up by a series of formulae such as:

$$S_{rc} = 0.9 * max(S_{rr}) + 0.1 * avg(S_{rr}),$$
$$S_{cc} = 0.9 * max(S_{rc}) + 0.1 * avg(S_{rc})$$

where S_{cc} is the similarity between a given pair of components a and b, S_{rc} the similarity between a given contour within component a and component b, and S_{rr} the similarity between a given pair of contours from a and b. Two aspects of Peng and Chen's technique render it potentially very successful in reflecting human similarity judgments; firstly, its ability to capture alternative human interpretations of scene contours (and hence model some of the observed variability in human similarity matching), and secondly, the way in which vertex angle values are quantized for string matching, which accords well with the experiments of Goldmeier [27] on the importance of singularity in human shape similarity judgments. Against this, there have to be concerns about the robustness of their string representation method in the presence of noise, and the problems of spurious contour generation. It is difficult to assess how effective Peng and Chen's technique will prove in practice, since their paper presents only preliminary test results.

Kim et al. (1999). Kim et al. have identified a number of shortcomings in the edge-based approaches described above, highlighting the lack of robustness of methods based on extraction of closed boundary contours (see Fig. 13.6), and the lack of sensitivity of edge direction histograms, illustrated by examples where very different shapes generate almost identical distributions of edge direction. Instead, they have developed a technique for trademark image retrieval based mainly on comparison of selected Zernike moment magnitudes [36, 37]. Their method involves normalizing each trademark image to fit within a given maximum extent circle, and then calculating magnitudes of all Zernike moments up to order 17 – a total of 90 moments for each image. Instead of using all 90 moments for search purposes, they use statistics of the distribution of each moment across the entire database to identify the most salient features for searching. Each query is analyzed to compute the values of all 90 moments. The moment for which $P(z_{nm} \geq Z_{nm})$ is smallest* is held to have the greatest discriminating power, and is therefore used for searching.

Kim et al. do not claim that their method is sufficient on its own for effective trademark retrieval. Nevertheless, experimental results presented in their paper suggest that it can be very effective in retrieving noisy or distorted versions of a given trademark image. Using a database of 3,000 trademark images, 10 query images were each subjected to a variety of different types of deformation or noise. On average, 92.5% of deformed images were retrieved within the first 30 images output by the system. The time taken to search the

* where z_{nm} is the value of the Zernike moment of order n and repetition m in a stored image, and Z_{nm} its value in the query.

database using a 200 MHz Pentium machine was around 0.6 s. The method's ability to retrieve similar images is less certain. An average precision figure of around 65% was reported when 10 independent observers were asked to judge the relevance of the first 30 results from similarity matching of the same 10 queries. However, no recall figures are given, so it is impossible to know how many similar images were missed.

Alwis and Austin (1998). Alwis and Austin [3] have taken the Gestalt approach used in ARTISAN one stage further, investigating ways in which the further application of Gestalt principles could solve problems of the kind illustrated in Fig. 13.6. Their approach has been to identify all significant line segments within the image, irrespective of their region of origin, and then cluster them into perceptually significant groups using techniques based on those of Sarkar and Boyer [58]. Line and arc segments exhibiting significant degrees of proximity, parallelism, or co-curvilinearity are assembled into closed boundaries to form a Gestalt image indicating how the image might actually be perceived. The process is shown in Fig. 13.8.

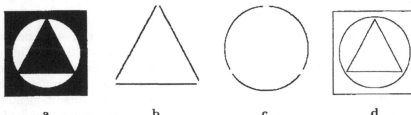

a   b   c   d

Fig. 13.8 An example of Gestalt image formation from the shape shown in Fig. 13.6: **a** original image, **b** lines grouped into triangle by end-point proximity, **c** arcs grouped into circle by end-point proximity and co-curvilinearity, **d** final Gestalt image.

Features are then extracted from both raw and Gestalt images for use in shape matching. The feature set used is similar to that chosen for ARTISAN [20], though the method of similarity estimation is quite different. Instead of computing a distance metric such as Euclidean distance between feature vectors, an evidence counting method is used. Histograms of feature values for each closed contour are computed, and two contours are held to match if the number of feature values falling into the same bins exceeds a given threshold. The strength of matching between any two images is then computed by summing normalized evidence counts for each pair of constituent closed contours. In comparative tests on a pilot database of 210 images, evidence counting consistently yielded higher retrieval effectiveness scores than matching based on Euclidean distance – a finding consistent with Tversky's experiments on human similarity estimation [67]. Preliminary experiments comparing the effectiveness of shape matching based on raw images, Gestalt images and a combination of the two suggest that, although Gestalt images are less effective on their own for shape matching than raw images, a combination of the two can yield better results than either. Different methods for combining similarity ratings from raw and Gestalt images were also investigated; evidence combination using the Dempster–Schafer method gave the best results [4].

Ravela and Manmatha (1999). A system capable of retrieving trademark images by appearance or by text descriptors has recently been described by Ravela and Manmatha [54]. This allows users to search either for representational or abstract trademark images. Queries are initially submitted in text form, and can then be refined by selecting any retrieved image for matching on the basis of visual appearance. Text retrieval facilities are based on the INQUERY probabilistic retrieval system [6]; image matching draws on the authors' earlier work on retrieval by appearance [53]. The feature vector used to represent image appearance for trademark retrieval is a histogram of local curvature ratios and gradient orientations computed from Gaussian derivatives at several different scales. The global similarity between two images is then given by the normalized cross-covariance of feature vectors. Image histograms are matched sequentially, the process taking only a few seconds for a database of 60,000 images.

The system's effectiveness in retrieving similar images was measured using six queries put to a test database of 2,048 images supplied by the US Patent and Trademark Office, and calculating precision scores for standard recall levels. A 50% recall was achieved with a precision of 75%, though for 100% recall, precision was only 9%. Unfortunately no independent relevance judgments were available for the experiments, so these figures have to be treated with caution. A larger evaluation study, using the same set of images, queries, and relevance judgments used in the evaluation of ARTISAN [20] is now under way.

13.4.4 Logo Recognition

Several groups have reported success in the related problem of logo recognition. Cesarini et al. [9] describe the development of a logo recognizer based on auto-associative neural networks. After training with low-resolution versions of company logos in the presence of noise and occlusion, their network could recognize noisy versions of all 44 logos in their test set. The paper makes no mention of scale or rotation invariance. Doermann et al. [15] use more conventional image processing technology in the form of hierarchical feature matching. Their system provides coarse screening by matching automatically extracted text characters and common shapes such as circles and rectangles, followed by more precise matching of a series of geometric invariants reflecting both global and local shape properties. It was reported to give reliable identification of scaled and rotated versions of 10 test logos from a 100-logo database. Francesconi et al. [25] have developed a method based on recursive neural networks which is capable of recognizing logos in the presence of rotation, blurring, and other types of distortion. Images are represented as a tree structure of components following the method of Cortelazzo [13], and each region boundary represented by a histogram of *curvature-runs* (boundary segments whose curvature falls within specified limits). The feature tree making up each image sequence is then used to train a recursive neural network. Experiments with 200 distorted versions of 40 test logos showed that the system could recognize the correct logo in over 99% of cases.

13.5 Conclusions

13.5.1 Effectiveness of Current Techniques

How effective are the trademark image retrieval systems described above? As far as is known, none is in routine use at any national trademark registry, suggesting that none can yet meet the exacting standards required for registration purposes. This is not surprising: the nature of the application is unusually demanding. Since retrieval failure could have serious financial implications, systems need to offer virtually 100% recall. They need to be able to match human similarity judgments based on a subtle and variable mix of shape, spatial and "perceptual" features, together with semantic interpretations in many cases. Since many trademark databases are large and growing rapidly, retrieval has to be provided reasonably efficiently.

Most current research has concentrated on techniques for shape retrieval. Fourier descriptor and moment invariants appear to be the most popular shape measures adopted, though other global measures such as edge direction histograms and Zernike moments have their adherents as well. In the absence of definitive studies comparing the effectiveness of these different techniques, it is impossible to tell which (if any) is superior to the others. It is in fact most unlikely that any one technique will be sufficient on its own. Most investigators who have studied the problem in detail have concluded that combinations of different techniques are generally more effective than any single method alone – particularly in databases where image type, style, and quality can vary considerably. There seems to be considerable scope here for the use of knowledge-based techniques to model human search behavior.

Work on other aspects of trademark image similarity has so far been more limited. STAR makes provision for retrieval by image structure and semantic content, but only via annotations added by human indexers. Virtually no-one has yet attempted to automate either of these processes. The relative scarcity of spatial matching facilities in experimental trademark image retrieval systems seems particularly surprising, given the wide variety of techniques already available. The lack of research into automatic retrieval by semantic content is more understandable.

13.5.2 Future Prospects

A number of promising ideas can be discerned in the research outlined above, suggesting that substantial improvements in retrieval effectiveness can be expected within the next few years. These include:

- Feature generation by multiresolution filtering [54], allowing matching at different levels of granularity
- The adoption of shape measures reflecting experimental findings on human shape perception [20, 49]
- The use of local shape features [34, 55], allowing shape matching by region as well as for complete images
- Methods for aggregating image components into perceptually significant groups [3, 55], facilitating retrieval by appearance
- The use of evidence combination methods rather than relying on distance metrics [3, 4], yielding results more likely to model human similarity judgments
- Multi-level matching strategies to improve search efficiency [34]
- Identification of the most salient features for a given query [36], improving discriminating power without sacrificing efficiency

However, many substantial problems remain. The prime task is to find better ways of modeling human similarity perception, not just using shape features, but by combining evidence from perceived shape, structure, and, where relevant, other aspects such as color and texture. New measures of shape and structure, and better methods of similarity estimation, are needed, taking into account experimental findings such as those of Goldmeier [27] and Tversky [67]. Since automatic retrieval by semantic content is a distant goal, improved methods of combining text keywords or Vienna classification codes (the only realistic way of capturing semantic content in the near future) with image features are also needed. Another relatively neglected area of research relates to device and word marks. Both text content and overall appearance can be significant determinants of their distinctiveness. Sophisticated character recognition techniques, far more advanced than those normally found in OCR systems, are needed to recognize what can be highly stylized or distorted text. Another aspect of a trademark's appearance which could potentially be captured in automated form is distinctiveness – the extent to which it differs from other marks for the same type of goods. Studies of the frequency distribution of different types of visual feature in collections of trademark images might well generate results which could be useful in refining searches.

Ways of improving retrieval efficiency are also urgently needed. As discussed above, multi-dimensional indexing has proved of limited use for most image retrieval applications, owing to its rapid loss of efficiency as the number of dimensions rises. Pre-clustering of similar images [69], or inverted-file indexing using visual codewords [44] seem more likely to provide answers. Finally, interface design for image retrieval systems (of any kind) is in its infancy, and research is needed into improved methods of query formulation and refinement. Particular problems include how to specify an initial query to the system, and how best to refine search parameters if system output fails to meet user requirements. Relevance feedback [56] is an obvious technique to apply here, but there may well be better methods.

The final difficulty limiting progress in trademark retrieval concerns system evaluation. Unless there are reliable and widely accepted ways of measuring the effectiveness of new techniques, it will be impossible to judge whether they represent any advance on existing methods. This will inevitably limit progress. The problem is not limited to trademark retrieval – in fact, it is even more acute in the general image retrieval field. It is possible to assemble standard collections of images, queries, and (most importantly) relevance judgments in the trademark retrieval field, because there is a body of users who know fairly accurately what they want. This is not necessarily true of image retrieval in general. It has been argued elsewhere [21] that one of the most useful developments in the image retrieval field would be the introduction of an equivalent of the TREC series of comparative trials of system effectiveness. Trademark images could well provide a good starting-point for such an exercise.

References

1. US Patent and Trademark Office: Basic Facts About Registering a Trademark US Patent and Trademark Office, Washington, DC, 1995. (http://www.uspto.gov/web/offices/tac/doc/basic/basic_facts.html).
2. World Intellectual Property Organization: International Classification of the Figurative Elements of Marks (Vienna Classification), Fourth Edition, ISBN 92-805-0728-1. World Intellectual Property Organization, Geneva, 1998.
3. Alwis, S and Austin, J, "A Novel Architecture for Trademark Image Retrieval Systems," presented at The Challenge of Image Retrieval research workshop, Newcastle upon Tyne, 1998. Available in BCS electronic Workshops in Computing series at http://www.ewic.org.uk/ewic/workshop/view.cfm/CIR-98.
4. Alwis, S and Austin, J, "Trademark Image Retrieval Using Multiple Features," presented at CIR-99: The Challenge of Image Retrieval, Newcastle, 1999.
5. Beckmann, N, "The R*-Tree: An Efficient And Robust Access Method for Points and Rectangles," ACM SIGMOD Rec, 19(2), pp. 322–331, 1990.
6. Callan, JP, Croft, WB, and Harding, SM, "The INQUERY Retrieval System," 3rd International Conference on Database and Expert System applications, pp. 78–83, 1992.
7. Canny, J, "A Computational Approach to Edge Detection," IEEE Trans Patt Anal Mach Intell, 8, pp. 679–698, 1986.
8. Carson C, Belongie, S, Greenspan, H, and Malik, J, "Region-Based Image Querying," IEEE Workshop on Content-Based Access of Image and Video Libraries, San Juan, Puerto Rico, 1997.

9. Cesarini, F, Francesconi, E, Gori, M, Marinai, S, Sheng, JQ, and Soda, G, "A Neural-Based Architecture for Spot-Noisy Logo Recognition," 4th International Conference on Document Analysis and Recognition, Ulm, pp. 175–179, 1997.
10. Chang, SF, Chen W, and Sundaram, H, "Semantic Visual Templates: Linking Visual Features to Semantics," IEEE International Conference on Image Processing (ICIP'98), Chicago, Illinois, pp. 531–53, 1998.
11. Chang, SK, Shi, QY, and Yan, CW, "Iconic Indexing by 2-D Strings," IEEE Trans Patt Anal Mach Intell, 9(3), pp. 413–427, 1987.
12. Cleverdon, CW and Keen, EM, "Factors Determining the Performance of Indexing Systems," Cranfield College of Aeronautics, Cranfield, UK, 1966.
13. Cortelazzo, G, Mian, GA, Vezzi, G, and Zamperoni, P, "Trademark Shape Description by String-Matching Techniques," Patt Recogn, 27(8), pp. 1005–1018, 1994.
14. Del Bimbo, A and Pala, P, "Visual Image Retrieval by Elastic Matching of User Sketches," IEEE Trans Patt Anal Mach Intell, 19(2), pp. 121-132, 1997.
15. Doermann, D, Rivlin, E, and Weiss, I, "Applying Algebraic and Differential Invariants for Logo Recognition," Mach Vision Applic, 9, pp. 73–86, 1996.
16. Dudani, SA, Breeding, KJ, and McGhee, RB, "Aircraft Identification by Moment Invariants," IEEE Trans Computers, 26, pp. 39–45, 1977.
17. Eakins, JP, "Design Criteria for a Shape Retrieval System," Computers Indust, 21, pp. 167–184, 1993.
18. Eakins, JP, Shields, K, and Boardman, JM, "ARTISAN – a Shape Retrieval System Based on Boundary Family Indexing," in Storage and Retrieval for Image and Video Databases IV, Proc SPIE 2670, pp. 17-28, 1996.
19. Eakins, JP, Graham, ME, and Boardman, JM, "Evaluation of a Trademark Retrieval System," 19th BCS IRSG Research Colloquium on Information Retrieval, Robert Gordon University, Aberdeen, 1997. Available in BCS electronic Workshops in Computing series (http://www.ewic.org.uk/ewic/workshop/view.cfm/IRR-97).
20. Eakins, JP, Boardman, JM, and Graham, ME, "Similarity Retrieval of Trademark Images," IEEE Multimed, 5(2), pp. 53–63, 1998.
21. Eakins, JP and Graham, M.E, "Content-Based Image Retrieval: A Report to the JISC Technology Applications Programme," JISC Technology Applications Programme Report 39, 1999.
22. Ellis, D, "The Dilemma of Measurement in Information Retrieval Research," J Am Soc Inform Sci, 47, pp. 23–36, 1996.
23. Faloutsos, C, Barber, R, Flickner, M, Hafner, J, Niblack, W, and Equitz, W, "Efficient and Effective Querying by Image Content," J Intell Inform Syst, 3, pp. 231–262, 1994.
24. Forsyth, DA, Malik, J, Fleck, M, Leung, T, Belongie, S, Carson, C, and Bregler, C, "Finding Pictures of Objects in Large Collections of Images," in Digital Image Access and Retrieval: 1996 Clinic on Library Applications of Data Processing (Heidorn, P and Sandore, B, eds), Graduate School of Library and Information Science, University of Illinois at Urbana-Champaign, pp. 118-139, 1997.
25. Francesconi, E, Frasconi, P, Gori, M, Marinai, S, Sheng, JQ, Soda, G, and Sperduti, A, "Logo Recognition by Recursive Neural Networks," 2nd International Workshop on Graphics Recognition: Algorithms and Systems, Lecture Notes in Computer Science, 1389, pp. 104–117, 1998.
26. Freeman, H, "Computer Processing of Line-Drawing Images," ACM Comput Surv, 6(1), pp. 57–97, 1974.

27. Goldmeier, E, "Similarity in Visually Perceived Forms," Psychol Issues 8(1), pp. 1–135, 1972.
28. Gudivada, VN and Raghavan, VV, "Design and Evaluation of Algorithms for Image Retrieval by Spatial Similarity," ACM Trans Inform Syst, 13(2), pp. 115–144, 1995.
29. Hou, YT, Hsu, A, Liu, P, and Chiu, MY, "A Content-Based Indexing Technique Using Relative Geometry Features," in Image Storage and Retrieval Systems, Proc SPIE 1662, pp. 59–68, 1992.
30. Hu, MK, "Visual Pattern Recognition by Moment Invariants," IRE Trans Inform Theory, IT-8, pp. 179–187, 1962.
31. Idris, F and Panchanathan, S, "Review of Image and Video Indexing Techniques," J Visual Commun Image Represent 8(2), pp. 146–166, 1997.
32. Jain, AK, Zhong, Y, and Lakshmanan, S, "Object Matching Using Deformable Templates," IEEE Trans Patt Anal Mach Intell, 18(3), pp. 267–277, 1996.
33. Jain, AK and Vailaya A, "Image retrieval Using Color and Shape," Patt Recogn, 29(8), pp. 1233–1244, 1996.
34. Jain, AK and Vailaya, A, "Shape-Based Retrieval: A Case Study With Trademark Image Databases" Patt Recogn, 31(9), pp. 1369–1390, 1998.
35. Kato, T, "Database Architecture for Content-Based Image Retrieval," in Image Storage and Retrieval Systems (Jambardino, A and Niblack, W, eds), Proc SPIE 2185, pp. 112–123, 1992.
36. Kim, YS and Kim, WY, "Content-Based Trademark Retrieval System Using a Visually Salient Feature," Image Vision Comput, 16, pp. 931–939, 1998.
37. Kim, YS, Kim, Y, Kim, W, Kim, M, "Development of Content-Based Trademark Retrieval System on the World-Wide Web," ETRI J, 21(1), pp. 39–53, 1999.
38. Kurniawati, R, Jin, J, and Shepherd, J, "The SS+ Tree: An Improved Index Structure for Similarity Searches in High-Dimensional Feature Space," in Storage and Retrieval for Image and Video Databases V, Proc. SPIE 3022, pp. 110–120, 1997.
39. Lee, CS, Ma, W and Zhang, H, "Information Embedding Based on Users' Relevance Feedback for Image Retrieval," in Multimedia Storage and Archiving Systems IV, Proc. SPIE 3846, pp. 294–304, 1999.
40. Levine, MD, Vision in Man and Machine, ch. 10, McGraw-Hill, New York, 1985.
41. Liang, KC and Kuo, C, "Implementation and Performance Evaluation of a Progressive Image Retrieval System," in Storage and Retrieval for Image and Video Databases VI, Proc. SPIE 3312, pp. 37–48, 1998.
42. Lin, KI, Jagadish, H, Faloutsos, C, "The TV-Tree: an Index Structure for High Dimensional Data," J Very Large Databases, 3(4), pp. 517–549, 1994.
43. Liu, F and Picard, RW, "Periodicity, Directionality and Randomness: Wold Features for Image Modelling and Retrieval," IEEE Trans Patt Anal Mach Intell, 18(7), pp. 722–733, 1996.
44. Ma, WY and Manjunath, BS, "A Texture Thesaurus for Browsing Large Aerial Photographs," J Am Soc Inform Sci, 49(7), pp. 633–648, 1998.
45. Manjunath, BS and Ma, WY, "Texture Features for Browsing and Retrieval of Large Image Data," IEEE Trans Patt Anal Mach Intell, 18, pp. 837–842, 1996.
46. Mehrotra, R and Gary, JE, "Similar-Shape Retrieval in Shape Data Management," IEEE Computer, 28(9), pp. 57–62, 1995.

47. Mehtre, BM, Kankanhalliet, MS, and Lee, FV, "Shape Measures for Content-Based Image Retrieval: A Comparison," Inform Process Manag, 33(3), pp. 319–337, 1997.
48. Oliva, A, and Torrlba, AB, "Global Semantic Classification of Scenes Using Power Spectrum Templates," presented at CIR-99, the Challenge of Image Retrieval, Newcastle, February, 1999.
49. Peng, HL and Chen, SY, "Trademark Shape Recognition Using Closed Contours," Patt Recogn Lett, 18, pp. 791–803, 1997.
50. Pentland, A, Picard, R, and Sclaroff, S, "Photobook: Content-Based Manipulation of Image Databases," Int J Computer Vision, 18(3), pp. 233–254, 1996.
51. Rauber, TW and Steiger, AS, "Shape Description by UNL Fourier Features – an Application to Handwritten Character Recognition," presented at 11th IAPR International Conference on Pattern Recognition, The Hague, 1992.
52. Ravela, S and Manmatha, R, "Retrieving Images by Appearance," IEEE International Conference on Computer Vision (ICCV98), Bombay, India, pp. 608–613, 1998.
53. Ravela, S and Manmatha, R, "On Computing Global Similarity in Images," IEEE Workshop on Applications of Computer Vision (WACV98), Princeton, NJ, pp. 82–87, 1998.
54. Ravela, S and Manmatha, R, "Multi-Modal Retrieval of Trademark Images Using Global Similarity," Internal Report, University of Massachusetts at Amherst, 1999.
55. Ren, M, Eakins, JP, Briggs, P, "Human Perception of Trademark Images: Implications for Retrieval System Design," in Multimedia Storage and Archiving Systems 1V, Proc SPIE 3846, pp. 114-125, 1999.
56. Rui, Y, Huang, TS, and Mehrota, S, "Relevance Feedback Techniques in Interactive Content-Based Image Retrieval," in Storage and Retrieval for Image and Video Databases VI, Proc SPIE 3312, pp. 25–36, 1997.
57. Rui, Y, Huang, TS, Chang, SF, "Image Retrieval: Current Techniques, Promising Directions, and open issues," J Visual Commun Image Represent, 10(1), pp. 39–62, 1999.
58. Sarkar, S and Boyer, KL, "A Computational Structure for Preattentive Perceptual Organization: Graphical Enumeration and Voting Methods," IEEE Trans Syst Man Cybern, 24, pp. 246–267, 1994.
59. Scassellati, B, Alexopolous S, and Flickner, M, "Retrieving Images by 2-D Shape: A Comparison of Computation Methods with Human Perceptual Judgements," in Storage and Retrieval for Image and Video Databases II (Niblack, WR and Jain, RC, eds), Proc SPIE 2185, pp. 2–14, 1994.
60. Smith, JR and Chang, SF, "Querying by Color Regions Using the VisualSEEk Content-Based Visual Query System," Intelligent Multimedia Information Retrieval (Maybury, ed), AAAI Press, Menlo Park, CA, pp. 23–41, 1997.
61. Sparck Jones, K, "Reflections on TREC," Inform Process Manag, 31(3), pp. 291–314, 1995.
62. Squire, D, and Pun, T, "A Comparison of Human and Machine Assessments of Image Similarity for the Organization of Image Databases," 10th Scandinavian Conference on Image Analysis, Lappeenranta, Finland, pp. 51–58, 1997.
63. Swain, MJ and Ballard, DH, "Color Indexing," Int J Computer Vision, 7(1), pp. 11–32, 1991.
64. Tamura, H, Mori, S, and Yamawaki, Y, "Textural Features Corresponding to Visual Perception," IEEE Trans Syst Man Cybern, 8(6), pp. 460–472, 1978.

65. Teague, MR, "Image Analysis by the General Theory of Moments," J Opt Soc Am, 70, pp. 920–930, 1980.
66. Teh, CH and Chin, RT, "Image Analysis by Methods of Moments," IEEE Trans Patt Anal Mach Intell, 10(4), pp. 496–513, 1988.
67. Tversky, A, "Features of Similarity," Psychol Rev, 84(4), pp. 327–352, 1977.
68. Umetani, Y and Taguchi, K, "Discrimination of General Shapes by Psychological Feature Properties," Dig Syst Indust Autom, 1(2-3), pp. 179–198, 1982.
69. Vellaikal, A and Kuo, C, "Hierarchical Clustering Techniques for Image Database Organization and Summarization," in Multimedia Storage and Archiving Systems III, Proc SPIE 3527, pp. 68–79, 1998.
70. Vleugels, J and Veltkamp, R, "Efficient Image Retrieval Through Vantage Objects," presented at VISUAL99: 3rd International Conference on Visual Information and Information Systems, Lecture Notes in Computer Science 1614, pp. 769–776, 1999.
71. Wertheimer, M, "Untersuchungen zur Lehre von der Gestalt," Psychologische Forschung, 4, pp. 301-350, 1923; Translated as "Laws of organization in perceptual forms," in A Sourcebook of Gestalt Psychology, Humanities Press, New York, 1950.
72. Whalen, TE, Lee, ES, and Safayeni, F, "The Retrieval of Images from Image Databases: Trademarks," Behav Inform Technol, 14(1), pp. 3–13, 1995.
73. Witkin, AP and Tenenbaum, JM, "On the Role of Structure in Vision," in Human and Machine Vision, pp. 481–543, Academic Press, New York, 1983.
74. Wu, JK, Lam, CP, Mehtre, BM, Gao, YJ, Narasimhalu, A, "Content-Based Retrieval for Trademark Registration," Multimed Tools Applic, 3, pp. 245–267, 1996.
75. Zahn, CT and Roskies, CZ, "Fourier Descriptor for Plane Closed Curves," IEEE Trans Computers, C-21, pp. 269–281, 1972.

Author Index

Subject Index